# Microprocessors *and* Interfacing Devices

**Rupender Singh**
BTech, MTech, PGDESD

Assistant Professor
World Institute of Technology
Gurgaon, Haryana

**Sarika Jain**
MSc (IIT Delhi), PhD Scholar (IIT Delhi)
Sr. Lecturer
Amity University
Gurgaon, Haryana

**CBS**

**CBS Publishers & Distributors** Pvt Ltd

New Delhi • Bengaluru • Chennai • Kochi • Mumbai • Pune
Hyderabad • Kolkata • Nagpur • Patna • Vijayawada

# Microprocessors
*and*
# Interfacing Devices

**ISBN:** 978-81-239-2539-4

Copyright © Authors and Publisher

**First Edition:** 2015

Published by Satish Kumar Jain and Produced by Varun Jain for

**CBS Publishers & Distributors** Pvt Ltd

4819/XI Prahlad Street, 24 Ansari Road, Daryaganj, New Delhi 110 002, India.
Ph: 23289259, 23266861, 23266867        Website: www.cbspd.com
Fax: 011-23243014        e-mail: delhi@cbspd.com; cbspubs@airtelmail.in.
*Corporate Office:* 204 FIE, Industrial Area, Patparganj, Delhi 110 092
Ph: 4934 4934        Fax: 4934 4935        e-mail: publishing@cbspd.com; publicity@cbspd.com

*Branches*

- **Bengaluru:** Seema House 2975, 17th Cross, K.R. Road,
  Banasankari 2nd Stage, Bengaluru 560 070, Karnataka
  Ph: +91-80-26771678/79        Fax: +91-80-26771680        e-mail: bangalore@cbspd.com
- **Chennai:** 7, Subbaraya Street, Shenoy Nagar, Chennai 600 030, Tamil Nadu
  Ph: +91-44-42032115        Fax: +91-44-42032115        e-mail: chennai@cbspd.com
- **Kochi:** 36/14 Kalluvilakam, Lissie Hospital Road, Kochi 682 018, Kerala
  Ph: +91-484-4059061-65        Fax: +91-484-4059065        e-mail: kochi@cbspd.com
- **Mumbai:** 83-C, Dr E Moses Road, Worli, Mumbai-400018, Maharashtra
  Ph: +91-22-24902340/41        Fax: +91-22-24902342        e-mail: mumbai@cbspd.com
- **Pune:** Bhuruk Prestige, Sr. No. 52/12/2+1+3/2 Narhe, Haveli
  (Near Katraj-Dehu Road Bypass), Pune 411 041, Maharashtra
  Ph: +91-20-64704058, 64704059, 32392277 Fax: +91-20-24300160        e-mail: pune@cbspd.com

*Representatives*

- **Hyderabad**  0-9885175004    • **Kolkata**  0-9831437309, 0-9051152362
- **Nagpur**        0-9021734563    • **Patna**      0-9334159340
- **Vijayawada** 0-9000660880

*Printed at* Magic International, Greater Noida, UP

# Preface

This book *Microprocessors and Interfacing Devices* has been written for the students of BE/BTech, MSc (electronics), MSc (IT), MTech, MCA, BSc (electronics) and Diploma in Electronics of all Indian universities. This book uses plain, lucid language to explain fundamentals of the subject. The book provides logical method of explaining various complicated concepts and stepwise methods to explain the important topics. Each chapter is well supported with necessary illustrations, practical examples and solved problems. All the chapters in the book are arranged in a proper sequence that permits each topic to build upon earlier studies. Many of the examples and problems have been selected from recent question papers of various university and engineering examinations.

Microprocessors are regarded as one of the most important devices in our computers. Microprocessors are also used in other advanced electronic systems such as computer printers, automobiles and jet airliners. The importance of microprocessors is well known in various engineering fields. This increasing importance has led to enhanced demand for courses dealing with microprocessors and interfacing devices. This book is an attempt to provide a suitable textbook which will meet the needs of the engineering students.

This book contains 10 chapters with comprehensive material discussed in a very systematic and elaborative manner. Chapter 1 explains the history of microprocessors and companies associated with them. Chapter 2 gives the details of basic digital concepts in a systematic manner. A large number of programming examples based on assembly language are given in Chapters 3 and 5 for both 8085 and 8086 microprocessors. Chapters 6 to 9 discuss interfacing devices with their architecture and pin configuration. The details of Pentium processors have been discussed in Chapter 10.

The first author gratefully acknowledges the moral support and encouragement provided by his family members Mr Resham Singh and Mr Hoshiyar Singh.

Thanks are due to CBS Publishers & Distributors for taking interest in the publication and marketing of the book.

Suggestions for improvement in the text shall be personally acknowledged and deeply appreciated which will help to make it an ideal book for all.

**Rupender Singh**
**Sarika Jain**

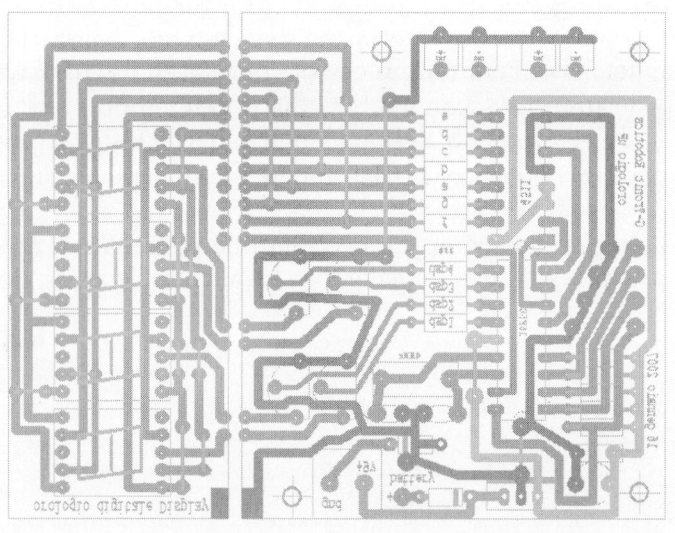

# Contents

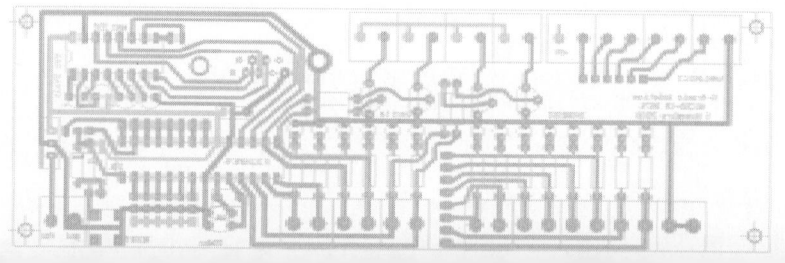

# 1

# Introduction to Microprocessors

## 1.1 INTRODUCTION

A microprocessor (μp) is a multipurpose, programmable logic device (IC) that reads binary instructions from a storage device called memory, accepts binary data as input and processes data according to those instructions, and provides results as output. A multipurpose device means it can be used to perform various sophisticated computing tasks or functions, as well as simple tasks. A programmable device means that it can be instructed to perform given tasks within its capability. Today's microprocessor is designed to understand and execute many binary instructions. Microprocessor is also called Central Processing Unit (CPU) since it is the functional centre of the computer system and it is used to process data. The first microprocessor was Intel 4004 (early 1970s) used in calculators. It was designed by Intel Corporation and become known as the 4-bit microprocessor. It was quickly replaced by the 8-bit microprocessor (Intel 8008), which in turn superseded by the Intel 8080. In the mid 1970s, the Intel 8080 was widely used in control applications, and small computers were also designed using the 8080 as the CPU.

In order to function as a programmable device the microprocessor must work in a complete system comprising of three components: microprocessor, memory and input/output. This system is called microprocessor-based system or microcomputer system. Figure 1.1 shows microcomputer system. These three components will work together or interact with each other to perform a given task.

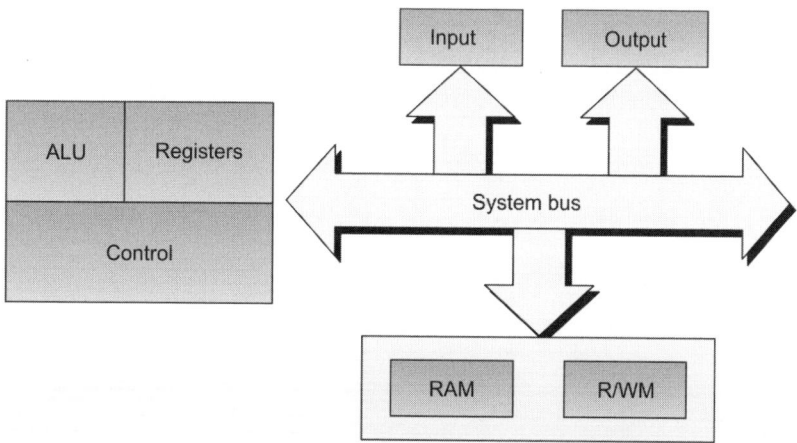

**Fig. 1.1:** Block diagram of microcomputer

## 1.2 HISTORICAL DEVELOPMENT OF THE MICROPROCESSORS

Microprocessors were categorized into five generations: first, second, third, fourth and fifth generations. Figure 1.2 shows generation of microprocessors yearwise on the basis of speed in Hz. Their characteristics are described below.

### 1.2.1 First Generation

The microprocessors that were introduced in 1971 to 1972 were referred to as the first generation systems. First generation microprocessors processed their instructions serially—they fetched the instruction, decoded it, then executed it. When an instruction was completed, the microprocessor updated the instruction pointer and fetched the next instruction, performing this sequential drill for each instruction in turn.

### 1.2.2 Second Generation

By the late 1970s, enough transistors were available on the IC to usher in the second generation of microprocessor sophistication: 16-bit arithmetic and pipelined instruction processing. Motorola's MC68000 microprocessor, introduced in 1979, is an example. Another example is Intel's 8080. This generation is defined by overlapped fetch, decode and execute steps (Computer 1996). As the first instruction is processed in the execution unit, the second instruction is decoded and the third instruction is fetched. The distinction between the first and second generation devices was primarily the use of newer semiconductor technology to fabricate the chips. This new technology resulted in a five-fold increase in instruction, execution, speed and higher chip densities.

### 1.2.3 Third Generation

The third generation, introduced in 1978, was represented by Intel's 8086 and the Zilog Z8000, which were 16-bit processors with minicomputer-like performance. The third generation came about as IC transistor counts approached 250,000.

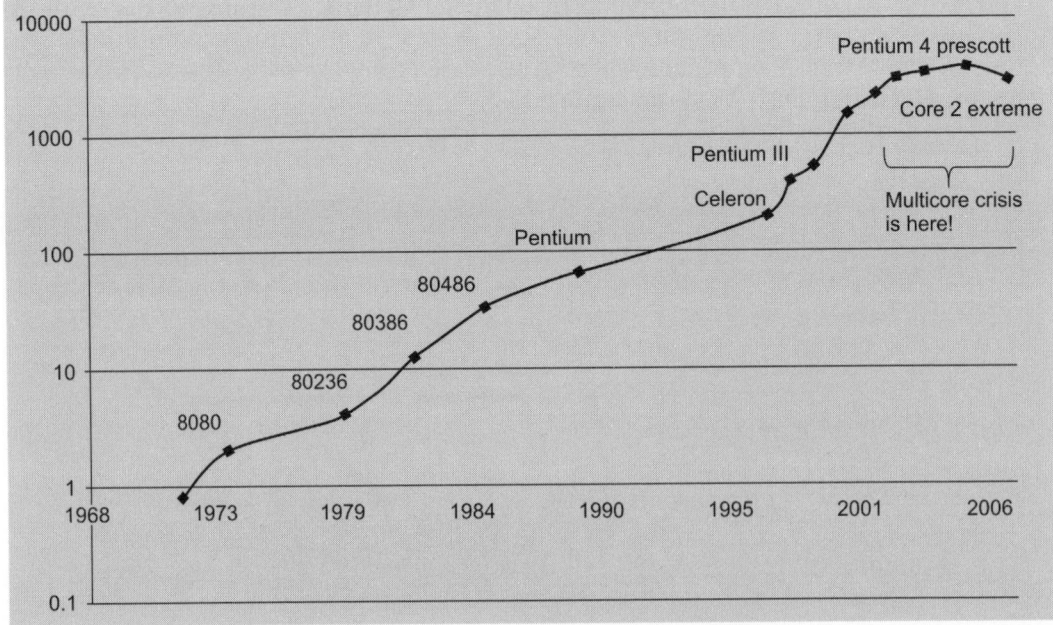

**Fig. 1.2:** Generation of Intel microprocessors

Motorola's MC68020, for example, incorporated an on-chip cache for the first time and the depth of the pipeline increased to five or more stages. This generation of microprocessors was different from the previous ones in that all major workstation manufacturers began developing their own RISC-based microprocessor architectures (Computer, 1996).

### 1.2.4 Fourth Generation

As the workstation companies converted from commercial microprocessors to in-house designs, microprocessors entered their fourth generation with designs surpassing a million transistors. Leading-edge microprocessors such as Intel's 80960CA and Motorola's 88100 could issue and retire more than one instruction per clock cycle.

### 1.2.5 Fifth Generation

Microprocessors in their fifth generation, employed decoupled super scalar processing, and their design soon surpassed 10 million transistors. In this generation, PCs are a low margin, high-volume-business dominated by a single microprocessor.

### 1.3 MICROPROCESSOR PROGRESS: INTEL

Table 1.1 helps us to understand the differences between different processors that Intel has introduced over the years.

Information about Table 1.1 is as follows:

- **The date** is the year that the processor was 1st introduced. Many processors are reintroduced at higher clock speeds for many years after the original release date.
- **Transistors mean** the number of transistors used on the chip. We can see that the number of transistors on a single chip has risen steadily over the years.
- **Microns** indicate the width, in microns of the thinnest wire on the chip. For comparison, a human hair is 100 microns thick. As the feature size on the chip goes down, the number of transistors rises.
- **Clock speed** stands for the maximum rate that the chip can be clocked at. Clock speed will make more sense later.

**Table 1.1** Comparison of different microprocessors

| Name | Date | Transistors | Microns | Clock speed | Data width | Address lines | Mips |
|------|------|-------------|---------|-------------|------------|---------------|------|
| 8080 | 1974 | 6 K | 6 | 2 MHz | 8 bits | 16 bits | 0.64 |
| 8086 | 1978 | 29 K | 6 | 8 MHz | 16 bits | 20 bits | 0.80 |
| 8088 | 1979 | 29 K | 3 | 5 MHz | 16 bits 8 bits bus | 20 bits | 0.33 |
| 80286 | 1982 | 134 K | 1.5 | 6 MHz | 16 bits | 24 bits | 1 |
| 80386 | 1985 | 275 K | 1.5 | 16 MHz | 32 bits | 32 bits | 5 |
| 80486 | 1989 | 1200 K | 1 | 25 MHz | 32 bits | 32 bits | 20 |
| Pentium | 1993 | 3100 K | 0.8 | 60 MHz | 32 bits/64 bits | 32 bits | 100 |
| Pentium II | 1997 | 7500 K | 0.35 | 233 MHz | 32 bits/64 bits | 32 bits | ~300 |
| Pentium III | 1999 | 9500 K | 0.25 | 450 MHz | 32 bits/64 bits | 32 bits | ~510 |
| Pentium IV | 2000 | 42000 K | 0.18 | 1.5 GHz | 32 bits/64 bits | 32 bits | ~1700 |
| Pentium IV Prescott | 2004 | 125000 K | 0.09 | 3.6 GHz | 32 bits/64 bits | 32 bits | ~7000 |

- **Data width** is the width of the ALU in bits. An 8-bit ALU can add/subtract/multiply, etc. two 8-bit numbers, while a 32-bit ALU can manipulate 32-bit numbers. An 8-bit ALU would have to execute 4-instructions to add two 32-bit numbers, while a 32-bit ALU can do it in one instruction. In many cases, the external data bus is of the same width as the ALU, but not always. The 8088 had a 16-bit ALU and an 8-bit external data bus, while the modern Pentiums fetch data 64-bits at a time for their 32-bit ALUs.

- **MIPS** stands for 'millions of instructions per second' and is a rough measure of the performance of a CPU. Modern CPUs can do so many different things that MIPS ratings lose a lot of their meaning, but we can get a general sense of the relative power of the CPU from this column.

## 1.4  HOW MICROPROCESSOR WORKS

To understand how a microprocessor works, it is helpful to look inside and learn about the Logic used to create one. In the process, we can also learn about **assembly language** → the native language of a microprocessor → and many of the things that engineers can do to boost the speed of a processor.

A microprocessor executes a collection of machine instructions that tell the processor what to do. Based on the instructions, a microprocessor does three basic things:

- Using its ALU (Arithmetic Logic Unit), a microprocessor can perform mathematical operations like addition, subtraction, multiplication and division. Modern micro-processors contain complete floating-point processors that can perform extremely sophisticated operations on large floating-point numbers.

- A microprocessor can move data from one memory location to another.

- A microprocessor can make decisions and jump to a new set of instructions based on those decisions.

There may be very sophisticated things that a microprocessor does, but those are its three basic activities. The diagram shown in Fig. 1.3 is an extremely simple representation of the microprocessor capable of doing those three things.

This is about as simple as a microprocessor gets. This microprocessor has:

- **An address bus** (it may be 8, 16, 32 or 64-bits wide) that sends an address to memory or I/O devices.

- **A data bus** (it may be 8, 16, 32 or 64-bits wide) that can send data to memory or receive data from memory.

- **An RD (read bar) and WR (write bar) line** to tell the memory whether it wants to receive or deliver to it.

- **A clock line** that sequences the processor clock pulses.

- **A reset line** that resets the program counter to zero (or whatever) and restarts execution.

Let us assume that both the address and data buses are 8-bits wide in this example. Here are the components of this simple microprocessor:

- Registers A, B and C are simply registers (latches) made out of 8-flip-flops.

- The address registers (latch) is just like registers A, B and C.

- The program counter is a register with the extra ability to increment by 1 and also to reset to zero when told to do so.

- The ALU could be as simple as an 8-bit adder or it might be able to add, subtract, multiply and divide 8-bit data.

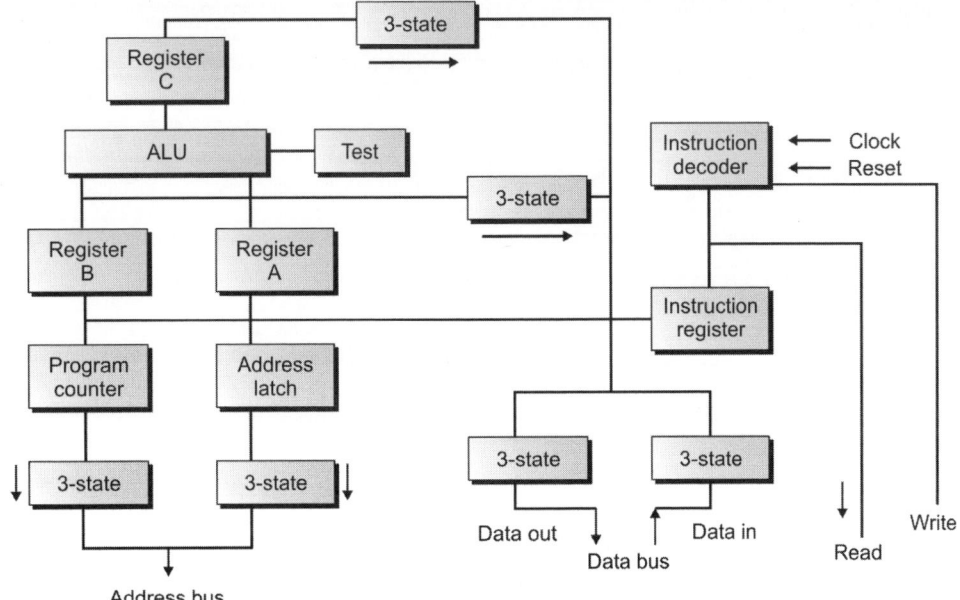

**Fig. 1.3:** Block diagram of microprocessor

- The test register is a special latch that can hold values from comparisons performed in the ALU. An ALU can normally compare 2-numbers and determine if they are equal, if one is greater than the other, etc. The test register can also normally hold a carry bit from the last stage of the adder. It stores these values in flip-flops and then the instruction decoder can use the values to make decisions.
- There are 6-boxes marked '3-state' in the diagram. These are **tri-state buffers.** A tri-state buffer can pass a 1, a 0 or it can essentially disconnect its output (imagine a switch that totally disconnects the output line from the wire that the output is heading toward). A tri-state buffer allows multiple outputs to connect to a wire, but only one of them to actually drive a 1 or a 0 onto the line.
- The instruction register and instruction decoder are responsible for controlling all of the other components.

## 1.5 MICROPROCESSOR PERFORMANCE

The number of **transistors** available has a huge effect on the performance of a processor. Figure 1.4 shows the number of transistors yearwise on the basis of Moore's law who stated that number of transistors doubles every two years. As seen earlier, a typical instruction in a processor like an 8088 took 15-clock cycles to execute. Because of the design of the multiplier, it took approximately 80-cycles just to do one 16-bit multiplication on the 8088. With more transistors, much more powerful multipliers capable of single-cycle speeds become possible. More transistors also allow for a technology called **pipelining.** In a pipelined architecture, instruction execution overlaps, so even though it might take 5-clock cycles to execute each instruction, there can be 5-instructions in various stages of execution simultaneously. That way it looks like 1-instruction completes every clock cycle. Many modern processors have multiple instruction decoders, each with their own pipeline. This allows for multiple instruction streams, which means that more than 1-instruction can complete during each clock cycle. This technique can be quite complex to implement, so it takes a lot of transistors.

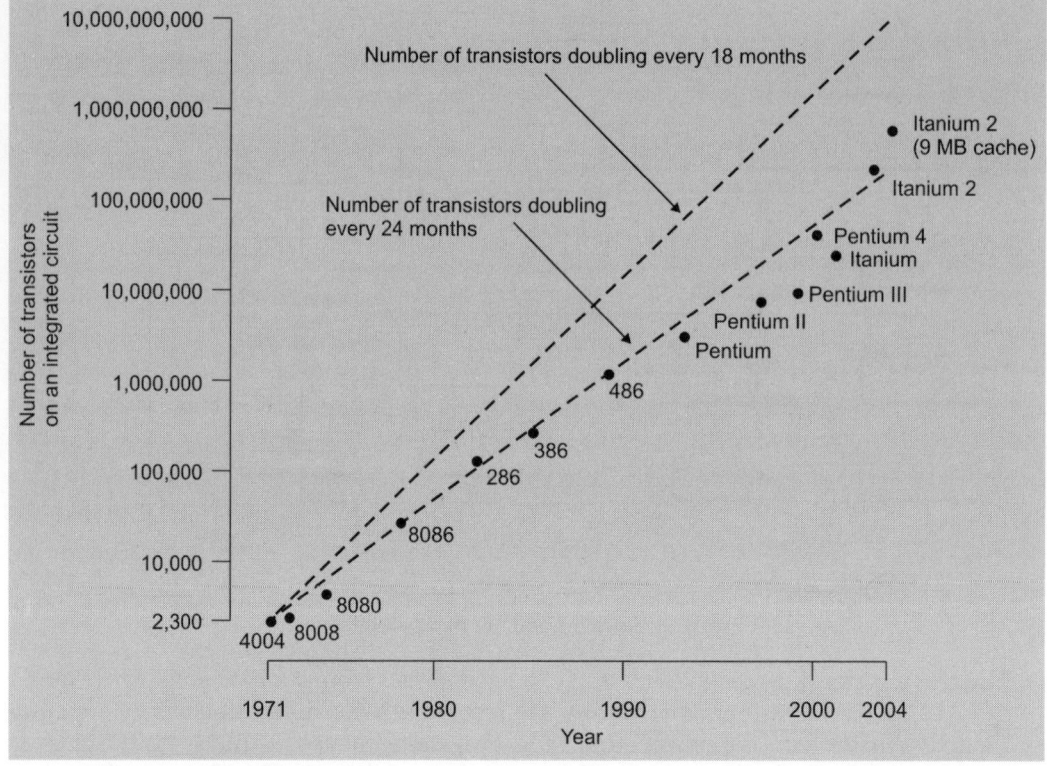

**Fig. 1.4:** Number of transistors in different microprocessors on the basis of Moore's law

## 1.6 INTEL'S MICROPROCESSORS

### 1.6.1 4-Bit Microprocessors

#### 1.6.1.1 Microprocessor 4004 (1971)

It was a 4-bit microprocessor. Historically, the 4-bit microprocessor was the first general purpose microprocessor introduced in the market. The basic design of the early microprocessors was derived from that of the desk calculator. The Intel 4004, a 4-bit design, was the grandfather of microprocessors. Introduced in late 1971, the 4004 was originally designed for a Japanese manufacturer as the processing element of a desk calculator; it was not designed as a general-purpose computer. The shortcomings of the 4004 were recognized as soon as it was introduced, but it was the first general purpose computing device on a chip to be placed on the market. Many of the chips introduced at about the same time by other companies were, in fact, mere calculator chips.

*Features*

- Introduced: November 15, 1971
- Clock Rate: 740 kHz
  - 0.07 MIPS
- Bus Width: 4 bits (multiplexed address/data due to limited pins)
- PMOS technology
- Number of Transistors: 2,300 at 10 μm
- Addressable Memory: 640 bytes

- Program Memory: 4 KB (4 KB)
- Originally designed to be used in Busicom calculator.

### 1.6.2 8-Bit Microprocessors

#### 1.6.2.1 Microprocessor 8008 (1972)

The 8008 increased the 4004's word length from four to eight bits, and doubled the volume of information that could be processed. It was still an invention in search of a market, however, as the technology world was just beginning to view the microprocessor as a solution to many needs.

*Features*

- Introduced: April 1, 1972
- Clock Rate: 500 kHz
  - 0.05 MIPS
- Bus Width: 8 bits (multiplexed address/data due to limited pins)
- Enhancement Load: PMOS logic
- Number of Transistors: 3,500 at 10 µm
- Addressable Memory: 16 KB
- Typical in dumb terminals, general calculators, bottling machines.

#### 1.6.2.2 Microprocessor 8080 (1974)

The 8080 were 20 times as fast as the 4004 and contained twice as many transistors. The 8080, designed as a successor to Intel's 8008, was the first powerful microprocessor introduced in the market. Several others microprocessors of similar performance were introduced in the market within a year after the 8080 appeared and several additional powerful designs were introduced later. Technically, however, the 8080 long remained the most powerful product in the market. Furthermore, Intel was the first company to invest in the development of support chips and software for its products. This ensured the continued success of the 8080 because its performance was then sufficient for many applications. The early 8080 competitors were introduced with at least a nine-month delay and failed to dislodge it. The 8080 is still sold today thought it has been largely eclipsed by successor products—most notably the 8085 microprocessor. The 8080 is available in three versions: the standard 8080A with a 2 MHz clock, the 8080A-2, and the 8080A-1 with a 3 MHz clock.

*Features*

- Introduced: April 1, 1974
- Clock Rate: 2 MHz
  - 0.64 MIPS
- Bus Width: 8 bits data, 16 bits address
- Enhancement Load: NMOS logic
- Number of Transistors: 6,000
- Assembly language downwards compatible with 8008
- Addressable Memory: 64 KB
- Up to 10X the performance of the 8008
- Used in the Altair 8800, traffic light controller, cruise missile
- Required six support chips versus 20 for the 8008.

### 1.6.2.3 Microprocessor 8085 (1976)

The **8085** is an 8-bit microprocessor introduced by Intel in 1976. The 8085 was the follow-on processor to the very successful Intel 8080A processor. The 8085 got its name because it was Intel's first 5 volt microprocessor. The 8085 was 100% software compatible with the 8080A with increased systems performance. The initial 8085s were based on NMOS technology and the later "H" versions were based on HMOS technology. The 8085 used 6,500 transistors.

The 8085 incorporated all the features of the 8224 (clock generator) and the 8228 (system controller) increasing the level of system integration.

*Features*

- Introduced: March 1976
- Clock Rate: 5 MHz
  - 0.37 MIPS
- Bus Width: 8 bits data, 16 bits address
- Depletion Load: NMOS logic
- Number of Transistors: 6,500 at 3 μm
- Binary compatible downwards with the 8080
- Used in Toledo scale and also used as a computer peripheral controller-modems, hard disk, printers, etc.
- High level of integration, operating for the first time on a single 5 volt power supply from 12 volts previously. Also featured serial I/O, 3 maskable interrupts, 1 non-maskable interrupt, 1 externally expandable interrupt [8259], status, DMA.

## 1.6.3 16-Bit Microprocessors

### 1.6.3.1 Microprocessor 8086 (1978)

Intel's 8086 microprocessor is a *first member* of x 86 families of processors. Advertised as a "source-code compatible" with Intel 8080 and Intel 8085 processors, the 8086 was not object code compatible with them. The 8086 has complete 16-bit architecture, 16-bit internal registers, 16-bit data bus and 20-bit address bus (1 MB of physical memory). Since the processor has 16-bit index registers and memory pointers, it can effectively address only 64 KB of memory.

*Features*

- Introduced: June 8, 1978
- Clock Rates: 5 MHz with 0.33 MIPS
  - 8 MHz with 0.66 MIPS
  - 10 MHz with 0.75 MIPS
- The memory is divided into odd and even banks. It accesses both the banks simultaneously in order to read 16 bit of data in one clock cycle.
- Bus Width: 16 bits data, 20 bits address
- Number of Transistors: 29,000 at 3 μm
- Addressable Memory: 1 megabyte
- Up to 10X the performance of 8080 (typically lower)

- Used in portable computing, and in the IBM PS/2 Model 25 and Model 30. Also used in the AT&T PC6300/Olivetti M24, a popular IBM PC-compatible (predating the IBM PS/2 line.)
- Used segment registers to access more than 64 KB of data at once, which many programmers complained made their work excessively difficult.

### 1.6.3.2 Microprocessor 8088 (1979)

Created as a cheaper version of Intel's 8086, the 8088 was a 16-bit processor with an 8-bit external bus. This chip became the most ubiquitous in the computer industry when IBM chose it for its first PC. The success of the IBM PC and its clones gave Intel a dominant position in the semiconductor industry.

*Features*

- Introduced: June 1, 1979
- Clock Rates: 4.77 MHz with 0.33 MIPS
  - 9 MHz with 0.75 MIPS
- Internal Architecture: 16 bits
- External Bus Width: 8 bits data, 20 bits address
- Number of Transistors: 29,000 at 3 μm
- Addressable Memory: 1 megabyte
- Identical to 8086 except for its 8 bit external bus (hence an 8 instead of a 6 at the end)
- Used in IBM PCs and PC clones.

### 1.6.3.3 Microprocessor 80186 (1982)

*Features*

- Introduced: 1982
- Included two timers, a DMA controller and an interrupt controller on the chip in addition to the processor (These were at fixed addresses which differed from the IBM PC, making it impossible to build a 100% PC-compatible computer around the 80186.)
- Added a few opcodes and exceptions to the 8086 design; otherwise identical instruction set to 8086 and 8088.
- Used mostly in embedded applications: Controllers, point-of-sale systems, terminals.
- Used in several non-PC-compatible MS-DOS computers including RM Nimbus, Tandy 2000.

### 1.6.3.4 Microprocessor 80286 (1982)

With 16 MB of addressable memory and 1 GB of virtual memory, this 16-bit chip is referred to as the first "modern" microprocessor. Many novices were introduced to desktop computing with a "286 machine" and it became the dominant chip of its time. It contained 130,000 transistors and packed serious compute power (12 MHz) into a tiny footprint.

*Features*

- Introduced: February 1, 1982
- Clock Rates: 6 MHz with 0.9 MIPS
  - 8 MHz, 10 MHz with 1.5 MIPS

- 12.5 MHz with 2.66 MIPS
- 16 MHz, 20 MHz and 25 MHz available
- Bus Width: 16 bit data, 24 bit address
- Included memory protection hardware to support multitasking operating systems with per-process address space
- Number of Transistors: 34,000 at 1.5 μm
- Addressable Memory: 16 MB (16 MB)
- Added protected-mode features to 8086 with essentially the same instruction set
- 3–6X the performance of the 8086
- Widely used in IBM-PC AT and AT clones contemporary to it.

### 1.6.4 32-Bit Microprocessors

#### 1.6.4.1 Microprocessor 80386 (1985), 80486 (1989)

The price/performance curve continued its steep climb with the 386 and later the 486—32-bit processors that brought real computing to the masses. The 386, which became the best-selling microprocessor in history, featured 275,000 transistors; the 486 had more than a million.

*Features of 80386*

- Introduced: October 17, 1985
- Clock Rates:
  - 16 MHz with 5 to 6 MIPS
  - 20 MHz with 6 to 7 MIPS, introduced on February 16, 1987
  - 25 MHz with 8.5 MIPS, introduced on April 4, 1988
  - 33 MHz with 11.4 MIPS (9.4 SPECint92 on Compaq/i 16K L2), introduced on April 10, 1989
- Bus Width: 32 bit data, 32 bit address
- Number of Transistors: 275,000 at 1 μm
- Addressable Memory: 4 GB (4 GB)
- Virtual Memory: 64 TB (64 TiB)
- First X86 chip to handle 32-bit data sets
- Reworked and expanded memory protection support including paged virtual memory and virtual 86 mode features required by Linux, Windows 95 and OS/2 Warp
- Used in Desktop computing.

*Features of 80486*

- Introduced: April 10, 1989
- Clock Rates:
  - 25 MHz with 20 MIPS
  - 33 MHz with 27 MIPS
  - 50 MHz with 41 MIPS
- Bus Width: 32 bits
- Number of Transistors: 1.2 million at 1 μm; the 50 MHz was at 0.8 μm

- Addressable Memory: 4 GB
- Virtual Memory: 1 TB
- Level 1 cache of 8 KB on chip
- Math coprocessor on chip
- 50X performance of the 8088
- Used in Desktop computing and servers
- Family 4 model 3.

### 1.6.5 64-Bit Microprocessors

#### 1.6.5.1 Pentium I (1993)

Adding systems-level characteristics to enormous raw compute power, the Pentium supports demanding I/O, graphics and communications-intensive applications with more than 3 million transistors.

*Features*

- Bus Width: 64 bits
- System Bus Clock Rate: 60 or 66 MHz
- Address Bus: 32 bits
- Addressable Memory: 4 GB
- Virtual Memory: 64 TB
- Superscalar architecture
- Runs on 5 volts
- Used in desktops
- 16 KB of L1 cache

P5: 0.8 μm process technology

- Introduced: March 22, 1993
- Number of Transistors: 3.1 million
- Socket 4 273 pin PGA processor package
- Package Dimensions: 2.16″ × 2.16″
- Family 5 model 1.

#### 1.6.5.2 Pentium Pro (1995)

The newest Pentium has dynamic instruction execution and other performance enhancing features such as a large L2 cache in the chip package, in addition to its more than 5.5 million transistors.

*Features*

- Introduced: November 1, 1995
- Primarily used in server systems
- Socket 8 processor package (387 pins) (Dual SPGA)
- Number of Transistors: 5.5 million
- Family 6 model 1

0.6 μm process technology
- 16 KB L1 cache
- 256 KB integrated L2 cache
- System Bus Clock Rate: 60 MHz
- Variants: 150 MHz

0.35 μm process technology or 0.35 μm CPU with 0.6 μm L2 cache
- Number of Transistors: 5.5 million
- 512 KB or 256 KB integrated L2 cache
- System Bus Clock Rate: 60 or 66 MHz
- Variants
  - 166 MHz (66 MHz bus clock rate, 512 KB 0.35 μm cache) Introduced on November 1, 1995
  - 180 MHz (60 MHz bus clock rate, 256 KB 0.6 μm cache) Introduced on November 1, 1995
  - 200 MHz (66 MHz bus clock rate, 256 KB 0.6 μm cache) Introduced on November 1, 1995
  - 200 MHz (66 MHz bus clock rate, 512 KB 0.35 μm cache) Introduced on November 1, 1995
  - 200 MHz (66 MHz bus clock rate, 1 MB 0.35 μm cache) Introduced on August 18, 1997.

### 1.6.5.3 Pentium II (1997)

The 7.5 million-transistor Pentium II processor incorporates Intel MMXTM technology, which is designed specifically to process video, audio and graphics data efficiently.

*Features*
- Introduced: May 7, 1997
- Pentium Pro with MMX and improved 16-bit performance
- 242-pin Slot 1 (SEC) Processor Package
- Number of Transistors: 7.5 million
- 32 KB L1 cache
- 512 KB ½ bandwidth external L2 cache

Klamath: 0.35 μm process technology (233, 266, 300 MHz)
- System Bus Clock Rate: 66 MHz
- Family 6 model 3
- Variants
  - 233 MHz Introduced on May 7, 1997
  - 266 MHz Introduced on May 7, 1997
  - 300 MHz Introduced on May 7, 1997.

### 1.6.5.4 Pentium II Xeon (1998)

The Pentium II Xeon processors are designed to meet the performance requirements of mid-range and higher servers and workstations. Consistent with Intel's strategy to deliver unique processor products targeted for specific market segments, the Pentium II Xeon processors feature technical innovations specifically designed for workstations and servers that utilize demanding business applications, such as Internet services, corporate data warehousing, digital content creation, and electronic and mechanical design automation. Systems based on the processor can be configured to scale to four or eight processors and beyond.

### 1.6.5.5 Celeron (PII Based, 1998)

Continuing Intel's strategy of developing processors for specific market segments, the Intel Celeron processor is designed for the value PC market segment. It provides consumers great performance at an exceptional value, and it delivers excellent performance for uses, such as gaming and educational software.

*Features*

- Covington: 0.25 μm process technology
- Introduced: April 15, 1998
- 242-pin Slot 1 SEPP (Single Edge Processor Package)
- Number of Transistors: 7.5 million
- System Bus Clock Rate: 66 MHz
- 32 KB L1 cache
- No L2 cache
- Variants
  - 266 MHz Introduced on April 15, 1998
  - 300 MHz Introduced on June 9, 1998

### 1.6.5.6 Pentium III (1999)

The Pentium III processor features 70 new instructions. It was designed to significantly enhance Internet experiences, allowing users to do such things as browse through realistic online museums and stores and download high-quality videos. The processor incorporates 9.5 million transistors and was introduced using 0.25-micron technology.

*Features*

Katmai: 0.25 μm process technology
- Introduced: February 26, 1999
- Improved PII, i.e. P6-based core, now including Streaming SIMD Extensions (SSE)
- Number of Transistors: 9.5 million
- 512 KB ½ bandwidth L2 External cache
- 242-pin Slot 1 SECC2 (Single Edge Contact Cartridge 2) processor package
- System Bus Clock Rate: 100 MHz, 133 MHz (B-models)
- Family 6 model 7
- Variants
  - 450 MHz Introduced on February 26, 1999
  - 500 MHz Introduced on February 26, 1999
  - 550 MHz Introduced on May 17, 1999
  - 600 MHz Introduced on August 2, 1999
  - 533 MHz (133 MHz bus clock rate) Introduced on September 27, 1999
  - 600 MHz (133 MHz bus clock rate) Introduced on September 27, 1999

Coppermine: 0.18 μm process technology
- Introduced: October 25, 1999
- Number of Transistors: 28.1 million
- 256 KB Advanced Transfer L2 Cache (Integrated)

- 242-pin Slot-1 SECC2 (Single Edge Contact Cartridge 2) Processor Package, 370-pin FC-PGA (Flip-chip Pin Grid Array) Package
- System Bus Clock Rate: 100 MHz (E-models), 133 MHz (EB models)
- Slot 1, Socket 370
- Family 6 model 8.

### 1.6.5.7 Pentium III Xeon (1999)

The Pentium III Xeon processor extends Intel's offerings to the workstation and server market segments, providing additional performance for e-Commerce applications and advanced business computing. The processors incorporate the Pentium III processor's 70 SIMD instructions, which enhance multimedia and streaming video applications. The Pentium III Xeon processor's advance cache technology speeds information from the system bus to the processor, significantly boosting performance. It is designed for systems with multiprocessor configurations.

*Features*
- Introduced: October 25, 1999
- Number of Transistors: 9.5 million at 0.25 $\mu$m or 28 million at 0.18 $\mu$m
- L2 cache is 256 KB, 1 MB or 2 MB Advanced Transfer Cache (Integrated)
- Processor Package Style is Single Edge Contact Cartridge (SECC2) or SC330
- System Bus Clock Rate: 133 MHz (256 KB L2 cache) or 100 MHz (1–2 MB L2 cache)
- System Bus Width: 64 bit
- Addressable Memory: 64 GB
- Used in two-way servers and workstations (256 KB L2) or 4- and 8-way servers (1–2 MB L2)
- Family 6 model 10
- Variants
  - 500 MHz (0.25 $\mu$m process) Introduced on March 17, 1999
  - 550 MHz (0.25 $\mu$m process) Introduced on August 23, 1999
  - 600 MHz (0.18 $\mu$m process, 256 KB L2 cache) Introduced on October 25, 1999
  - 667 MHz (0.18 $\mu$m process, 256 KB L2 cache) Introduced on October 25, 1999
  - 733 MHz (0.18 $\mu$m process, 256 KB L2 cache) Introduced on October 25, 1999.

### 1.6.5.8 Pentium IV

*Features*
- 0.18 $\mu$m process technology (1.40 and 1.50 GHz)
- Introduced: November 20, 2000
- L2 cache was 256 KB Advanced Transfer Cache (Integrated)
- Processor Package Style was PGA423, PGA478
- System Bus Clock Rate: 400 MHz
- SSE2 SIMD Extensions
- Number of Transistors: 42 million
- Used in desktops and entry-level workstations

### 1.6.5.9 Intel Core 2

*Features*

- Conroe: 65 nm process technology
- Desktop CPU (SMP support restricted to 2 CPUs)
- Two cores on one die
- Introduced: July 27, 2006
- SSSE3 SIMD Instructions
- Number of Transistors: 291 Million
- 64 KB of L1 cache per core (32 + 32 KB 8-way)
- Intel VT, multiple OS support
- TXT, Enhanced Security Hardware Extensions
- Execute Disable Bit
- EIST (Enhanced Intel Speed Step Technology)
- iAMT2 (Intel Active Management Technology), remotely manage computers
- LGA 775
- Variants
  - Core 2 Duo E6850 - 3.00 GHz (4 MB L2, 1333 MHz FSB)
  - Core 2 Duo X6800 - 2.93 GHz (4 MB L2, 1066 MHz FSB)
  - Core 2 Duo X6850 - 2.54 GHz (4 MB L2, 1066 MHz FSB)
  - Core 2 Duo E6750 - 2.67 GHz (4 MB L2, 1333 MHz FSB)
  - Core 2 Duo E6700 - 2.67 GHz (4 MB L2, 1066 MHz FSB)
  - Core 2 Duo E6600 - 2.40 GHz (4 MB L2, 1066 MHz FSB)
  - Core 2 Duo E6550 - 2.33 GHz (4 MB L2, 1333 MHz FSB)
  - Core 2 Duo E6420 - 2.13 GHz (4 MB L2, 1066 MHz FSB)
  - Core 2 Duo E6400 - 2.13 GHz (2 MB L2, 1066 MHz FSB)
  - Core 2 Duo E6320 - 1.86 GHz (4 MB L2, 1066 MHz FSB)
  - Core 2 Duo E6300 - 1.86 GHz (2 MB L2, 1066 MHz FSB)

### 1.6.5.10 Pentium Dual Core

*Features*

- Allendale: 65 nm process technology
- Desktop CPU (SMP support restricted to 2 CPUs)
- Two CPUs on one die
- Introduced: January 21, 2007
- SSSE3 SIMD Instructions
- Number of Transistors: 167 million
- TXT, Enhanced Security Hardware Extensions
- Execute Disable Bit
- EIST (Enhanced Intel Speed Step Technology)
- Variants
  - Intel Pentium E2220 - 2.40 GHz (1 MB L2, 800 MHz FSB)
  - Intel Pentium E2200 - 2.20 GHz (1 MB L2, 800 MHz FSB)

– Intel Pentium E2180 - 2.00 GHz (1 MB L2, 800 MHz FSB)

– Intel Pentium E2160 - 1.80 GHz (1 MB L2, 800 MHz FSB)

– Intel Pentium E2140 - 1.60 GHz (1 MB L2, 800 MHz FSB)

• Wolfdale: 45 nm process technology

– Intel Pentium E6500 - 2.93 GHz (2 MB L2, 1066 MHz FSB)

– Intel Pentium E6300 - 2.80 GHz (2 MB L2, 1066 MHz FSB)

– Intel Pentium E5400 - 2.70 GHz (2 MB L2, 800 MHz FSB)

– Intel Pentium E5300 - 2.60 GHz (2 MB L2, 800 MHz FSB)

– Intel Pentium E5200 - 2.50 GHz (2 MB L2, 800 MHz FSB)

– Intel Pentium E2210 - 2.20 GHz (1 MB L2, 800 MHz FSB)

## 1.7 MANUFACTURING OF MICROPROCESSORS

Economical manufacturing of microprocessors requires mass production. Microprocessors are constructed by depositing and removing thin layers of conducting, insulating and semiconducting materials in hundreds of separate steps. Nearly every layer must be patterned accurately into the shape of transistors and other electronic elements. Usually this is done by photolithography, which projects the pattern of the electronic circuit onto a coating that changes when exposed to light. Since these patterns are smaller than the shortest wavelength of visible light, short wavelength ultraviolet radiation must be used. Microprocessor features are so small and precise that a single speck of dust can destroy the microprocessor. Microprocessors are made in filtered clean rooms where the air may be a million times cleaner than in a typical home (PC World, 2000).

## 1.8 ARCHITECTURE OF MICROPROCESSOR

The microprocessor has three major components, namely Registers, Arithmetic and Logic Unit (ALU) and Control Unit (CU) as shown in Fig 1.5. Registers are used to store the data temporarily during the execution of the instruction. There are two types of registers, namely general purpose registers and special purpose registers.

Arithmetic and logic unit is responsible for the mathematical operation, such as addition, subtraction, etc. and logical operation, such as AND, OR, etc. The control unit provides the various control and timing signal to all the operations.

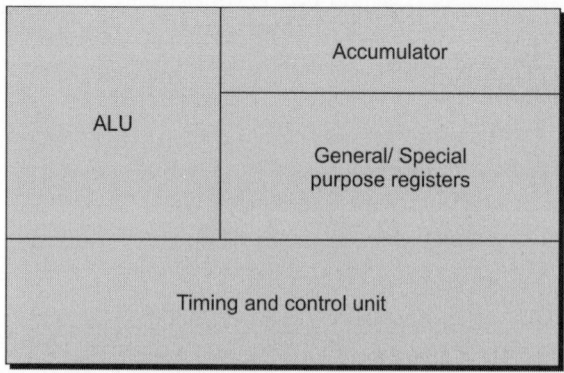

**Fig. 1.5:** Basic architecture of microprocessor

## 1.9 MICROCOMPUTER STRUCTURE

Figure 1.6 shows microcomputer structure which is discussed below.

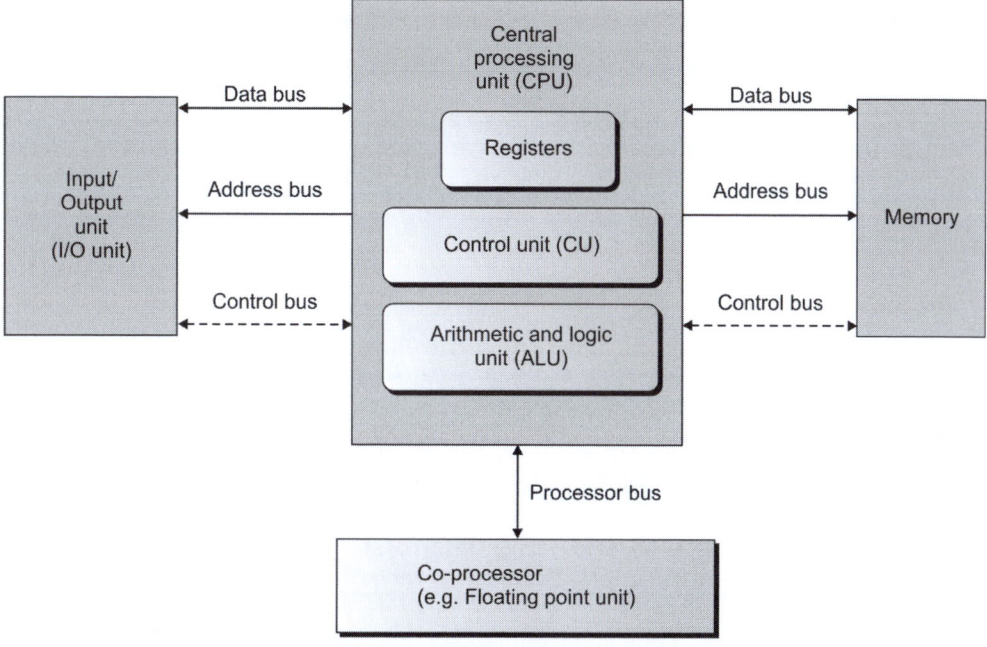

**Fig. 1.6:** Microcomputer structure

### 1.9.1 Clocks and Reset Circuits

Microprocessors use a fixed frequency clock to synchronize the operation of their internal logic. This clock signal is usually generated by a quartz-crystal controlled oscillator. The processor must be initialized to a known state when power is first applied or when the system "crashes." Most microprocessors require external circuits to detect the power-on condition or external resets.

### 1.9.2 Buffers

In larger systems, the microprocessor must be connected to more chips than the microprocessor chip's electrical specifications allow. ICs called buffers, bus drivers or transceivers are used between the microprocessor and the other chips.

### 1.9.3 Latches

Some microprocessors reduce the number of pins required on the chip by using the same pin for two purposes, such as using the same pins first as part of the address bus and then as part of the data bus. An external chip (a latch) is required to temporarily hold the address bus value.

### 1.9.4 Address Decoders

Since microprocessors can usually address more memory than an individual memory chips contains, there has to be external logic to select the appropriate memory chip for a given address. This function is called address decoding.

### 1.9.5 Wait State Generators

The access, hold, and setup-times of memory and I/O devices often exceed the values provided by the microprocessor's read and write cycles. Circuits are used to force the processor to "wait" one or more clock cycles to extend the durations of read or write cycles.

### 1.9.6 Interrupt Controllers

I/O devices that require immediate attention can interrupt a program's normal execution by asserting an interrupt signal to the microprocessor. When a system has several sources of interrupts, it is useful to be able to distinguish between them and to ensure that higher-priority interrupts are recognized first.

### 1.9.7 Timers and Counters

It is often useful for programs to be able to measure elapsed time, so that they can compute the time of day and measure intervals between events. Timers are counters driven from a clock and whose value can be read by the CPU. Timers are often used to generate periodic interrupts to allow an operating system to switch between tasks.

### 1.9.8 I/O Interfaces

I/O interfaces are used to read and write data to storage and I/O devices, such as keyboards, printers, disk drives, etc. Serial interfaces transfer one bit at a time while parallel interfaces transfer several (typically 8) bits at a time. A typical serial interface is the "RS-232" serial interface. Typical parallel interfaces include the "Centronics" (printer), IDE, SCSI and GPIB interfaces.

### 1.9.9 DMA Controllers and Bus Arbitration

DMA (Direct Memory Access) allows peripherals to directly read or write memory or other peripherals on the system bus independently of the CPU. Special logic is required to ensure that only one device at time tries to use the bus.

### 1.9.10 Memory

Since microprocessors alone cannot accommodate the large amount of memory required to store program instructions and data, such as the text in a word-processing program, transistors can be used as memory elements in combination with the microprocessors.

#### 1.9.10.1 Random Access Memory

Integrated circuits, called random-access memory (RAM), which contain large numbers of transistors, are used in conjunction with the microprocessor to provide the needed memory. RAM is a volatile memory. There are different kinds of random-access memory. Static RAM (SRAM) holds information as long as power is turned on and is usually used as cache memory because it operates very quickly.

Another type of memory, dynamic RAM (DRAM), is slower than SRAM and must be periodically refreshed with electricity or the information it holds is lost. DRAM is more economical than SRAM and serves as the main memory element in most computers. Dynamic RAM stores data as electric charge in a capacitor. DRAM requires that every memory cell's content be "refreshed" periodically by recharging the capacitor. Circuits are required to ensure that the DRAM contents are periodically refreshed.

### 1.9.10.2 Read Only Memory

It is used to store the data permanently. It is a non-volatile memory, i.e. the contents of the memory cannot be changed by the programmer. There are different types of ROMs, such as masked ROM, PROM, EPROM, EEPROM and flash memory.

### 1.9.10.3 Cache Memory

Many modern microprocessors require faster access to memory than is possible with inexpensive memory devices. Fast auxiliary memories called cache memories are used to store the contents of frequently used memory locations and thus improve the overall performance of the system.

## 1.9.11 Buses

A bus is a group of related signals. Basically there are three types of buses, i.e. data bus, address bus and control bus. Most microprocessor systems include several buses. The signals appearing on the pins on the microprocessor chip are called the processor bus. Many microcomputers allow peripherals and memory to be placed on physically separate PC cards which plug into connectors on a "motherboard" or "backplane." The signals on these connectors are called the system bus. Examples of system buses include the ISA, VME and PCI buses. Peripherals (such as modems or disk drives) are connected to the computer using different connectors. The signals on these connectors are called the peripheral bus. Some common peripheral buses include the RS-232 serial interface and the SCSI (Small Computer Systems Interface) bus. These buses are often further divided into smaller buses. For example, the pins on the microprocessor and system bus can be grouped into a data bus, an address bus, a control bus, and a utility bus. A large part of this course is devoted to the study of typical processor, system and peripheral buses.

## 1.10 TYPES OF MICROPROCESSORS

Microprocessors are classified by the semiconductor technology of their design (TTL-transistor-transistor logic; CMOS-complementary-metal-oxide semiconductor; or ECL-emitter-coupled logic), by the width of the data format (4-bit, 8-bit, 16-bit, 32-bit or 64-bit) they process; and by their instruction set (CISC, complex-instruction-set computer or RISC, reduced instruction-set computer). TTL technology is most commonly used, while CMOS is favored for portable computers and other battery-powered devices because of its low power consumption. ECL is used where the need for its greater speed offsets the fact that it consumes the most power. Four-bit devices, while inexpensive, are good only for simple control applications; in general, the wider the data format, the faster and more expensive the device. CISC processors, which have 70 to several hundred instructions, are easier to program than RISC processors, but are slower and more expensive microprocessors, have been described in many different ways. They have been compared with the brain and the heart of humans. Their operation is to a switch board just as to the nervous system in an animal. They have often been called microcomputers. The purpose of the microprocessor was to control memory. Specifically, a microprocessor is "a component that implements memory".

## 1.11 MICROPROCESSOR PROGRAMMING LANGUAGES

*A programming language is a notation for writing programs, which are specifications of a computation or algorithm.* There is only one programming language that any computer can actually understand and execute its own native binary machine code. This is the lowest possible level of language in which it is possible to write a computer program. All other languages are said to be *high level* or *low level* according to how closely they can be said to resemble machine code.

In this context, a low-level language corresponds closely to machine code, so that a single low-level language instruction translates to a single machine-language instruction. A high-level language instruction typically translates into a series of machine-language instructions.

*Low-level languages have the advantage that they can be written to take advantage of any peculiarities in the architecture of the central processing unit (CPU) which is the "brain" of any computer.* Thus, a program written in a low-level language can be extremely efficient, making optimum use of both computer memory and processing time. However, to write a low-level program takes a substantial amount of time, as well as a clear understanding of the inner workings of the processor itself. Therefore, low-level programming is typically used only for very small programs or for segments of code that are highly critical and must run as efficiently as possible.

*High-level languages permit faster development of large programs.* The final program which is executed by the computer is not efficient, but the savings in programmer time.

This is because the cost of writing a program is nearly constant for each line of code, regardless of the language. Thus, a high-level language where each line of code translates to 10 machine instructions costs only one tenth as much in program development as a low-level language where each line of code represents only a single machine instruction.

In addition to the distinction between high-level and low-level languages, there is a further distinction between *compiler languages* and *interpreter languages*.

## 1.11.1 Machine Language

The very lowest possible level at which one can program a computer is in its own native machine code, consisting of strings of 1's and 0's and stored as binary numbers. The main problems with using machine code directly are that it is very easy to make a mistake, and very hard to find it once you realize the mistake has been made.

## 1.11.2 Assembly Language

*Assembly language is the symbolic representation of machine code*, which also allows symbolic designation of memory locations. Thus, an instruction to add the contents of a memory location to an internal CPU register called the *accumulator* might add a number instead of a string of binary digits (*bits*). No matter how close assembly language is to machine code, the computer still cannot understand it.

The assembly-language program must be translated into machine code by a separate program called an *assembler*. The assembler program recognizes the character strings that make up the symbolic names of the various machine operations, and substitutes the required machine code for each instruction. At the same time, it also calculates the required address in memory for each symbolic name of a memory location, and substitutes those addresses for the names. The final result is a machine-language program that can run on its own at any time; the assembler and the assembly language program are no longer needed.

To help distinguish between the "before" and "after" versions of the program, the original assembly language program is also known as the *source code*, while the final machine language program is designated as the *object code*.

If an assembly-language program needs to be changed or corrected, it is necessary to make the changes to the source code and then re-assemble it to create a new object program.

### 1.11.3 Compiler Language

Compiler languages are the high-level equivalent of assembly language. Each instruction in the compiler language can correspond to many machine instructions. Once the program has been written, it is translated to the equivalent machine code by a program called a *compiler*. Once the program has been compiled, the resulting machine code is saved separately, and can be run on its own at any time.

As with assembly-language programs, updating or correcting a compiled program requires that the original (source) program be modified appropriately and then recompiled to form a new machine-language (object) program.

Typically, the compiled machine code is less efficient than the code produced when using assembly language. This means that it runs a bit more slowly and uses a bit more memory than the equivalent assembled program. To offset this drawback, however, we also have the fact that it takes much less time to develop a compiler-language program, so it can be ready to go sooner than the assembly-language program.

### 1.11.4 Interpreter Language

*An interpreter language, like a compiler language*, is considered to be high level. However, it operates in a totally different manner from a compiler language. Rather, the interpreter program resides in memory, and directly executes the high-level program without preliminary translation to machine code.

This uses an interpreter program to directly execute the user's program, but it has both advantages and disadvantages. The primary advantage is that one can run the program to test its operation, make a few changes and run it again directly. There is no need to recompile because no new machine code is ever produced. This can enormously speed up the development and testing process.

On the down side, this arrangement requires that both the interpreter and the user's program reside in memory at the same time. In addition, because the interpreter has to scan the user's program *one line at a time* and execute internal portions of itself in response; execution of an interpreted program is much slower than for a compiled program.

### 1.12 MICROPROCESSOR APPLICATIONS

When microprocessors appeared, they were first used in computer systems for a negative reason. In the early 1970s, there were a few support chips and microprocessors were programmed to perform functions that are now done by a wide variety of hardware chips. For this reason, assembling a complete microprocessor-based system required both hardware and software expertise. Only five years later in 1976, companies realized that microprocessors could be used to build inexpensive personal computers. It then took several more years to manufacture computers that were adequate for business and professional purposes. Yet the technology had been there all along. Many of the early microprocessor applications found markets by accident rather than by design.

New product development had generally been a direct result of the dissemination of technical information. In the early 1970s, the necessary combination of hardware and software expertise was rarely found outside the computer manufacturing industry. This was not perceived as a problem because when microprocessors were introduced, the computer establishment saw them only as low-cost processors for simple control applications. In fact, the first 8-bit microprocessor, the Intel 8008, was designed for direct control of a CRT display. *Microprocessors are now used for controlling virtually every computer peripheral that does not require bipolar speeds.* Initially, such applications were limited by the relatively low speed of early microprocessors, but now, with the faster microprocessors coupled with specialized peripheral controller chips, such as CRT and floppy disk controllers, it is possible to control fast devices, such as CRTs and disks.

With microprocessors, we have now entered the era of distributed systems. In distributed systems, intercommunication between a numbers of processors is reduced to a minimum because they do not interact in real-time but exchange data words or block. Each processor is then a direct process controller that completely controls a process. Such network may involve multiple microprocessors. Traditionally, a multiprocessor system is one in which several processors interact with each other in real-time for control purposes. Most systems involving networks of microprocessors do not interact so closely and therefore do not qualify as *"multi-microprocessor systems."*

The widespread use of microprocessors to replace random logic has dramatically increased since the early 1980s. Microprocessors afford flexibility which is not available in conventional "hardwired" circuitry. Design and production costs of a single high-volume system can be amortized by using different programs to tailor the system to meet the diverse needs of several specific applications. Incorporating last-minute design changes is normally quicker and easier in software to that in hardware. Finally, many inexpensive microprocessors are now capable of speeds that are more than adequate for many products.

Microprocessors are utilized in computer systems ranging from notebook computers to small personal computers to supercomputer-class workstations. Programs include word processing, *electronic mail, spreadsheets, animation, graphics, and database processing.* Owing to their low cost and flexibility, microprocessors appear in many everyday household products, such as microwave ovens, handheld electronic games, washing machines, programmable video cassette recorders (VCRs) and programmable thermostats. Newer cars incorporate microprocessor controlled ignition and emission systems to improve engine operation, increasing fuel economy while reducing pollution. With the continuing progress of LSI technology, most microprocessor systems actually use multiple processors distributed over several chips. Processors can often be found in the peripheral chips of the system, i.e. the PIO, the UART or other system chips. This makes the programming tasks more difficult than with traditional systems; however, it does result in standardized systems, all of the traditional chips that were merely interface devices in the past are now fully programmable. Programmed instructions are sent to these devices by the microprocessor. These processors, residing in peripheral devices, should be considered as slaves.

### 1.12.1 Some Important Applications

1. Calculators
2. Accounting system
3. Games machine
4. Complex industrial controllers
5. Traffic light control
6. Data acquisition systems
7. Multi user, multi-function environments

8. Military applications

9. Communication systems

## 1.13 COMPANIES ASSOCIATED WITH MICROPROCESSORS

Overall, Intel Corporation dominated the microprocessor area even though other companies like Texas Instruments, Motorola, etc. also introduced some microprocessors. Listed below are the microprocessors that each company created.

### (A) Intel

As indicated previously, Intel Corporation dominated the microprocessor technology and is generally acknowledged as the company that introduced the microprocessor successfully into the market.

Its first microprocessor was the 4004 in 1971. The 4004 took the integrated circuit one step further by locating all the components of a computer (CPU, memory and input and output controls) on a minuscule chip. It evolved from a development effort for a calculator chip set. Previously, the IC had to be manufactured to fit a special purpose, now only one microprocessor could be manufactured and then programmed to meet any number of demands.

The 4004 microprocessor was the central component in a four-chip set, called the 4004 Family: 4001–2,048-bit ROM, a 4002–320-bit RAM and a 4003–10-bit I/O shift register. The 4004 had 46 instructions, using only 2,300 transistors in a 16-pin DIP. It ran at a clock rate of 740 kHz (eight clock cycles per CPU cycle of 10.8 microseconds)—the original goal was 1 MHz to allow it to compute BCD arithmetic as fast (per digit) as a 1960s era IBM 1620.

Following in 1972 was the 4040 which was an enhanced version of the 4004, with an additional 14 instructions, 8K program space and interrupt abilities (including shadows of the first 8 registers). In the same year, the 8008 was introduced. It had a 14-bit PC. The 8008 was intended as a terminal controller and was quite similar to the 4040. The 8008 increased the 4004's word length from four to eight bits, and doubled the volume of information that could be processed.

In April 1974, 8080, the successor to 8008 was introduced. It was the first device with the speed and power to make the microprocessor an important tool for the designer. It quickly became accepted as the standard 8-bit machine. It was the first Intel microprocessor announced before it was actually available. It represented such an improvement over existing designs that the company wanted to give customers adequate lead time to design the part into new products. The use of 8080 in personal computers and small business computers was initiated in 1975 by MITS's Altair microcomputer. A kit selling for $395 enabled many individuals to have computers in their own homes (Computer, 1996). Following closely, in 1976, was 8048, the first 8-bit single-chip microcomputer. It was also designed as a microcontroller rather than a microprocessor—low cost and small size was the main goal. For this reason, data was stored on-chip, while program code was external. The 8048 was eventually replaced by the very popular but bizarre 8051 and 8052 (available with on-chip program ROMs). While the 8048 used 1-byte instructions, the 8051 had a more flexible 2-byte instruction set, eight 8-bit registers plus an accumulator A. Data space was 128 bytes and could be accessed directly or indirectly by a register, plus another 128 above that in the 8052 which could only be accessed indirectly (usually for a stack).

In 1978, Intel introduced its high-performance, 16-bit MOS processor—the 8086. This microprocessor offered power, speed and features far beyond the second-generation machines of the mid-70's. It is said that the personal computer revolution did not really start until the 8088 processor was created. This chip became the most ubiquitous in the computer industry when IBM chose it for its first PC.

In 1982, the 80286 (also known as 286) was next and was the first Intel processor that could run all the software written for its predecessor, the 8088. Many novices were introduced to desktop computing with a "286 machine" and it became the dominant chip of its time. It contained 130,000 transistors.

In 1985, the first multi-tasking chip, the 386 (80386) was created. This multi-tasking ability allowed Windows to do more than one function at a time. This 32-bit microprocessor was designed for applications requiring high CPU performance. In addition to providing access to the 32-bit world, the 80386 addressed two other important issues: It provided system-level support to system designers, and it was object-code compatible with the entire family of 8086 microprocessors (Computer, 1996). The 80386 was made up of six functional units: (i) execution unit (ii) segment unit (iii) page unit (iv) decode unit (v) bus unit and (vi) prefetch unit. The 80386 had registers divided into such categories as general-purpose registers, debug registers and test registers. It had 275,000 transistors.

The 486 (80486) generation of chips really advanced the point-and-click revolution. It was also the first chip to offer a built-in math coprocessor, which gave the central processor the ability to do complex math calculations. The 486 had more than a million transistors. In 1993, when Intel lost a bid to trademark the 586, to protect its brand from being copied by other companies, it coined the name Pentium for its next generation of chips and there began the Pentium series—Pentium Classic, Pentium II, III and currently, IV.

### (B) Motorola

The MC68000 was the first 32-bit microprocessor introduced by Motorola in early 1980s. This was followed by higher levels of functionality on the microprocessor chip in the MC68000 series. For example, MC68020, introduced later, had three times as many transistors, was about three times as big and was significantly faster. Motorola 68000 was one of the second generation systems that were developed in 1973. It was known for its graphics capabilities. The Motorola 88000 (originally named the 78000) is a 32-bit processor, one of the first load-store CPUs based on a Harvard Architecture (Noyce, 1981).

### (C) Digital Equipment Corporation (DEC)

In March 1974, Digital Equipment Corporation (DEC) announced that it would offer a series of microprocessor modules built around the Intel 8008.

### (D) Texas Instruments (TI)

A precursor to these microprocessors was the 16-bit Texas Instruments 1900 microprocessor which was introduced in 1976. The Texas Instruments TMS370 is similar to the 8051, another of TI's creations. The only difference between the two was the addition of a B accumulator and some 16-bit support.

## PROBLEMS

1. What is microprocessor?

2. Explain microcomputer system.

3. What is the difference between RAM and ROM?

4. Explain the generation of microprocessor.

5. Write short note on Cache memory.

6. Explain the various applications of microprocessor based systems.

7. Define programming language. Explain different types of programming languages.

8. What are compiler, assembler and interpreter?

9. What is bus? Describe different types of buses.

10. What are the common support circuits of microprocessor?

# Digital System

## 2.1 DIGITAL AND ANALOG SIGNALS

Signals are form of energy. They can carry information and they can be any physical quantity that varies with time, space or any other independent variable. For example, alternating signal or sin wave whose amplitude varies with respect to time or the motion of a particle with respect to space can be considered as signals.

A system is a combination of different elements that performs an operation on a signal. For example, a rectifier circuit is used to convert AC signal in Pulsating DC. In this case, the rectifier performs some operation(s) on the signal, which has the effect of changing negative part of AC into positive part to get rectification.

Signals can be categorized in various ways; for example, discrete- and continuous-time domains. Discrete-time signals are defined only on a discrete set of times. Continuous-time signals are often referred to as continuous signals even when the signal functions are not continuous; an example is a square-wave signal.

Another category of signals is discrete-valued and continuous-valued or otherwise known as digital and analog signals. Digital signals are discrete-valued and analog signals are continuous electrical signals that vary in time as shown in Fig. 2.1. Analog devices and systems process signals whose voltages or other quantities vary in a continuous manner. They can take on any value across a continuous range of voltage, current or other metric.

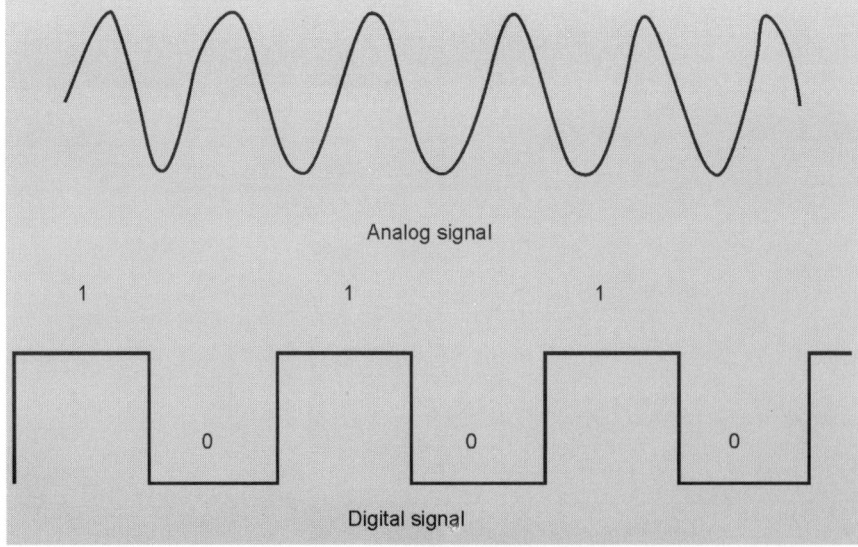

**Fig. 2.1:** Analog and digital signals

The analog signals can have an infinite number of values. Analog systems can be called wave systems. They have a value that changes steadily over time and can have any one of an infinite set of values in a range. Analog signals represent some physical quantity and they can be a model of the real quantity. Most of the time, the variations correspond to that of the non-electric (original) signal. For example, the telephone transmitter converts the sounds into an electrical voltage signal. The intensity of the voice causes electric current variations, therefore, the two are analogous, hence, the name analog. At the receiving end, the signal is reproduced in the same proportion, hence, the electric current is a model and is an electrical representation of one's voice.

Not all analog signals vary as smoothly as the waveform shown in Fig. 2.1. Digital signals are non-continuous, they change in individual steps. They consist of pulses or digits with discrete levels or values. The value of each pulse is constant, but there is an abrupt change from one digit to the next. Digital signals have two amplitude levels, the value of which are specified as one of two possibilities, such as 1 or 0, HIGH or LOW, TRUE or FALSE and so on. In reality, the values are anywhere within specific ranges and we define values within a given range.

A digital system is the one that handles only discrete values or signals. Any set that is restricted to a finite number of elements contains discrete information. The word digital describes any system based on discontinuous data or events. Digital is the method of storing, processing and transmitting information through the use of distinct electronic pulses that represent the binary digits 0 and 1. Examples of discrete sets are the 10 decimal digits, the 26 letters of the alphabet, etc. A digital system would be to flick the light switch on and off. There is no 'in between' values.

## 2.2 ADVANTAGES AND DISADVANTAGES OF DIGITAL SIGNALS

The usual advantages of digital circuits when compared to analog circuits are:

1. **Noise Margin (resistance to noise/robustness):** Digital circuits are less affected by noise. If the noise is below a certain level (the noise margin), a digital circuit behaves as if there was no noise at all. The stream of bits can be reconstructed into a perfect replica of the original source. However, if the noise exceeds this level, the digital circuit cannot give correct results.

2. **Error Correction and Detection:** Digital signals can be regenerated to achieve lossless data transmission within certain limits. Analog signal transmission and processing, by contrast, always introduces noise.

3. **Easily Programmable:** Digital systems interface well with computers and are easy to control with software. It is often possible to add new features to a digital system without changing hardware, and to do this remotely, just by uploading new software. Design errors or bugs can be worked-around with a software upgrade, after the product is in customer hands. A digital system is often preferred because of (re-) programmability and ease of upgrading without requiring hardware changes.

4. **Cheap Electronic Circuits:** More digital circuitry can be fabricated per square millimeter of integrated-circuit material. Information storage can be much easier in digital systems than in analog ones. In particular, the great noise-immunity of digital systems makes it possible to store data and retrieve it later without degradation. In an analog system, aging and wear and tear will degrade the information in storage, but in a digital system, as long as the wear and tear is below a certain level, the information can be recovered perfectly. Theoretically, there is no data-loss when copying digital

data. This is a great advantage over analog systems, which faithfully reproduce every bit of noise that makes its way into the signal.

Disadvantages of digital circuits in real world are:

1. The world in which we live is analog and signals from this world, such as light, temperature, sound, electrical conductivity, electric and magnetic fields, and phenomena, such as the flow of time, are for most practical purposes continuous and thus analog quantities rather than discrete digital ones. For a digital system to do useful things in the real world, translation from the continuous realm to the discrete digital realm must occur, resulting in quantization errors. This problem can usually be mitigated by designing the system to store enough digital data to represent the signal to the desired degree of fidelity. The Nyquist-Shannon sampling theorem provides an important guideline as to how much digital data is needed to accurately portray a given analog signal.

2. Digital systems can be fragile in that if a single piece of digital data is lost or misinterpreted, the meaning of large blocks of related data can completely change. This problem can be diminished by designing the digital system for robustness. For example, a parity bit or other error-detecting or error-correcting code can be inserted into the signal path so that minor data corruptions can be detected and possibly corrected.

3. Digital circuits use more energy than analog circuits to accomplish the same calculations and signal processing tasks, thus producing more heat as well. In portable or battery-powered systems, this can be a major limiting factor.

4. Digital circuits are made from analog components, and care has to be taken to all noise and timing margins, to parasitic inductances and capacitances, to proper filtering of power and ground connections, to electromagnetic coupling amongst data lines. Inattention to these can cause problems, such as "glitches"; pulses do not reach valid switching (threshold) voltages or unexpected (undecoded) combinations of logic states.

5. A corollary of the fact that digital circuits are made from analog components, is the fact that digital circuits are slower to perform calculations than analog circuits that occupy a similar amount of physical space and consume the same amount of power. However, the digital circuit will perform the calculation with much better repeatability, due to the high noise immunity of digital circuitry.

## 2.3 INTRODUCTION TO NUMBER SYSTEM

Number systems provide the basis for all operations in information processing systems. In a number system, the information is divided into a group of symbols; for example, 26 English letters, 10 decimal digits, etc. In conventional arithmetic, a number system based upon ten units (0 to 9) is used. However, arithmetic and logic circuits used in computers and other digital systems operate with only 0's and 1's because it is very difficult to design circuits that require ten distinct states. The number system with the basic symbols 0 and 1 is called binary, i.e. a binary system uses just two discrete values. The binary digit (either 0 or 1) is called a bit.

A group of bits which is used to represent the discrete elements of information is a symbol. The mapping of symbols to a binary value is known as a binary code. This mapping must be unique. For example, the decimal digits 0 through 9 are represented in a digital system with a code of four bits. Thus, a digital system is a system that manipulates discrete elements of information that is represented internally in binary form.

## 2.4 DECIMAL NUMBERS

The invention of decimal number system has been the most important factor in the development of science and technology. The decimal number system uses positional number representation, which means that the value of each digit is determined by its position in a number.

The base, also called the radix of a number system, is the number of symbols that the system contains. The decimal system has ten symbols: 0, 1, 2, 3, 4, 5, 6, 7, 8, 9. In other words, it has a base of 10. Each position in the decimal system is 10 times more significant than the previous position. The numeric value of a decimal number is determined by multiplying each digit of the number by the value of the position in which the digit appears and then adding the products. Thus, the number 4567 is interpreted as

$$4 \times 1000 + 5 \times 100 + 6 \times 10 + 7 \times 1 = 4000 + 500 + 60 + 7$$

Here 7 is the least significant digit (LSD) and 4 is the most significant digit (MSD).

In general, in a number system with a base or radix $r$, the digits used are from 0 to $r - 1$ and the number can be represented as

$$N = a_n r^n + a_{n-1} r^{n-1} + \ldots + a_1 r^1 + a_0 r^0 \quad \text{where, for } n = 0, 1, 2, 3, \ldots$$

$r$ = base or radix of the system
$a$ = number of digits having values between 0 and $r - 1$.

Above relation is for all integers and for the fractions (numbers between 0 and 1), the following equation holds:

$$N = a_{-1} r^{-1} + a_{-2} r^{-2} + \ldots + a_{-n+1} r^{-n+1} + a_{-n} r^{-n}$$

Thus, for decimal fraction 0.9248

$$N = 0.9000 + 0.0200 + 0.0040 + 0.0008$$

Where    $a - 1 = 9$
$a - 2 = 2$
$a - 3 = 4$
$a - 4 = 8$

## 2.5 BINARY NUMBERS

The binary number has a radix of 2. As $r = 2$, only two digits are needed, and these are 0 and 1. Like the decimal system, binary is a positional system, except that each bit position corresponds to a power of 2 instead of a power of 10. In digital systems, the binary number system and other number systems closely related to it are used almost exclusively. Hence, digital systems often provide conversion between decimal and binary numbers. The decimal value of a binary number can be formed by multiplying each power of 2 by either 1 or 0 followed by adding the values together.

**Example:** The decimal equivalent to the binary number 101100.

$$N = 101100$$
$$= 1 \times 2^5 + 0 \times 2^4 + 1 \times 2^3 + 1 \times 2^2 + 0 \times 2^1 + 0 \times 2^0$$
$$= 44$$

In binary, $r$ bits can represent $n = 2^r$ symbols, e.g. 3 bits can represent up to 8 symbols, 4 bits for 16 symbols, etc. For $N$ symbols to be represented, the minimum number of bits required is the lowest integer '$r$' that satisfies the relationship.

$$2^r > N$$

e.g. if $N = 9$, minimum $r$ is 5 since $2^3 = 8$ and $2^4 = 16$.

## 2.6 OCTAL NUMBERS

Digital systems operate only on binary numbers. Since binary numbers are often very long, two shorthand notations, octal and hexadecimal, are used for representing large binary numbers. Octal systems use a base or radix of 8. Thus, it has digits from 0 to 7 ($r - 1$). As in the decimal and binary systems, the positional value of each digit in a sequence of numbers is fixed. Each position in an octal number is a power of 8, and each position is 8 times more significant than the previous position.

**Example:** The decimal equivalent of the octal number 207.

$$N = 207$$
$$= 2 \times 8^2 + 0 \times 8^1 + 7 \times 8^0$$
$$= 135$$

## 2.7 HEXADECIMAL NUMBERS

The hexadecimal numbering system has a base of 16. There are 16 symbols. The decimal digits 0 to 9 are used as the first ten digits as in the decimal system, followed by the letters A, B, C, D, E and F, which represent the values 10, 11, 12, 13, 14 and 15, respectively. Table 2.1 shows the relationship among decimal, binary, octal and hexadecimal number systems.

Hexadecimal numbers are often used in describing the data in computer memory. A computer memory stores a large number of words, each of which is a standard size collection of bits. An 8-bit word is known as a **Byte.** A hexadecimal digit may be considered as half of a byte. Two hexadecimal digits constitute one byte, the rightmost

**Table 2.1:** Relationship among decimal, binary, octal and hexadecimal number systems

| Decimal | Binary | Octal | Hexadecimal |
|---------|--------|-------|-------------|
| 0 | 0000 | 0 | 0 |
| 1 | 0001 | 1 | 1 |
| 2 | 0010 | 2 | 2 |
| 3 | 0011 | 3 | 3 |
| 4 | 0100 | 4 | 4 |
| 5 | 0101 | 5 | 5 |
| 6 | 0110 | 6 | 6 |
| 7 | 0111 | 7 | 7 |
| 8 | 1000 | 10 | 8 |
| 9 | 1001 | 11 | 9 |
| 10 | 1010 | 12 | A |
| 11 | 1011 | 13 | B |
| 12 | 1100 | 14 | C |
| 13 | 1101 | 15 | D |
| 14 | 1110 | 16 | E |
| 15 | 1111 | 17 | F |

4 bits corresponding to half a byte, and the leftmost 4 bits corresponding to the other half of the byte. Often a half-byte is called nibble.

## 2.8 SIGNED NUMBERS

If "word" size is $n$ bits, there are $2n$ possible bit patterns, so only $2n$ possible distinct numbers can be represented. It implies that all possible numbers cannot be represented and some of these bit patterns (half?) to represent negative numbers. The negative numbers are generally represented with sign magnitude, i.e. reserve one bit for the sign and the rest of bits are interpreted directly as the number. For example, in a 4 bit system, 0000 to 0111 can be used to represent positive numbers from $+0$ to $+2^{n-1}$ and 1000 to 1111 can be used for negative numbers from $-0$ to $-2^{n-1}$. The two possible zeros are redundant and also it can be seen that such representations are arithmetically costly.

Another way to represent negative numbers are by radix and radix-1 complement (also called $r's$ and $(r-1)'s$). For example, $-k$ is represented as $R^n - k$. In the case of base 10 and corresponding 10's complement with $n = 2$, 0 to 99 are the possible numbers. In such a system, 0 to 49 is reserved for positive numbers and 50 to 99 are for positive numbers.

**Example:**

$$+8 = +8$$
$$-8 = 10^2 - 8 = 92$$

2's complement is a special case of complement representation. The negative number $-k$ is equal to $2n - k$. In 4 bits system, positive numbers 0 to $2^{n-1}$ is represented by 0000 to 0111 and negative numbers $-2^{n-1}$ to $-1$ is represented by 1000 to 1111. Such a representation has only one zero and arithmetic is easier. To negate a number, complement all bits and add 1.

**Example:**

$$44_{10} = 101100_2$$

Complementing bits will result

$$010011$$
$$\underline{+1} \qquad \text{add 1}$$
$$010100$$

that is $010100_2 = -44_{10}$

### 2.8.1 Properties of Two's Complement Numbers

1. $X$ plus the complement of $X$ equals 0.
2. There is one unique 0.
3. Positive numbers have 0 as their leading bit (MSB); while negatives have 1 as their MSB.
4. The range for an $n$-bit binary number in 2's complement representation is from $-2^{(n-1)}$ to $2^{(n-1)} - 1$.
5. The complement of the complement of a number is the original number.
6. Subtraction is done by addition to the 2's complement of the number.

### 2.8.2 Value of Two's Complement Numbers

For an n-bit 2's complement number, the weights of the bits are the same as for unsigned numbers except of the MSB. For the MSB or sign bit, the weight is $-2^{n-1}$. The value of the $n$-bit 2's complement number is given by:

$$A_{\text{2'S-complement}} = (a^{n-1}) \times (-2^{n-1}) + (a^{n-2}) \times (2^{n-1}) + ... + (a_1) \times (a^1) + a_0$$

For example, the value of the 4-bit 2's complement number 1010 is given by:

$$= 1 \times -2^3 + 0 \times 2^2 + 1 \times 2^1 + 0$$
$$= -8 + 0 + 2 + 0$$
$$= -6$$

An $n$-bit 2's complement number can be converted to an $m$-bit number where $m > n$ by appending $m-n$ copies of the sign bit to the left of the number. This process is called sign extension.

**Example:** To convert the 4-bit 2's complement number 1100 to an 8-bit representation, the sign bit (here = 1) must be extended by appending four 1's to left of the number:

$$1100_{\text{4-bit 2's-complement}} = 11111100_{\text{8-bit 2's-complement}}$$

To verify that the value of the 8-bit number is still –4; value of 8-bit number

$$= -2^7 + 2^6 + 2^5 + 2^4 + 2^3 + 2^2$$
$$= -128 + 64 + 32 + 16 + 8 + 4$$
$$= -128 + 124 = -4$$

Similar to decimal number addition, two binary numbers are added by adding each pair of bits together with carry propagation. An addition example is illustrated below:

| | |
|---|---|
| $X$ | 190 |
| $Y$ | 141 |
| $X + Y$ | 331 |

| | |
|---|---|
| 1 0 1 1 1 1 0 0 0 | Carry |
|   1 0 1 1 1 1 1 0 | $X$ |
| + 1 0 0 0 1 1 0 1 | $Y$ |
| 1 0 1 0 0 1 0 1 1 | |

Similar to addition, two binary numbers are subtracted by subtracting each pair of bits together with borrowing, where needed. For example:

| | |
|---|---|
| $X$ | 229 |
| $Y$ | 46 |
| $X - Y$ | 183 |

| | |
|---|---|
| 0 0 1 1 1 1 1 0 0 | Borrow |
| 1 1 1 0 0 1 0 1 | $X$ |
| 0 0 1 0 1 1 1 0 | $Y$ |
| 1 0 1 1 0 1 1 1 | $X - Y$ |

Two's complement addition/subtraction example

| | | | |
|---|---|---|---|
| 4 | 0100 | –2 | 1110 |
| –7 | 1001 | –6 | 1010 |
| –3 | 1101 | –8 | 11000 |

Overflow occurs if signs (MSBs) of both operands are the same and the sign of the result is different. Overflow can also be detected if the carry in the sign position is different from the carry out of the sign position. Ignore carry out from MSB.

## 2.9 NUMBER BASE CONVERSION

This section describes the conversion of numbers from one number system to another. Radix Divide and Multiply Method is generally used for conversion. There is a general procedure for the operation of converting a decimal number to a number in base $r$. If the number includes a radix point, it is necessary to separate the number into an integer part and a fraction part, since each part must be converted differently. The conversion of a decimal integer to a number in base $r$ is done by dividing the number and all successive quotients by $r$ and accumulating the remainders. The conversion of a decimal fraction is done by repeated multiplication by $r$ and the integers are accumulated instead of remainders.

Integer part—repeated divisions by $r$ yield LSD to MSD

Fractional part—repeated multiplications by $r$ yield MSD to LSD

**Example:** Conversion of decimal 29 to binary is by dividing decimal value by 2 (the base) until the value is 0.

Integer

| | | |
|---|---|---|
| 29 | | |
| 14 | 1 | LSB |
| 7 | 0 | |
| 3 | 1 | |
| 1 | 1 | |
| 0 | 1 | MSB |

The answer is $29_{10} = 11101_2$.

Divide number by 2; keep track of remainder; repeat with dividend equal to quotient until zero; first remainder is binary LSB and last is MSB.

The conversion from decimal integers to any base-$r$ system is similar to this above example, except that division is done by $r$ instead of 2.

**Example:** Convert $(0.7854)_{10}$ to binary.

$0.7854 \times 2 = 1.5708; a_{-1} = 1$

$0.5708 \times 2 = 1.1416; a_{-2} = 1$

$0.1416 \times 2 = 0.2832; a_{-3} = 0$

$0.2832 \times 2 = 0.5664; a_{-4} = 0$

The answer is $(0.7854)_{10} = (0.1100)_2$

Multiply fraction by two; keep track of integer part; repeat with multiplier equal to product fraction; first integer is MSB , last is the LSB; conversion may not be exact; a repeated fraction. The conversion from decimal fraction to any base-$r$ system is similar to this above example, except the multiplication is done by $r$ instead of 2.

The conversion of decimal numbers with both integer and fraction parts is done by converting the integer and the fraction separately and then combining the two answers.

$$\text{Thus, } (29.7854)_{10} = (11101. 1100)_2$$

For converting a binary number to octal, the following two-step procedure can be used.
1. Group the number of bits into 3's starting at least significant symbol. If the number of bits is not evenly divisible by 3, then add 0's at the most significant end.
2. Write the corresponding 1 octal digit for each group.

**Example:**

| 100 | 010 | 111 | (binary) |
|-----|-----|-----|----------|
| 4   | 2   | 7   | (octal)  |

| 10 | 101 | 110 | (binary) |
|----|-----|-----|----------|
| 2  | 5   | 6   | (octal)  |

Similarly, for converting a binary number to hex, the following two-step procedure can be used.
1. Group the number of bits into 4's starting at least significant symbol. If the number of bits is not evenly divisible by 4, then add 0's at the most significant end.
2. Write the corresponding 1 hex digit for each group.

**Example:**

| 1001 | 1110 | 0111 | 0000 | (binary) |
|------|------|------|------|----------|
| 9    | e    | 7    | 0    | (hex)    |

| 1 | 1111 | 1010 | 0011 | (binary) |
|---|------|------|------|----------|
| 1 | f    | a    | 3    | (hex)    |

The hex to binary conversion is very simple; just write down the 4 bit binary code for each hexadecimal digit.

**Example:**

| 3    | 9    | c    | 8    | (hex)    |
|------|------|------|------|----------|
| 0011 | 1001 | 1100 | 1000 | (binary) |

Similarly, for octal to binary conversion, write down the 8 bit binary code for each octal digit.

The hex to octal conversion can be carried out in two steps; first the hex to binary followed by the binary to octal. Similarly, decimal to hex conversion is completed in two steps; first the decimal to binary and from binary to hex as described above.

## 2.10 BOOLEAN ALGEBRA AND BASIC OPERATORS

Due to historical reasons, digital circuits are called switching circuits, digital circuit functions are called switching functions and the algebra is called switching algebra. The algebraic system known as Boolean algebra named after the mathematician George Boole. George Boole invented multi-valued discrete algebra (1854) and E.V. Huntington developed its postulates and theorems (1904). Historically, the theory of switching networks (or systems) is credited to Claude Shannon, who applied mathematical logic to describe relay circuits (1938). Relays are controlled electromechanical switches and they have been replaced by electronic controlled switches called logic gates. A special case of Boolean algebra known as Switching Algebra is a useful mathematical model for describing the combinational circuits. In this section, we will briefly discuss how the Boolean algebra is applied to the design of digital systems.

Examples of Huntington's postulates are given below:
1. **Closure**
   If $X$ and $Y$ are in set $(0, 1)$, then operations $X + Y$ and $X \cdot Y$ are also in set $(0, 1)$.
2. **Identity**
$$X + 0 = X \qquad X \cdot 1 = X$$
3. **Distributive**
$$X \cdot (Y + Z) = (X + Y) + (Y \cdot Z)$$
$$X + (Y \cdot Z) = (X + Y) \cdot (Y + Z)$$

## 4. Complement

$$X + \bar{X} = 1$$
$$X \cdot \bar{X} = 0$$

Note that for each property, one form is the dual of the other; (zeros to ones, ones to zeros, '.' operations to '+' operations, '+' operations to '.' operations).
From the above postulates, the following theorems could be derived.

## 5. Associative

$$X + (Y + Z) = (X + Y) + Z$$
$$X \cdot (Y \cdot Z) = (X \cdot Y) \cdot Z$$

## 6. Idempotence

$$X \cdot X = X$$
$$X + X = X$$

## 7. Absorption

$$X + (X \cdot Y) = X$$
$$X \cdot (X + Y) = X$$

## 8. Simplification

$$X + \left( \bar{X} \cdot Y \right) = X + Y$$
$$X \cdot \left( \bar{X} + Y \right) = X \cdot Y$$

## 9. Consensus

$$X \cdot Y + \bar{X} \cdot Z + Y \cdot Z = X \cdot Y + \bar{X} \cdot Z$$
$$(X + Y) \cdot \left( \bar{X} + Z \right) \cdot (Y + Z) = (X + Y) \cdot \left( \bar{X} + Z \right)$$

## 10. Adjacency

$$X \cdot Y + X \cdot \bar{Y} = X$$
$$(X + Y) \cdot \left( X + \bar{Y} \right) = X$$

## 11. Demorgans

$$\overline{X + Y} = \bar{X} \cdot \bar{Y}$$
$$\overline{X \cdot Y} = \bar{X} + \bar{Y}$$

In general form

$$\overline{F \left( \cdot, +, X_1, \dots X_n \right)} = G \left( +, \cdot, \overline{X_1}, \dots \overline{X_n} \right)$$

Very useful for complementing function expressions; for example

$$F = X + Y \cdot Z; \quad \bar{F} = \overline{X + Y \cdot Z}$$
$$\bar{F} = \bar{X} \cdot \overline{Y \cdot Z}; \quad F = \bar{X} \cdot \left( \bar{Y} + \bar{Z} \right)$$
$$\bar{F} = \bar{X} \cdot \bar{Y} + \bar{X} \cdot \bar{Z}$$

## 2.11 LOGICAL OPERATIONS

A set is a collection of objects (or elements) and for example, a set Z {0, 1} means that Z is a set containing two elements distinguished by the symbols 0 and 1. There are three primary operations AND, OR and NOT.

## NOT

It is unary complement or inversion operation, usually shown as over bar ( $\overline{X}$ ), other forms are $X'$ and $\sim X$ (Fig. 2.2).

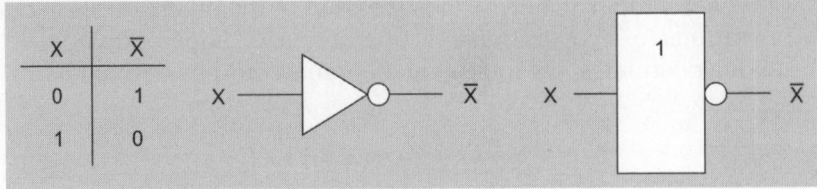

**Fig. 2.2:** NOT gate

## AND

Also known as the conjunction operation; output is true (1) only if all inputs are true. Algebraic operators are '.', 'and', '$\wedge$' (Fig. 2.3).

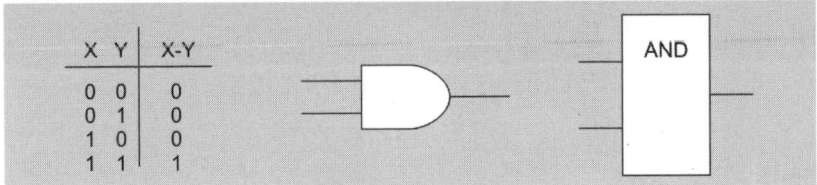

**Fig. 2.3:** AND gate

## OR

Also known as the disjunction operation; output is true (1) if any input is true. Algebraic operators are '+', '|', '$\vee$' (Fig. 2.4).

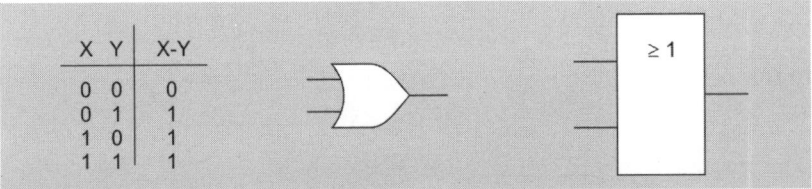

**Fig. 2.4:** OR gate

AND and OR are called binary operations because they are defined on two operands $X$ and $Y$. NOT is called a unary operation because it is defined on a single operand $X$. All of these operations are closed. That means if one applies the operation to two elements in a set $Z$ {0, 1}, the result will be always an element in the set B and not something else.

Like standard algebra, switching algebra operators have a precedence of evaluation. The following rules are useful in this regard:

1. NOT operations have the highest precedence.

2. AND operations are next.

3. OR operations are the lowest.

4. Parentheses explicitly define the order of operator evaluation and it is a good practice to use parentheses especially for situations which can cause doubt.

Note that in Boolean algebra, the operators AND and OR are not linear group operations; so one cannot solve equations by "adding to" of "multiplying" on both sides of the equal sign as is done with real, complex numbers in standard algebra.

## 2.11.1 Additional Logic Operations

For two inputs, there are 16 ways through which we can assign output values. Besides AND and OR, there are five other operations which are useful.

### BUFFER

The unary Buffer operation is useful in the real world (Fig. 2.5).

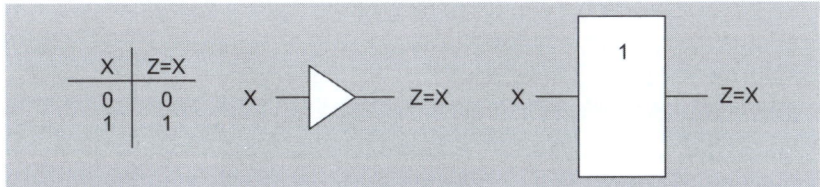

**Fig. 2.5:** Buffer

### NAND

NAND (NOT - AND ) is the complement of the AND operation (Fig. 2.6).

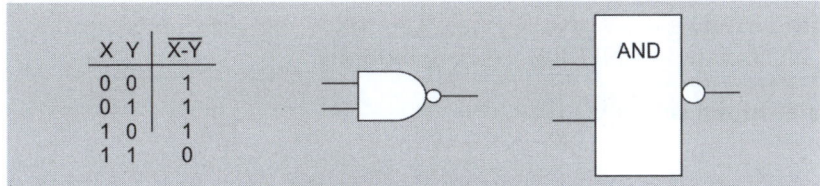

**Fig. 2.6:** NAND gate

### NOR

NOR (NOT - OR) is the complement of the OR operation (Fig. 2.7).

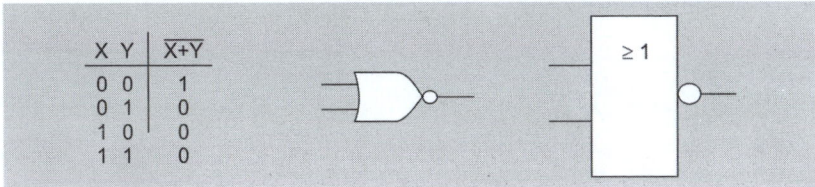

**Fig. 2.7:** NOR gate

### XOR

Exclusive OR is similar to the inclusive OR except output is 0 for 1. It is stated in other words as the output is 1 when modulo 2 input sum is equal to 1 (Fig. 2.8).

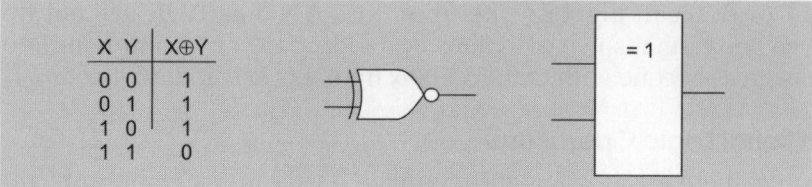

**Fig. 2.8:** XOR gate

## XNOR

Exclusive NOR is the complement of the XOR operation. Alternatively, the output is 1 when modulo 2 input sum is not equal to 1 (Fig. 2.9).

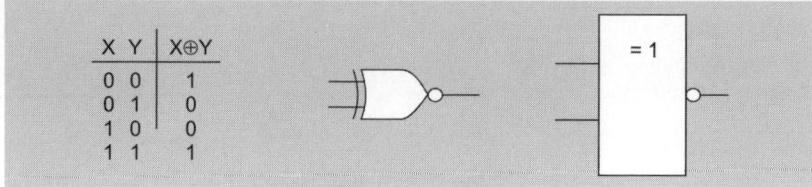

**Fig. 2.9:** XNOR gate

## 2.12 MINIMAL LOGIC OPERATOR SETS

AND, OR, NOT are all that is needed to express any combinational logic function as switching algebra expression. However, two other minimal logic operator sets are also possible with NAND gates or NOR gates. The following is a demonstration of how just NANDs or NORs can do AND, OR, NOT operations.

**NAND as a minimal set** (Fig. 2.10)

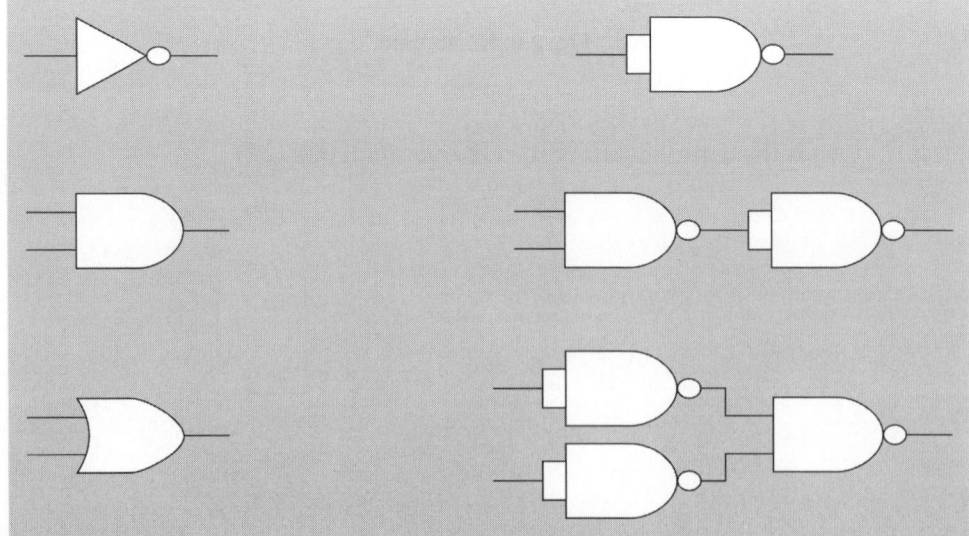

**Fig. 2.10:** Implementation of various gates using NAND

**NOR as a minimal set** (Fig. 2.11)

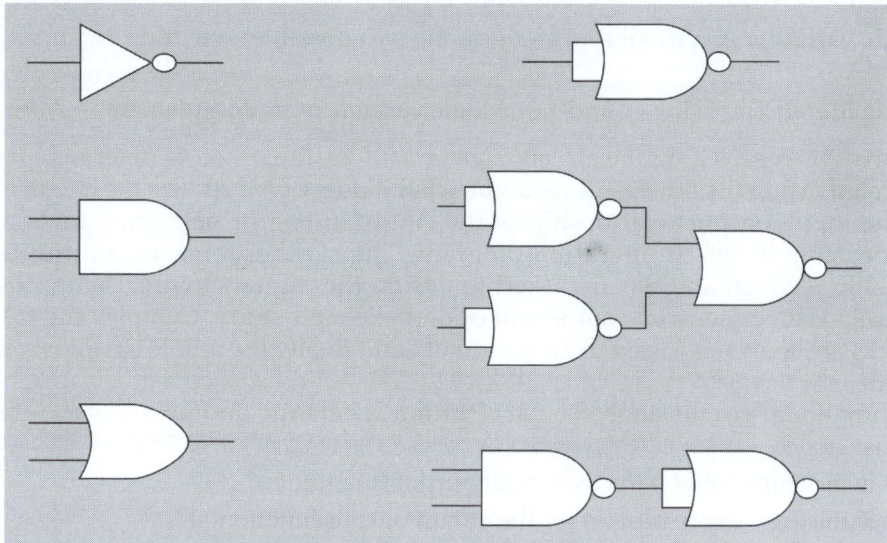

Fig. 2.11: Implementation of various gates using NOR

## 2.13 THREE STATE OUTPUTS

Standard logic gate outputs have only two states—high and low. Outputs are effectively either connected to +V or ground, that means there is always a low impedance path to the supply rails. Certain applications require a logic output that we can "turn off" or disable. It means that the output is disconnected (high impedance state). This is the three-state output and can be implemented by a standalone unit (a buffer) or part of another function output. This circuit is so-called tri-state because it has three output states: high (1), low (0) and high impedance (Z).

In the logic circuit of Fig. 2.12a, there is an additional switch to a digital buffer, which is called enabled input denoted by EN. When EN is low, the output is disconnected from the input circuit. When EN is high, the switch is connected and the circuit behaves like a digital buffer. All these states are listed in Truth table in Fig. 2.12b. Figure 2.12c depicts the symbol of a Tri-state buffer.

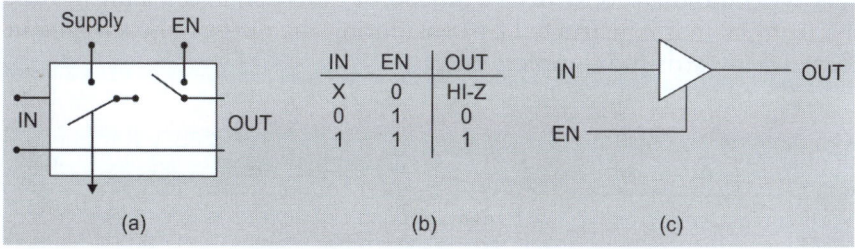

Fig. 2.12: (a) Switch configuration (b) Truth table and (c) Symbol of a tri-state buffer

### 2.13.1 Analyses and Synthesis of Combinational Logic Circuits

The important terms we are discussing in this section are as follows:

a. **Logic expression:** A mathematical formula consisting of logical operators and variables.

b. **Logic operator:** A function that gives a well-defined output according to switching algebra.

c. **Logic variable:** A symbol representing the two possible switching algebra values of 0 and 1.

d. **Logic literal:** The values 0 and 1 or a logic variable or its complement.

The analysis means a digital circuit is given and we are asked to determine its input–output relationship (its purpose, operation, what it does). One studies the circuit and then states the input–output relationship of the circuit in text or on a truth table or on an operation table or on an operation diagram. The synthesis means an input–output relationship is given and we are asked to design the digital circuit. The input–output relationship is a very crucial component of digital circuit study. Complex digital circuits and blocks are analyzed/designed individually and finally the whole circuit is analyzed/designed.

Combinational circuit analysis starts with a schematic and answers the following questions:

What is the truth table(s) for the circuit output function(s)?

What is the logic expression(s) for the circuit output function(s)?

Two types of analyses are possible: literal as well as symbolic analysis. Literal analysis is a process of manually assigning a set of values to the inputs, tracing the results and recording the output values. For '$n$' inputs, there are $2n$ possible input combinations. From input values, gate outputs are evaluated to form next set of gate inputs and evaluation continues until gate outputs are circuit outputs. The literal analysis only gives us the truth table.

Symbolic analysis also starts with the circuit diagram like literal analysis, but instead of assigning values, gate output expressions are determined. Intermediate expressions are combined in the following gates to form complex expressions. Symbolic analysis requires more work but gives us complete information of both the truth table and logic expression.

Now we will consider an example for the analysis of combinational logic circuit shown in Fig. 2.13.

Analyzing this circuit, it can be seen that
- Output of Gate $G1 = AB$
- Output of Gate $G2 = CD$
- Output of Gate $G3 = AB + CD$

From this, we could then construct a truth table (Table 2.2) to calculate the output of the circuit. The truth table is constructed by considering the output of each gate in turn and then building up towards the complete output.

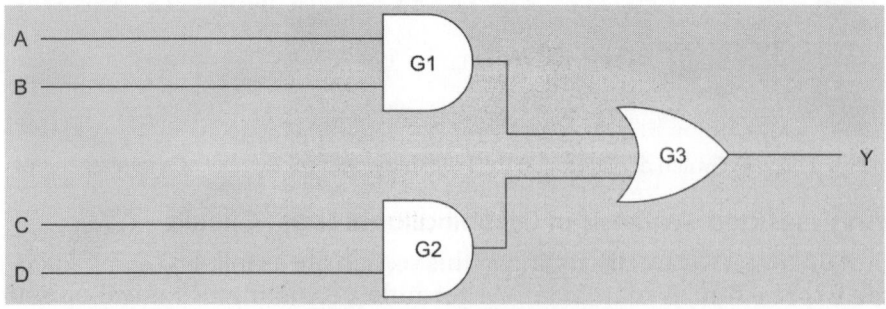

**Fig. 2.13:** Combinational circuit

Truth Table for given Boolean function is given in Table 2.2.

**Table 2.2:** Truth table

| A | B | C | D | AB | CD | AB + CD |
|---|---|---|---|----|----|---------|
| 0 | 0 | 0 | 0 | 0 | 0 | 0 |
| 0 | 0 | 1 | 0 | 0 | 1 | 0 |
| 0 | 1 | 0 | 0 | 1 | 0 | 0 |
| 0 | 1 | 1 | 0 | 1 | 1 | 0 |
| 1 | 0 | 0 | 1 | 0 | 0 | 1 |
| 1 | 0 | 1 | 1 | 0 | 1 | 1 |
| 1 | 1 | 0 | 1 | 1 | 0 | 1 |
| 1 | 1 | 1 | 1 | 1 | 1 | 1 |
| 0 | 0 | 0 | 0 | 0 | 0 | 0 |
| 0 | 0 | 1 | 0 | 0 | 1 | 0 |
| 0 | 1 | 0 | 0 | 1 | 0 | 0 |
| 0 | 1 | 1 | 0 | 1 | 1 | 0 |
| 1 | 0 | 0 | 1 | 0 | 0 | 1 |
| 1 | 0 | 1 | 1 | 0 | 1 | 1 |
| 1 | 1 | 0 | 1 | 1 | 0 | 1 |

Alternatively, the output of the circuit can be evaluated by substituting values directly into the logic equation.

For example, when $A = 1, B = 1, C = 1, D = 0$
then $Y = AB + CD = 1.1 + 1.0 = 1 + 0 = 1$

This can then be repeated for all other input combinations.

The analysis is followed by synthesis, i.e. we will consider how to design and implement a logic circuit to enable it to perform the desired specified operation. In this instance, we start with the equation and determine circuit to implement. For example, consider the logic function

$$X = AB + CDE$$

This is composed of two terms, $AB$ and $CDE$. The first term is formed by ANDing $A$ and $B$ and the second term is formed by ANDing together $C, D$ and $E$. These two terms are then ORed together. This can then be implemented using the AND and OR gates, as shown in Fig. 2.14. Generally, as the number of levels is increased, the overall delay is increased due to the contribution of propagation delays at each gate.

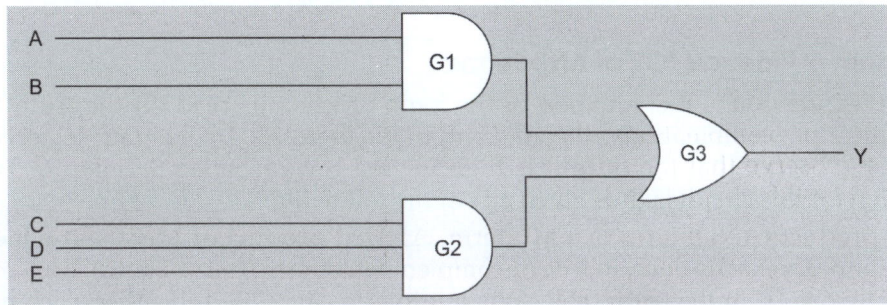

**Fig. 2.14:** Implementation of $X = AB + CDE$ using gates

Analysis can also be categorized into the functional analysis (determine what is computed) and the timing analysis (determine how long it takes to compute it). The logic expression is manipulated using Boolean (or switching) algebra and optimized to minimize the number of gates needed or to use specific type of gates.

## 2.14 CANONICAL AND STANDARD FORMS

A binary variable may be either in its true form $A$ or its complement $\bar{A}$. For $n$ variables, the maximum number of input variable combinations is given by $N = 2^n$. Then considering the AND gate, each of the $N$ logic expressions formed is called a standard product or minterm. As indicated in Table 2.3, binary digits '1' and '0' are taken to represent a given variable, for example, $A$ or its complement $\bar{A}$, respectively. Also from Table 2.3, note that each minterm is assigned a symbol $(P_j)$ each where $j$ is the decimal equivalent to the binary number of the minterm designated. Similarly, if we consider an OR gate, each of the $N$ logic expressions formed is called a standard sum or maxterm. In this case, binary digits '1' and '0' are taken to represent a given complemented variable $\bar{A}$ and its true form $A$, respectively. As shown in Table 2.3, a symbol $(S_j)$ is assigned to each maxterm where $j$ is the decimal equivalent to the binary number of the maxterm designated. Also observe that each maxterm is the complement of its corresponding minterm and vice versa.

The minterms and maxterms may be used to define the two standard forms for logic expressions, namely the sum of products (SOP) or sum of minterms, and the product of sums (POS) or product of maxterms. These standard forms of expression aid the logic circuit designer by simplifying the derivation of the function to be implemented. Boolean functions expressed as a **sum** of products or a product of sums are said to be in canonical form. Note that the POS is not the complement of the SOP expression.

**Table 2.3:** Input, minterms and maxterms

| Input | | | Minterms | | Maxterms | |
|---|---|---|---|---|---|---|
| $A$ | $B$ | $C$ | Terms | Designation | Terms | Designation |
| 0 | 0 | 0 | $\bar{A}\bar{B}\bar{C}$ | $P_0$ | $A+B+C$ | $S_0$ |
| 0 | 0 | 1 | $\bar{A}\bar{B}C$ | $P_1$ | $A+B+\bar{C}$ | $S_1$ |
| 0 | 1 | 0 | $\bar{A}B\bar{C}$ | $P_2$ | $A+\bar{B}+C$ | $S_2$ |
| 0 | 1 | 1 | $\bar{A}BC$ | $P_3$ | $A+\bar{B}+\bar{C}$ | $S_3$ |
| 1 | 0 | 0 | $A\bar{B}\bar{C}$ | $P_4$ | $\bar{A}+B+C$ | $S_4$ |
| 1 | 0 | 1 | $A\bar{B}C$ | $P_5$ | $\bar{A}+B+\bar{C}$ | $S_5$ |
| 1 | 1 | 0 | $AB\bar{C}$ | $P_6$ | $\bar{A}+\bar{B}+C$ | $S_6$ |
| 1 | 1 | 1 | $ABC$ | $P_7$ | $\bar{A}+\bar{B}+\bar{C}$ | $S_7$ |

### 2.14.1 Sum of Products (OR of AND Terms)

The SOP expression is the equation of the logic function as read off the truth table to specify the input combinations when the output is a logical 1. To illustrate, let us consider Table 2.4. Observe that the output is high for the rows labelled 3, 5 and 6. The SOP expression for this circuit is thus given any of the following:

Each product (AND) term is a Minterm. ANDed product of literals in which each variable appears exactly once, in true or complemented form (but not both). Each minterm has exactly one '1' in the truth table. When minterms are ORed together, each minterm contributes a '1' to the final function. Note that all product terms are not minterms.

**Table 2.4:** Sum of products and product of sums

| Input | | | Minterms | | Maxterms | | Output Y |
|---|---|---|---|---|---|---|---|
| A | B | C | Terms | Designation | Terms | Designation | |
| 0 | 0 | 0 | $\bar{A}\bar{B}\bar{C}$ | $P_0$ | $A+B+C$ | $S_0$ | 0 |
| 0 | 0 | 1 | $\bar{A}\bar{B}C$ | $P_1$ | $A+B+\bar{C}$ | $S_1$ | 0 |
| 0 | 1 | 0 | $\bar{A}B\bar{C}$ | $P_2$ | $A+\bar{B}+C$ | $S_2$ | 0 |
| 0 | 1 | 1 | $\bar{A}BC$ | $P_3$ | $A+\bar{B}+\bar{C}$ | $S_3$ | 1 |
| 1 | 0 | 0 | $A\bar{B}\bar{C}$ | $P_4$ | $\bar{A}+B+C$ | $S_4$ | 0 |
| 1 | 0 | 1 | $A\bar{B}C$ | $P_5$ | $\bar{A}+B+\bar{C}$ | $S_5$ | 1 |
| 1 | 1 | 0 | $AB\bar{C}$ | $P_6$ | $\bar{A}+\bar{B}+C$ | $S_6$ | 1 |
| 1 | 1 | 1 | $ABC$ | $P_7$ | $\bar{A}+\bar{B}+\bar{C}$ | $S_7$ | 0 |

1. $F = \bar{A}\cdot B\cdot C + A\cdot\bar{B}\cdot C + A\cdot B\cdot\bar{C}$
2. $F = P_3 + P_5 + P_6$
3. $F = (A,B,C) = \Sigma(3,5,6)$

## 2.14.2 Product of Sums (AND of OR Terms)

The POS expression is the equation of the logic function as read off the truth table to specify the input combinations when the output is a logical 0. To illustrate, let us again consider Table 2.4. Observe that the output is low for the rows labeled 0, 1, 2, 4 and 7. The POS expression for this circuit is thus given by any of the following:

## 2.15 MINIMIZATION

The goal of logic expression minimization is to find an equivalent of an original logic expression that has fewer variables per term, has fewer terms and needs less logic to implement. There are three main manual methods used for logic expression minimization: algebraic minimization, Karnaugh Map minimization and Quine-McCluskey (tabular) minimization.

### 2.15.1 Algebraic Minimization

The algebraic minimization process is the application of switching algebra postulates, laws and theorems to transform the original expression. It is hard to recognize when a particular law can be applied and difficult to know if resulting expression is truly minimal. The incorrect implementation or dropped variables, etc. can easily lead to a mistake.

The following are two examples of the algebraic minimization process by exploiting the adjacency theorem. Look for two terms that are identical except for one variable in the following expression:

$$A\cdot B\cdot C\cdot\bar{D} + A\cdot B\cdot C\cdot D$$

Application removes one term and one variable from the remaining term.

$$A\cdot B\cdot C\cdot\bar{D} + A\cdot B\cdot C\cdot D = A\cdot B\cdot C$$

$$(A\cdot B\cdot C)\cdot\bar{D} + (A\cdot B\cdot C)\cdot C\cdot D = A\cdot B\cdot C$$

$$(A\cdot B\cdot C)\cdot(\bar{D}+D) = (A\cdot B\cdot C)\cdot 1 = A\cdot B\cdot C$$

In the following example, one can look for the adjacency

$$F = \overline{A_3}A_2\overline{A_1}A_0 + \overline{A_3}A_2A_1\overline{A_0} + \overline{A_3}A_2A_1A_0 + A_3\overline{A_2}A\overline{A_0} + A_3\overline{A_2}\,\overline{A_1}A_0$$

The first and third term differ only $A_1$ and $\overline{A}$

The third and fourth term differ only $A_0$ and $A_0$

The second and third term differ only $A_0$ and $A_0$

Duplicate third term and rearrange

$$F = \overline{A_3}A_2\overline{A_1}A_0 + \overline{A_3}A_2A_1A_0 + \overline{A_3}A_2A_1\overline{A_0} + \overline{A_3}A_2A_1A_0 + A_3\overline{A_2}\,\overline{A_1}A_0 + A_3\overline{A_2}\,\overline{A_1}A_0$$

Apply adjacency on term pairs

$$F = \overline{A_3}A_2A_0 + \overline{A_3}A_2A_1 + A_3\overline{A_2}\,\overline{A_1}$$

## 2.15.2 Karnaugh Map (K-Map) Minimization

The Karnaugh map provides a systematic method for simplifying a Boolean expression or a truth table function. The $K$-map can produce the simplest SOP or POS expression possible. $K$-map procedure is actually an application of adjacency and guarantees a minimal expression. It is easy to use, visual, fast and familiarity with Boolean laws is not required. The $K$-map is a table consisting of $N = 2^n$ cells, where $n$ is the number of input variables. Assuming the input variables are $A$ and $B$, then the $K$-map illustrating the four possible variable combinations is shown in Fig. 2.15.

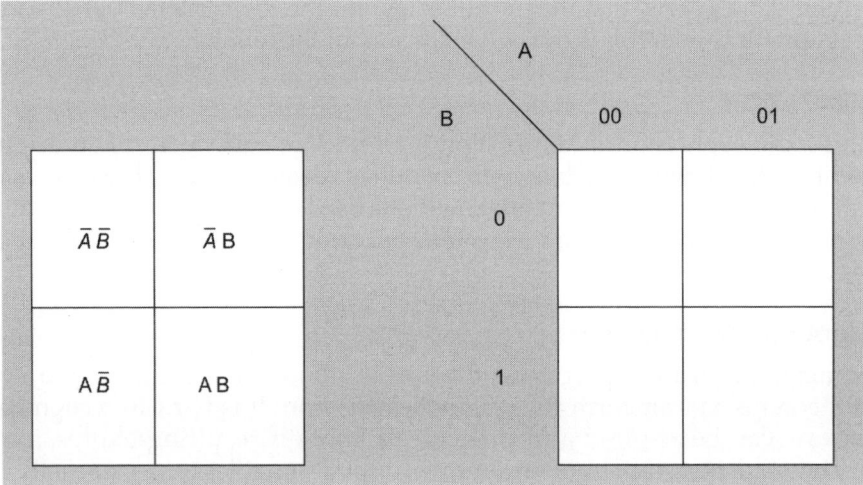

**Fig. 2.15:** Two variable $K$-maps

Similarly, three variable and four variable $K$-maps can be constructed as shown in Fig. 2.16.

For a SOP expression, each cell represents one particular combination of the variables in product form. The table format is such that there is a single variable change between any adjacent cells. This is the characteristic that will determine adjacency. This method is typically applicable to limited number of variables (4 ~ 8) and for $n > 5$, the $K$-map technique becomes impractical unless implemented on computer. Manual errors are possible in translation from Truth Table to $K$-map or when grouping of cells not done correctly.

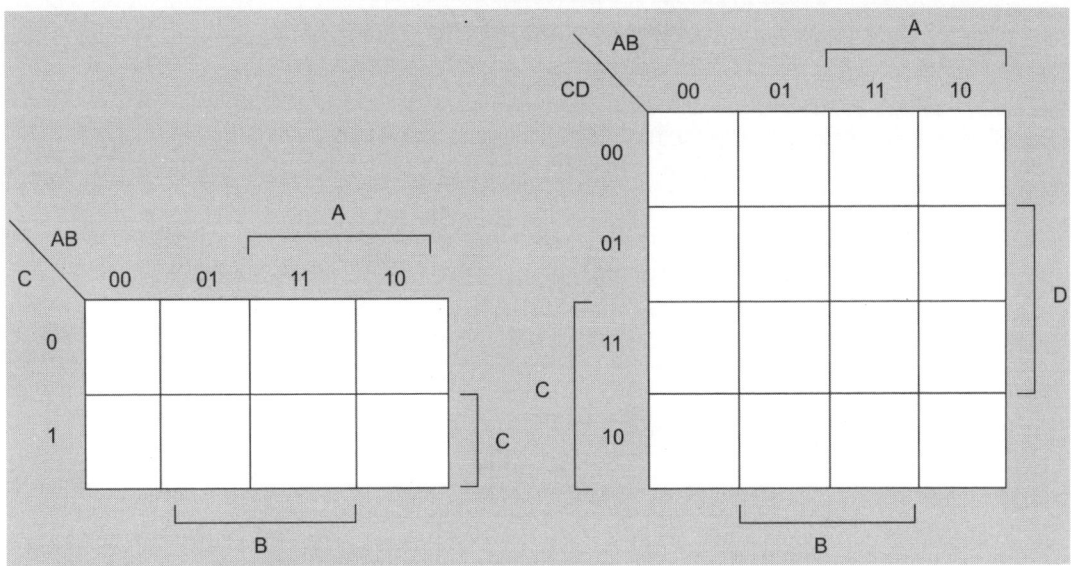

**Fig. 2.16:** Three variable and four variable *K*-maps

Basic K-map is a 2-D rectangular array of cells, each K-map represents one bit column of output and each cell contains one bit of output function. The arrangement of cells in array facilitates recognition of adjacent terms and adjacent terms differ in one variable value; equivalent to difference of one bit of input row values, e.g. m5 (101) and m7 (111). The standard Truth Table ordering does not show adjacency. One uses gray code for row order, however, it is still hard to see all possible adjacencies. For any cell in 2-D array, there are four direct neighbors (top, bottom, left, right). The 2-D array can therefore show adjacencies of up to four variables. One should not forget that cells are adjacent, top to bottom and side to side. The number of TT rows must match number of K-map cells. Watch out for ordering of 10 and 11 rows and columns.

To simplify a SOP for a Boolean expression using a *K*-map, first identify all the input combinations that produce an output of logic level 1 and place them in their appropriate *K*-map cell. Consequently, all other cells must contain zero (0). Second, group the adjacent cells that contain 1 in a manner that maximizes the size of the groups but also minimizes the total number of groups. All 1's in the output must be included in a group even if the group is only one cell. Third, as each SOP term represents an AND expression, each (AND) grouping is written with only the input variables that are common to the group. Finally, the simplified expression is formed by ORing each of the (AND) groups.

When the input combinations are irrelevant or cannot occur, the output states are in the Truth Table and the *K*-maps are filled with an X and are referred to as don't care states. The don't cares can work to our advantage during minimization; we can assign either 0 or 1 as needed. When simplifying *K*-maps with do not care states, the contents of the undefined cells (1 or 0) are chosen according to preference. The aim is to enlarge group sizes, thereby eliminating as many input variables from the simplified expression as possible. Only those X's that assist in simplifying the function should be included in the groupings. No additional X's should be added that would result in additional terms in the expression. To illustrate, let us consider the function specified by Table 2.5 and its corresponding *K*-map shown in Fig. 2.17. Note that the two groupings determine that the simplified expression is expressed as

$$F = C + \overline{A}\,\overline{B}$$

**Table 2.5:** Truth table of the function

| Input | | | Output |
|---|---|---|---|
| A | B | C | F |
| 0 | 0 | 0 | 1 |
| 0 | 0 | 1 | 1 |
| 0 | 1 | 0 | 0 |
| 0 | 1 | 1 | 1 |
| 1 | 0 | 0 | 0 |
| 1 | 0 | 1 | × |
| 1 | 1 | 0 | × |
| 1 | 1 | 1 | × |

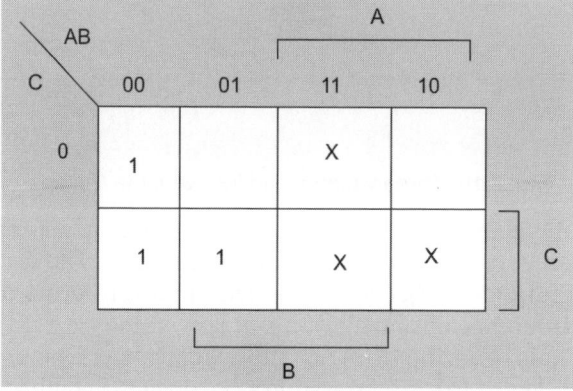

**Fig. 2.17:** $K$-Map for function $F = C + \bar{A}\bar{B}$

The minimum SOP expression for a function consists of some (but not necessarily all) of the prime implicants of a function. In other words, a SOP expression containing a term, which is not a prime implicant, cannot be the minimum. This is true because if a non-prime term were present, the expression could be simplified by combining the non-prime term with additional minterms. Any set of implicants that encloses (covers) all values is "sufficient", i.e. the associated logical expression represents the desired function. For example, all minterms or maxterms are sufficient. However, the smallest set of prime implicants that covers all values, forms a minimal expression for the desired function. There may be more than one minimal set.

### 2.15.3 Five Variable $K$-Map

A five variable $K$-map can be constructed using two 4 variable maps side-by-side. The groups spanning both maps occupy the same place in both maps (Figs 2.18 and 2.19).

When checking for adjacencies, each term should be checked against the five possible adjacent squares.

### 2.16 FLIP-FLOPS

The memory elements in a sequential circuit are called flip-flops, which are capable of storing 1-bit of information. A flip-flop is a synchronous version of the latch. "Flip-flop" is the common name given to two-state devices, which offer basic memory for sequential logic operations. Flip-flops are heavily used for digital data storage and transfer and are commonly used in banks called "registers" for the storage of binary numerical data.

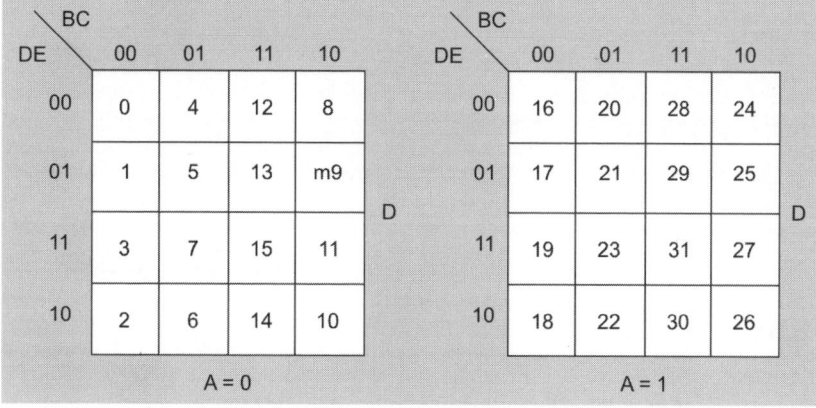

**Fig. 2.18:** 5 variable $K$-map

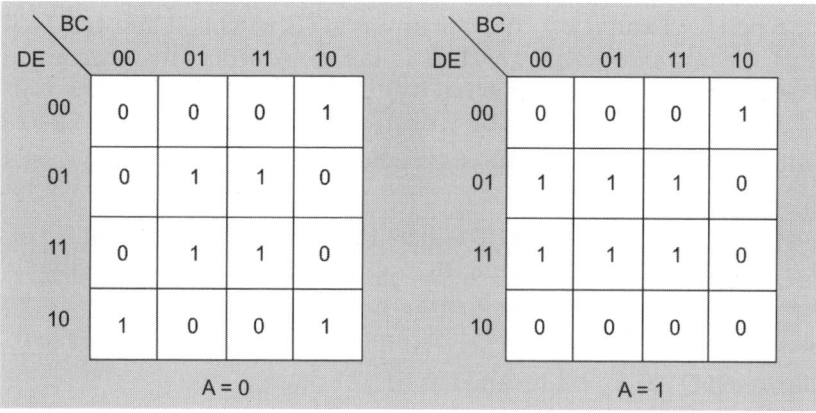

**Fig. 2.19:** $K$-map of $f(A, B, C, D, E) = \Sigma m$ (2, 5, 7, 8, 10, 13, 15, 17, 19, 21, 23, 24, 29, 31)

A flip-flop circuit has two outputs, one for the normal value and one for the complement value of the stored bit. Binary information can enter a flip-flop in a variety of ways and gives rise to different types of flip-flops. A flip-flop is usually controlled by one or two control signals and/or a gate or clock signal. The main difference between latches and flip-flops is the method used to change their states. Latches are level sensitive or level-triggered. This means that the outputs are dependent on the voltage level applied, not on any signal transition. Flip-flops are edge-triggered, that is, that they depend on the transition of a signal. This may either be a LOW-to-HIGH (rising edge) or a HIGH-to-LOW (falling edge) transition.

### 2.16.1 SR Flip-Flop

$S$-$R$ (Set-Reset) flip-flop can be constructed from NAND gates as shown in Fig. 2.20.

In the circuit shown in Fig. 2.20, there are three inputs, $S$ (Set), $R$ (Reset) and $Clk$ (Clock). In this circuit, $Clk$ input acts as an enable signal for the other two inputs, i.e. $S$ and $R$ inputs. The circuit operates normally when $Clk$ input is 1 and when it is 0, the circuit remains in its previous state. The information from $S$ and $R$ input is allowed to reach the output only when the $Clk$ input is 1.

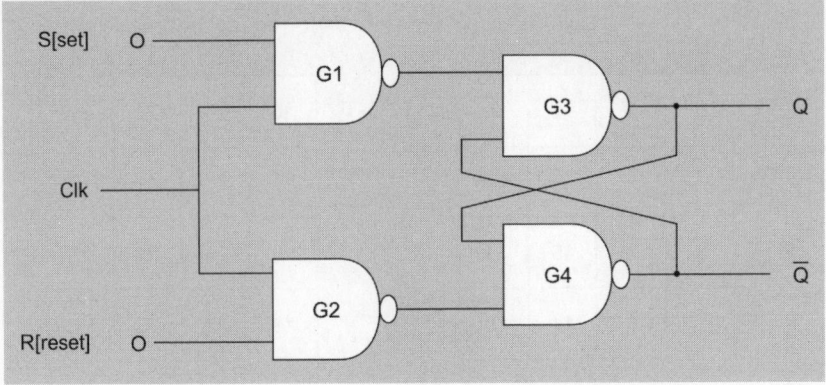

**Fig. 2.20:** *SR* flip-flop constructed from NAND gates

When $Clk = 1$, we can analyze the working of $SR$ flip-flop shown in Fig. 2.20 as follows:

1. **Case (i):** When $S = 1$ and $R = 0$, then the inputs at $G1$ will be 11 and at $G2$ will be 10 and as a result, the output of $G1$ is 0 and $G2$ is 1. These outputs then become the input to $G3$ and $G4$, respectively. We know that if any one input to NAND gate is 0, the output will be 1, so the output of $G3$ will be 1, i.e. $Q = 1$. Now the input at $G4$ becomes 11 and its output goes to 0, i.e. $\bar{Q} = 0$. This state when $Q = 1$, $\bar{Q} = 0$ is called as SET State.

2. **Case (ii):** Now when the input changes from $S = 1$ and $R = 0$ to $S = 0$ and $R = 1$, then the inputs at $G1$ will be 01 and at $G2$ will be 11 and as a result, the output of $G1$ is 1 and $G2$ is 0. These outputs then become the input to $G3$ and $G4$, respectively. We know that if any one input to NAND gate is 0, the output will be 1, so the output of $G3$ will be 1, i.e. $Q = 0$. Now the input at $G4$ becomes 11 and its output goes to 0, i.e. $\bar{Q} = 1$. This state when $Q = 0$ $\bar{Q} = 1$ is called as RESET State.

3. **Case (iii):** When from any of the two states explained above, the input changes to $S = 0$ and $R = 0$, then we see that the output remains the same as was in previous state, i.e. there is no change in the output.

When inputs change to $S = R = 1$, then an indeterminate condition occurs. In this condition, both the outputs $Q$ and $\bar{Q}$ are 0, which is not possible.

Table 2.6 shows the truth table or characteristic table of $S$-$R$ flip-flop.

| **Table 2.6:** Characteristic table of *S-R* flip-flop | | | |
|---|---|---|---|
| $S$ | $R$ | $Q$ | $\bar{Q}$ |
| 1 | 0 | 1 | 0 (SET) |
| 0 | 0 | 1 | 0 (No change) |
| 0 | 1 | 0 | 1 (RESET) |
| 0 | 0 | 0 | 1 (No change) |
| 1 | 1 | Not allowed | |

The standard truth table is given in Table 2.7.

In Table 2.7, output $Q(t + 1)$ represents next state and $Q(t)$ represents present state. When inputs are 00, then the output in the next state will be the same as present state, i.e. $Q(t +1) = Q(t)$ and so on.

**Table 2.7:** Standard truth table

| S | R | Q (t + 1) |
|---|---|-----------|
| 0 | 0 | Q(t) |
| 1 | 0 | 1 |
| 0 | 1 | 0 |
| 1 | 1 | Not allowed |

### 2.16.2 Present and Clear

Initially when the device is switched ON, then the state of the device is undefined and so is the states of the memory element, i.e. flip-flop, so to avoid this condition, we assign the initial state to the flip-flop using two asynchronous inputs Preset ($Pr$) and Clear ($Cr$). These inputs can be used along with the Clock but are not in synchronization with it. When both $Pr$ and $Cr$ inputs are High, i.e. $Pr = Cr = 1$, then the flip-flop operates normally. When $Pr = 1$ and $Cr = 0$, then the flip-flop is RESET. When $Pr = 0$ and $Cr = 1$ then the flip-flop is SET. The condition $Pr = Cr = 0$ must be avoided as in this condition, the state of the flip-flop will be uncertain (Table 2.8).

**Table 2.8:** State of flip-flop

| Pr | Cr | Operation of flip-flop |
|----|----|-----------------------|
| 1 | 1 | Normal operation |
| 1 | 0 | Reset  ($Q = 0$) |
| 0 | 1 | Set    ($Q = 1$) |

### 2.16.3 S-R Flip-Flop with Present and Clear

Block diagram of clocked $S$-$R$ flip-flop is shown in Fig. 2.21. It consists of Clock ($Clk$), Preset ($Pr$) and Clear ($Cr$) inputs along with the Set ($S$) and Reset ($R$) inputs.

Clock signal is used for synchronization in a circuit. The circuit operates when the clock signal is HIGH, i.e. $Clk = 1$. Preset and Clear inputs are applied to the flip-flop with a

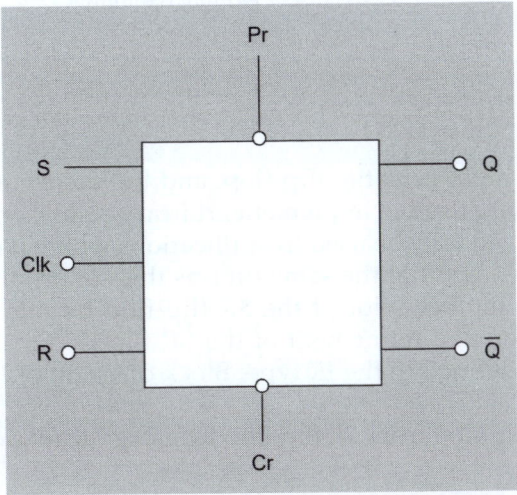

**Fig. 2.21:** Block diagram of clocked S-R flip-flop

bubble which signifies that these signals are active LOW, i.e. they are activated when their values are 0. A clocked *SR* flip-flop with *Clk*, *Cr* and *Pr* inputs is shown in Fig. 2.22.

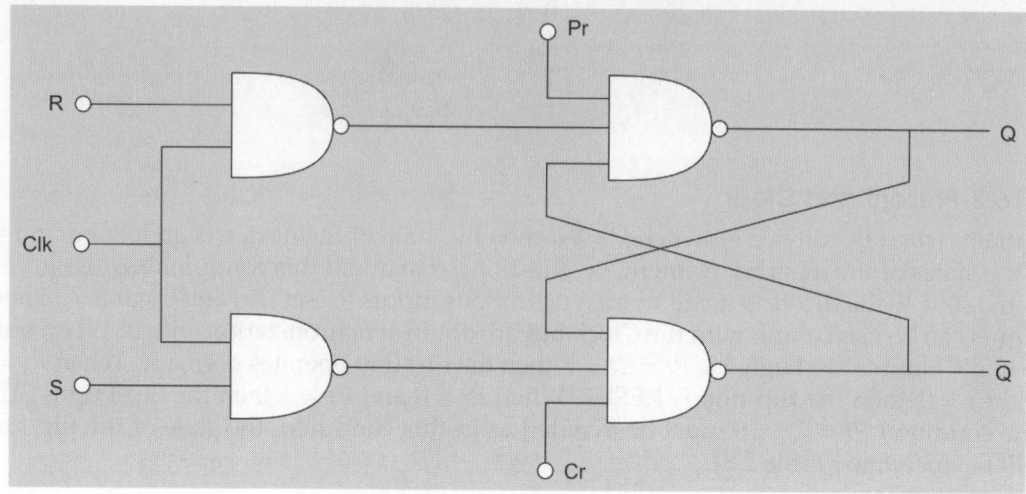

**Fig. 2.22:** Clocked SR flip-flop

*SR* flip-flop is activated, i.e. the flip-flop to the inputs only when the clock is high (*Clk* =1). The operation of *SR* flip-flop is similar to *SR* latch when *Clk* = 1 and *Pr* = *Cr* = 1. When *Pr* = 0 and *Cr* = 1 then the flip-flop is SET, i.e. *Q* = 1 and when *Pr* = 1 and *Cr* = 0 then the circuit is RESET, i.e. *Q* = 0. Pr and Cr signals are not in synchronization with the Clock pulse. They can operate both in the presence as well as in the absence of the Clock pulse. Truth table for *SR* flip-flop is given in Table 2.9.

In Table 2.9, the don't care condition (*X*) of *Clk* input represents that *Pr* and *Cr* inputs can override the *Clk* input, i.e. the circuit can be set or reset even in the presence of the *Clk* input.

**Table 2.9:** Truth table for *SR* flip-flop

| Clk | Pr | Cr | Output |
|-----|-----|-----|--------|
| 1 | 1 | 1 | Normal operation |
| X | 1 | 0 | 1(Preset) |
| X | 0 | 1 | 0 (Clear) |

### 2.16.4 *JK* Flip-Flop

The *JK* flip-flop is the most versatile flip-flop, and the most commonly used flip-flop when discrete devices are used to implement arbitrary state machines. The *SR* flip-flop has a limitation that it can only be used in applications where it can be guaranteed that both *S* and *R* cannot be logic 1 at the same time as this condition is indeterminate. The *JK* flip-flop augments the behavior of the *SR* flip-flop by interpreting the *S* = *R* = 1 condition. A *JK* flip-flop is a refinement of the *SR* flip-flop in that the indeterminate state of the *SR* type is defined in the *JK* type. Block diagram of *JK* flip-flop is shown in Fig. 2.23.

We can obtain a *JK* flip-flop from *SR* flip-flop by using the following equations;

$$S = J\overline{Q}$$

$$R = KQ$$

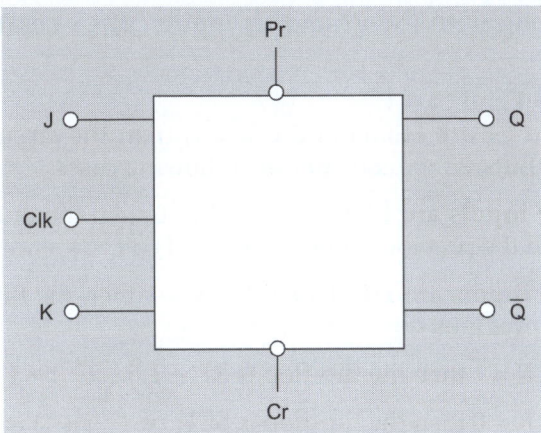

**Fig. 2.23:** Block diagram of *JK* flip-flop

A *JK* flip-flop obtained from such operation is shown in Fig. 2.24.

Figure 2.24(a) shows the basic block diagram of the conversion of SR flip-flop to *JK* flip-flop. In this block diagram, the inputs of *SR* flip-flop are ANDed with the outputs according to the equation defined. Figure 2.24(b) shows the logic circuit of the *JK* flip-flop

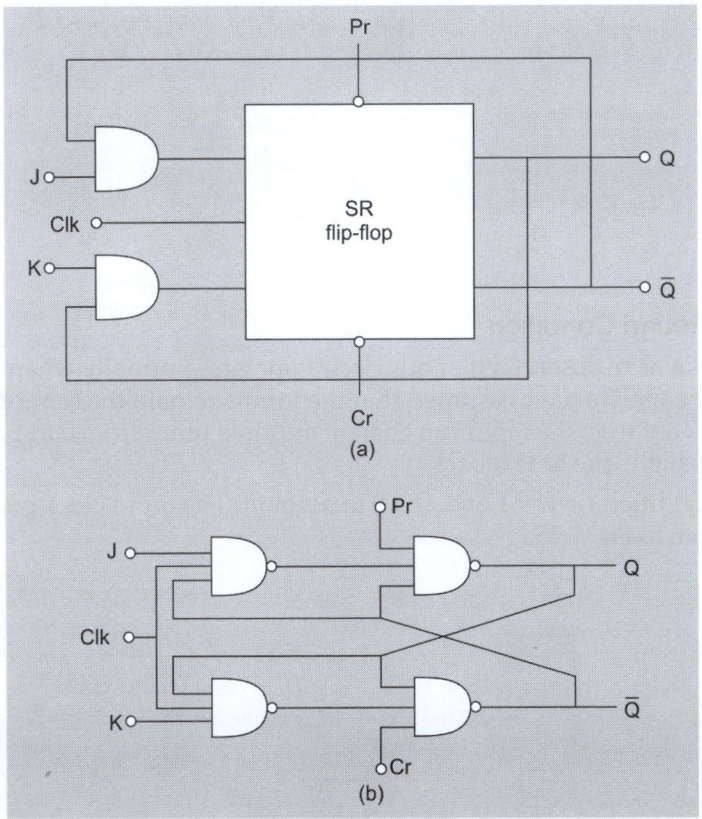

**Fig. 2.24:** A *JK* flip-flop obtained from *SR* flip-flop

using $SR$ flip-flop along with the $Pr$ and $Cr$ inputs. The operation of $JK$ flip-flop is summarized as follows:

1. **Case (i)** When $Pr = 1$ and $Cr = 1$

   When both $Pr$ and $Cr$ are HIGH and $Clk = 1$, then the circuit operates normally according to the inputs, so we consider the following cases:

   (a) **When both the inputs are LOW, i.e. $J = K = 0$** then the output of the flip-flop is same as it was in the previous state, i.e. $Q(t + 1) = Q(t)$.

   (b) **When both the inputs are HIGH, i.e. $J = K = 1$** then the flip-flop switches to the complement of previous state, i.e. $Q(t + 1) = \overline{Q(t)}$.

   (c) **When $J = 0$ and $K = 1$** then the flip-flop is RESET, i.e. $Q(t + 1) = 0$.

   (d) **When $J = 1$ and $K = 0$** then the flip-flop is SET, i.e. $Q(t + 1) = 1$.

2. **Case (ii)** When $Pr = 0$ and $Cr = 1$ then the flip-flop is SET, i.e. $Q(t + 1) = 1$ whatever be the values of the inputs $J$ and $K$.

3. **Case (iii)** Similarly, **When $Pr = 1$ and $Cr = 0$** then the flip-flop is cleared, i.e. $Q(t + 1) = 0$.

The condition $Pr = Cr = 0$ must be avoided as in this condition, the state of the flip-flop will be uncertain. Table 2.10 shows the truth table of $JK$ flip-flop.

| **Table 2.10:** Truth table of *JK* flip-flop | | | | | |
|---|---|---|---|---|---|
| *Clk* | *Pr* | *Cr* | *J(t)* | *K(t)* | *Q(t + 1)* |
| 1 | 0 | 1 | × | × | 1 |
| 1 | 1 | 0 | × | × | 0 |
| 1 | 1 | 1 | 0 | 0 | $Q(t)$ |
| 1 | 1 | 1 | 0 | 1 | 0 |
| 1 | 1 | 1 | 1 | 0 | 1 |
| 1 | 1 | 1 | 1 | 1 | $\overline{Q(t)}$ |

## 2.16.5 Race Around Condition

In $JK$ flip-flop, we have observed that our circuit operates normally when the $Clk$ signal is HIGH. In this theory, we have assumed that the inputs remain the same during the clock pulse. But this is not true, the input can change multiple times from 0 to 1 and 1 to 0 when the $Clk$ signal is high, due to feedback.

Let us take condition $J = K = 1$ and $Q = 0$ in account. The input clock pulse applied is of width $t_p$ as shown in Fig. 2.25.

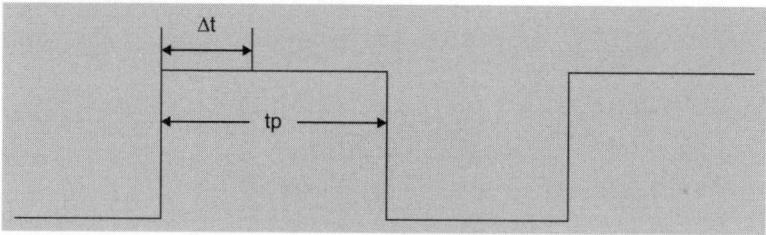

**Fig. 2.25:** Clock pulse of width $t_p$

The propagation delay of NAND gates is assumed to be $\Delta t$, so after $\Delta t$ time interval, the output changes from $Q = 0$ to $Q = 1$. Similarly, after $t$ time period, the output will again change from 1 to 0. If $t_p = 3 \times \Delta t$, then in a single clock pulse, the output will change thrice. Hence, at the end of the clock pulse, the output would be uncertain. This condition of uncertainty in the output is known as Race Around Condition.

Race around condition can be removed when $t_p < \Delta t$. This condition is quite difficult to achieve as the propagation delay itself is very small. There are two more ways of eliminating race around condition. They are:

(i) Master slave configuration

(ii) Edge triggering

### 2.16.6 Master Slave JK Flip-Flop

A master slave flip-flop is constructed from two separate JK flip-flops. One circuit serves as a master and the other as a slave. The block diagram of master slave JK flip-flop is shown in Fig. 2.26(a).

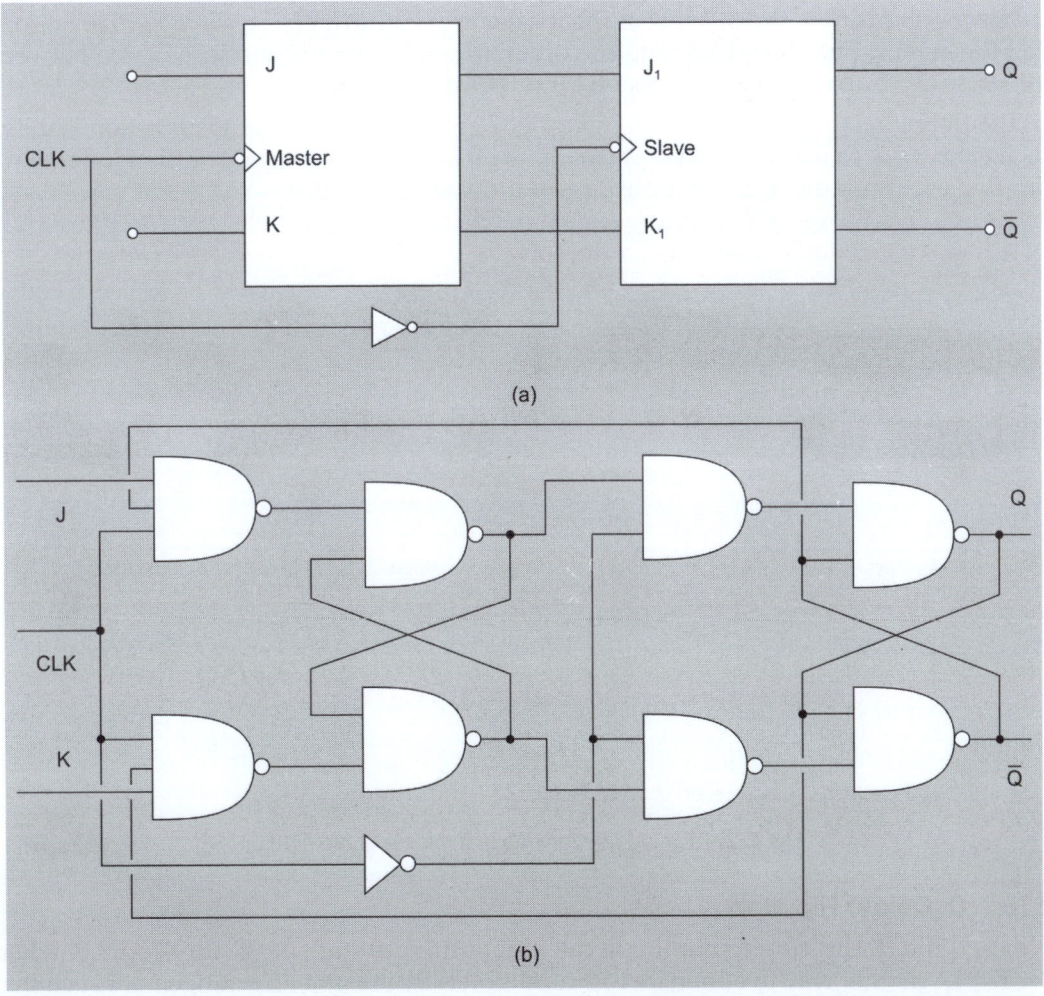

(a)

(b)

**Fig. 2.26: (a)** Block diagram of master slave JK flip-flop **(b)** Logic circuit of master slave JK flip-flop

The first flip-flop, i.e. master is enabled on the positive edge of the clock pulse *Clk* and the second flip-flop, i.e. slave is enabled on the negative edge of *Clk*. When *Clk* = 1 then the first flip-flop is enabled and produces the output, whereas at this time, for the second flip-flop *Clk* = 0 through the inverter and so is disabled. Again when *Clk* = 0, the first flip-flop is disabled and the output is the same as before, whereas now second flip-flop is enabled and accepts the output of the master flip-flop as input and accordingly produces the output. Hence, it is clear that the operation of first flip-flop depends on the primary inputs and of second flip-flop depends on first flip-flop, i.e. the second flip-flop simply follows the first flip-flop. Due to of this reason, the first flip-flop is called master and the second slave. The logic circuit of master slave *JK* flip-flop is shown in Fig. 2.26(b).

The working of master slave *JK* flip-flop overcomes the problem of race around condition that occurs with a *JK* flip-flop when $J = K = 1$. As any input to the master slave flip-flop at *J* and *K* is first seen by the master *FF* part of the circuit while *Clk* is High (= 1). This behavior effectively "locks" the input into the master *FF*. An important feature here is that the complement of the *Clk* pulse is fed to the slave *FF*. Therefore, the outputs from the master *FF* are only "seen" by the slave *FF* when *Clk* is Low (= 0). Hence, on the High-to-Low *Clk* transition, the outputs of the master are fed through the slave *FF*. This means that the at most one change of state can occur when $J = K = 1$ and so oscillation between the states $Q = 0$ and $Q = 1$ at successive *Clk* pulses does not occur.

The timing relationship is shown in Fig. 2.27 and is assumed that the flip-flop is in the clear state prior to the occurrence of the clock pulse. The output state of the master-slave flip-flop occurs on the negative transition of the clock pulse.

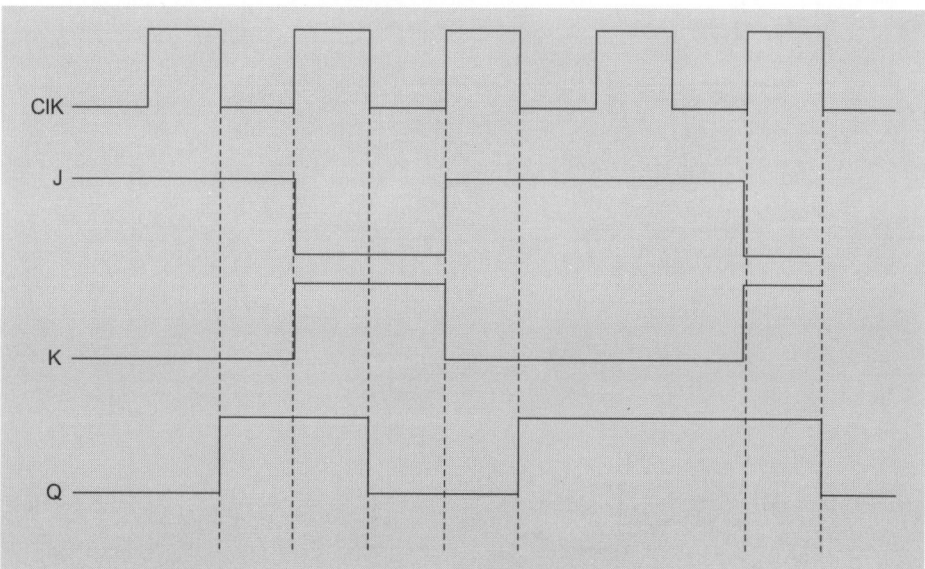

**Fig. 2.27:** Timing relationship in a master slave flip-flop

## 2.16.7 *D* (Delay) Flip-Flop

We used the *JK* flip-flop to eliminate the indeterminate state of *SR* flip-flop, i.e. when $S = R = 1$. Another way to eliminate this condition is by making an arrangement such that the two inputs are always complementary to each other. This condition is achieved in *D* flip-flop in which the $D$ input goes directly to the $S$ input and with an inverter to the $R$ input as shown in Fig 2.28.

The characteristic equation of $D$ flip-flop is $Q(t+1) = D$.

The operations of $D$ flip-flop are summarized in the truth table as shown in Table 2.11.

It is clear from the truth table as shown in Table 2.11 that the output at the end of the clock pulse, i.e. $Q(t+1)$, is equal to the input $D$ before the clock pulse. We can summarize this as data in $D$ flip-flop is transferred to the output after a certain delay and because of this reason, this flip-flop is known as Delay flip-flop (Fig. 2.29).

**Table 2.11:** Operations of $D$ flip-flop

| $D$ | $Q(t+1)$ |
|-----|----------|
| 0   | 0        |
| 1   | 1        |

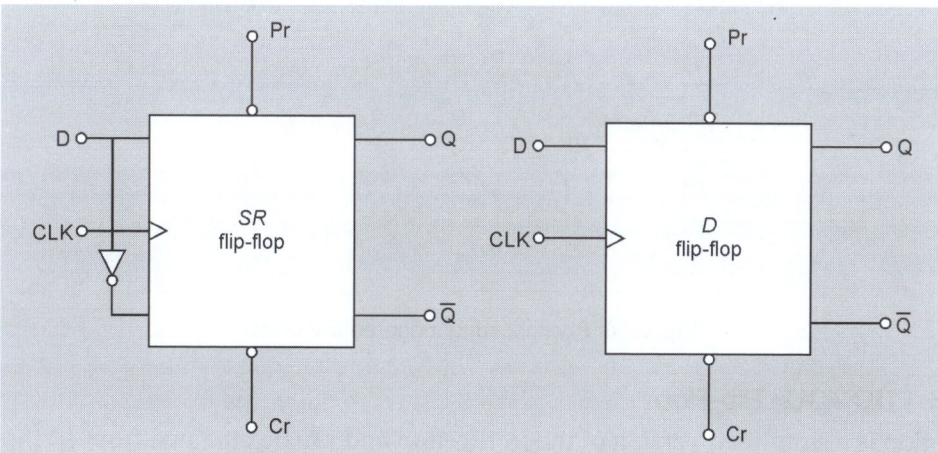

**Fig. 2.28:** Block diagram of $D$ flip-flop using $SR$

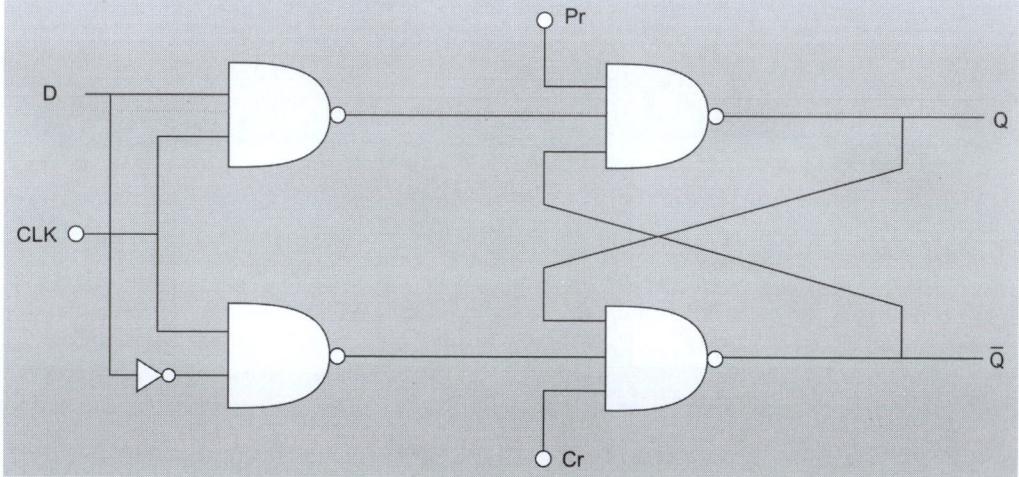

**Fig. 2.29:** $D$ flip-flop

A positive edge triggered $D$ flip-flop is shown in Fig. 2.30. In this figure, there are six NAND gates, which constitute three basic $SR$ flip-flops. The two $SR$ flip-flops in the beginning produces outputs, which are fed to third flip-flop at the end as inputs S and R and accordingly produces the final output.

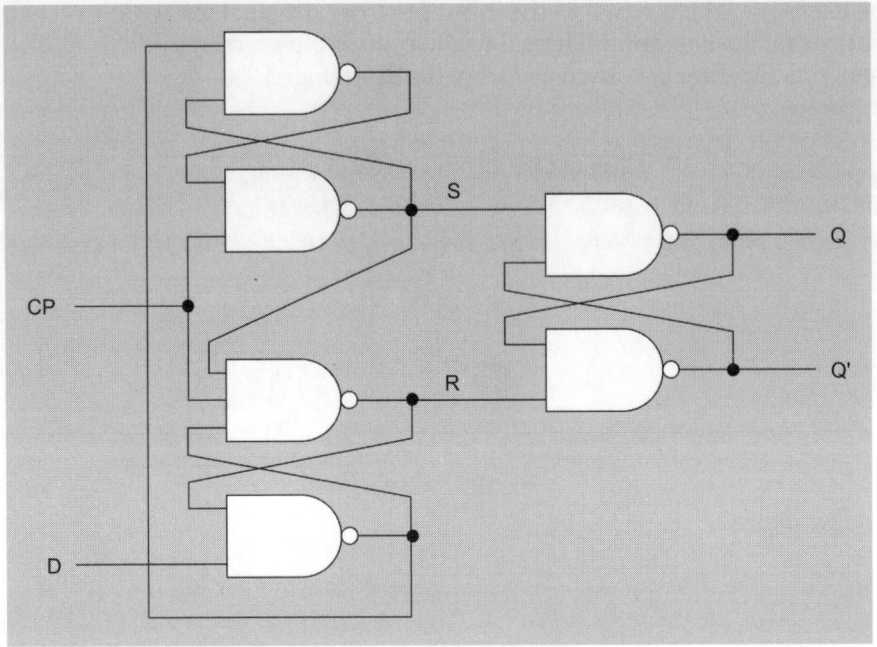

**Fig. 2.30:** Positive edge triggered $D$ flip-flop

### 2.16.8 T (TOGGLE) Flip-Flop

$T$ flip-flop is a simplified version of the $JK$ flip-flop and can be obtained from $JK$ flip-flop. When both the inputs of $JK$ flip-flop are tied together, then the resulting flip-flop is the $T$ flip-flop. Block diagram of $T$ flip-flop is shown in Fig. 2.31.

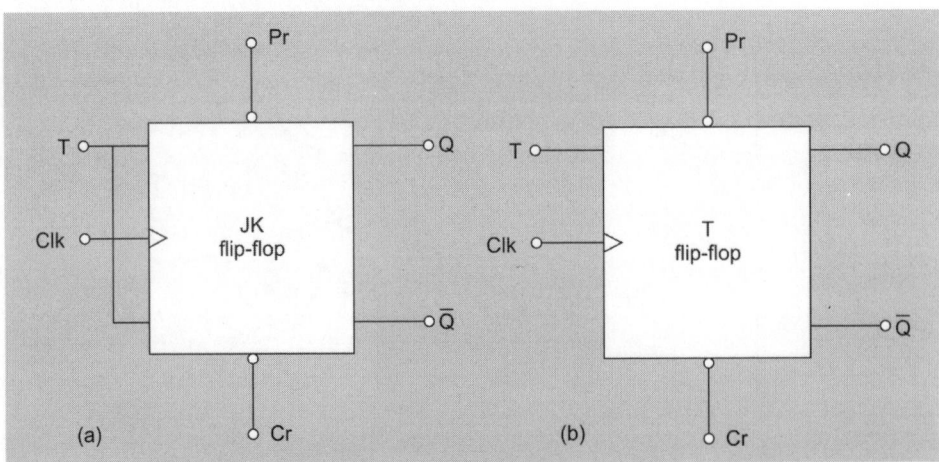

**Fig. 2.31:** Block diagram of $T$ flip-flop

Figure 2.31a shows the construction of $T$ flip-flop from $JK$ flip-flop and Fig. 2.31b shows the block diagram of $T$ flip-flop. If the toggle ($T$) input is high, the $T$ (Toggle) flip-flop changes state ("toggles"). If the $T$ input is low, it holds the previous value. The logic circuit of $T$ flip-flop is shown in Fig. 2.32.

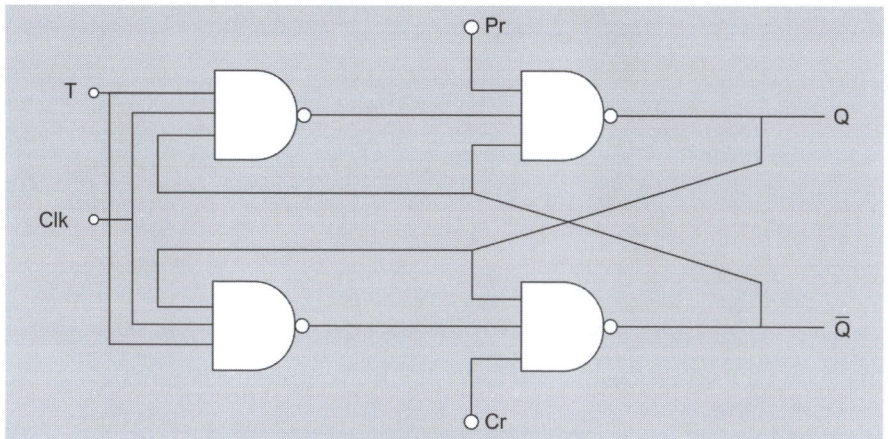

**Fig. 2.32:** Clocked $T$ flip-flop

The $T$ flip-flop operates normally when the clock input is HIGH, i.e. $Clk = 1$. The operation of $T$ flip-flop for $Clk = 1$ is summarized as:

When $T = 0$, the flip-flop does a hold. A hold means that the output, $Q$, is kept the same as it was before the clock edge, i.e. $Q(t + 1) = Q(t)$. When $T = 1$, the flip-flop does a toggle, which means the output $Q$ is negated after the clock edge, compared to the value before the clock edge, i.e. $Q(t + 1) = \overline{Q(t)}$.

The truth table of $T$ flip-flop is shown in Tables 2.12 and 2.13.

This behavior is described by the characteristic equation:

$$Q(t + 1) = T\ \overline{Q(t)} + \overline{T}\ Q(t)$$

**Table 2.12:** Truth table of $T$ flip-flop

| | | |
|---|---|---|
| 0 | 0 | 0(Q(t)) |
| 0 | 1 | 1($\overline{Q(t)}$) |
| 1 | 0 | 1(Q(t)) |
| 1 | 1 | 1($\overline{Q(t)}$) |

**Table 2.13:** Truth table of $T$ flip-flop

| $T$ | $Q(t + 1)$ |
|---|---|
| 0 | $Q(t)$ |
| 1 | $\overline{Q(t)}$ |

## 2.17 BINARY CODED DECIMAL

BCD represents each of the digits of an unsigned decimal as the 4-bit binary equivalents. As the largest decimal digit is 9, BCD of decimal greater than 9 can be represented by a combination of binary equivalent of each individual digit. Table 2.14 shows decimal to BCD conversion.

**Table 2.14:** Decimal to BCD conversion

| Decimal | Binary | BCD unpacked | Packed |
|---|---|---|---|
| 0 | 0000 0000 | 0000 0000 | 0000 0000 |
| 1 | 0000 0001 | 0000 0001 | 0000 0001 |
| 2 | 0000 0010 | 0000 0010 | 0000 0010 |
| 3 | 0000 0011 | 0000 0011 | 0000 0011 |
| 4 | 0000 0100 | 0000 0100 | 0000 0100 |
| 5 | 0000 0101 | 0000 0101 | 0000 0101 |
| 6 | 0000 0110 | 0000 0110 | 0000 0110 |
| 7 | 0000 0111 | 0000 0111 | 0000 0111 |
| 8 | 0000 1000 | 0000 1000 | 0000 1000 |
| 9 | 0000 1001 | 0000 1001 | 0000 1001 |
| 10 | 0000 1010 | 0000 0001 0000 0000 | 0001 0000 |
| 11 | 0000 1011 | 0000 0001 0000 0001 | 0001 0001 |
| 12 | 0000 1100 | 0000 0001 0000 0010 | 0001 0010 |
| 13 | 0000 1101 | 0000 0001 0000 0011 | 0001 0011 |
| 14 | 0000 1110 | 0000 0001 0000 0100 | 0001 0100 |
| 15 | 0000 1111 | 0000 0001 0000 0101 | 0001 0101 |
| 16 | 0001 0000 | 0000 0001 0000 0110 | 0001 0110 |
| 17 | 0001 0001 | 0000 0001 0000 0111 | 0001 0111 |
| 18 | 0001 0010 | 0000 0001 0000 1000 | 0001 1000 |
| 19 | 0001 0011 | 0000 0001 0000 1001 | 0001 1001 |
| 20 | 0001 0100 | 0000 0010 0000 0000 | 0010 0000 |

To illustrate the BCD code, $(234)_{10}$ can be represented as

| 2 | 3 | 4 | Decimal |
|---|---|---|---|
| ↓ | ↓ | ↓ | |
| 0010 | 0011 | 0100 | 4 bit binary of each digit |

### 2.17.1 Unpacked BCD

Unpacked BCD representation contains only one decimal digit per byte. The digit is stored in the least significant 4 bits; the most significant 4 bits are not relevant to the value of the represented number.

### 2.17.2 Packed BCD

Packed BCD representation packs two decimal digits into a single byte.

### 2.18 ASCII

ASCII stands for American Standard Code for Information Interchange. Computers can only understand numbers, so an ASCII code is the numerical representation of a character, such as 'a' or '@' or an action of some sort. ASCII was developed a long time ago and now the non-printing characters are rarely used for their original purpose. Table 2.15 shows the ASCII character table and this includes descriptions of the first 32 non-printing characters. ASCII was actually designed for use with teletypes and so the descriptions are

**Table 2.15:** ASCII character table

| Dec | Hex | Char | Dec | Hex | Char | Dec | Hex | Char | Dec | Hex | Char |
|---|---|---|---|---|---|---|---|---|---|---|---|
| 0 | 00 | Null | 32 | 20 | Space | 64 | 40 | @ | 96 | 60 | ` |
| 1 | 01 | Start of heading | 33 | 21 | ! | 65 | 41 | A | 97 | 61 | a |
| 2 | 02 | Start of text | 34 | 22 | " | 66 | 42 | B | 98 | 62 | b |
| 3 | 03 | End of text | 35 | 23 | # | 67 | 43 | C | 99 | 63 | c |
| 4 | 04 | End of transmit | 36 | 24 | $ | 68 | 44 | D | 100 | 64 | d |
| 5 | 05 | Enquiry | 37 | 25 | % | 69 | 45 | E | 101 | 65 | e |
| 6 | 06 | Acknowledge | 38 | 26 | & | 70 | 46 | F | 102 | 66 | f |
| 7 | 07 | Audible bell | 39 | 27 | ' | 71 | 47 | G | 103 | 67 | g |
| 8 | 08 | Backspace | 40 | 28 | ( | 72 | 48 | H | 104 | 68 | h |
| 9 | 09 | Horizontal tab | 41 | 29 | ) | 73 | 48 | I | 105 | 69 | i |
| 10 | 0A | Line feed | 42 | 2A | * | 74 | 4A | J | 106 | 6A | j |
| 11 | 0B | Vertical tab | 43 | 2B | + | 75 | 4B | K | 107 | 6B | k |
| 12 | 0C | Form feed | 44 | 2C | ' | 76 | 4C | L | 108 | 6C | l |
| 13 | 0D | Carriage return | 45 | 2D | – | 77 | 4D | M | 109 | 6D | m |
| 14 | 0E | Shift out | 46 | 2E | . | 78 | 4E | N | 110 | 6E | n |
| 15 | 0F | Shift in | 47 | 2F | / | 79 | 4F | O | 111 | 6F | o |
| 16 | 10 | Data link escape | 48 | 30 | 0 | 80 | 50 | P | 112 | 70 | p |
| 17 | 11 | Device control 1 | 49 | 31 | 1 | 81 | 51 | Q | 113 | 71 | q |
| 18 | 12 | Device control 2 | 50 | 32 | 2 | 82 | 52 | R | 114 | 72 | r |
| 19 | 13 | Device control 3 | 51 | 33 | 3 | 83 | 53 | S | 115 | 73 | s |
| 20 | 14 | Device control 4 | 52 | 34 | 4 | 84 | 54 | T | 116 | 74 | t |
| 21 | 15 | Neg acknowledge | 53 | 35 | 5 | 85 | 55 | U | 117 | 75 | u |
| 22 | 16 | Synchronous idle | 54 | 36 | 6 | 86 | 56 | V | 118 | 76 | v |
| 23 | 17 | End trans block | 55 | 37 | 7 | 87 | 57 | W | 119 | 77 | w |
| 24 | 18 | Cancel | 56 | 38 | 8 | 88 | 58 | X | 120 | 78 | x |
| 25 | 19 | End of medium | 57 | 39 | 9 | 89 | 59 | Y | 121 | 79 | y |
| 26 | 1A | Substitution | 58 | 3A | : | 90 | 5A | Z | 122 | 7A | z |
| 27 | 1B | Escape | 59 | 3B | ; | 91 | 5B | [ | 123 | 7B | { |
| 28 | 1C | File separator | 60 | 3C | < | 92 | 5C | \ | 124 | 7C | \| |
| 29 | 1D | Group separator | 61 | 3D | = | 93 | 5D | ] | 125 | 7D | } |
| 30 | 1E | Record separator | 62 | 3E | > | 94 | 5E | ^ | 126 | 7E | ~ |
| 31 | 1F | Unit separator | 63 | 3F | ? | 95 | 5F | _ | 127 | 7F | □ |

somewhat obscure. If someone says that they want your CV, however, in ASCII format, all this means is that they want 'plain' text with no formatting such as tabs, bold or underscoring—the raw format that any computer can understand. This is usually so, because they can easily import the file into their own applications without issues. Notepad.exe creates ASCII text or in MS Word you can save a file as 'text only'.

## PROBLEMS

1. What is the difference between Analog and Digital Signals?

2. What is hexadecimal number system? Convert decimal number 89 into hexadecimal.

3. Convert the following decimal numbers into binary numbers:
   (a) 20               (b) 49               (c) 98

4. Convert the following binary numbers into decimal numbers:
   (a) 1010111          (b) 011100          (c) 1000100

5. Convert the following decimal numbers into octal numbers:
   (a) 78               (b) 724             (c) 95

6. Convert the following octal numbers into decimal numbers:
   (a) 77               (b) 244             (c) 65

7. Convert the following hexadecimal numbers into binary numbers:
   (a) 7D               (b) 89              (c) A20

8. Convert the following binary numbers into hexadecimal numbers:
   (a) 110001101        (b) 100101          (c) 01111

9. Find 2's complement of given binary numbers:
   (a) 1001             (b) 011110          (c) 101010

10. Convert the following decimal numbers into BCD:
    (a) 13              (b) 30              (c) 43

11. Why are NAND and NOR gates known as universal gates?

12. Implement basic gates using NAND and NOR only.

13. Minimize the given Boolean function using K-map

$$Y = AB' + AB'C + A'BC + A'BC' + B'$$

14. What are flip-flops? What is the difference between flip-flops and latches?

15. What is the disadvantage in SR flip-flops?

16. What is race around condition?

17. What is ASCII code?

# Microprocessor 8085

## 3.1 INTRODUCTION

Intel 8085 is an 8-bit, N-channel Metal Oxide Semiconductor (NMOS) microprocessor (μp). It is a 40-pin IC package fabricated on a single Large Scale Integration (LSI) chip. The Intel 8085 uses a single +5V DC supply for its operation. Its clock speed is about 3MHz. The clock cycle is of 320 ns. The time for the clock cycle of the Intel 8085 is 200 ns. It has 80 basic instructions and 246 opcodes. The 8085 is an enhanced version of its predecessor, the 8080A; its instruction set is upward compatible with that of the 8080A, meaning that 8085 instruction set includes all the 8080A instructions plus some additional ones. Programs written for 8080A will be executed by 8085, but the 8085 and 8080A are not pin compatible.

8085 is an 8-bit microprocessor, it means that it can process 8-bits at a time. It has 16-bit address lines which means that it can address $2^{16}$ different memory locations through its address buses.

## 3.2 ARCHITECTURE OF 8085 MICROPROCESSOR

The architecture of Intel 8085 consists of three main sections: arithmetic and logic unit, timing and control unit and several registers. The functional block diagram of 8085 is shown in Fig. 3.1. These important sections are described in the subsequent sections.

### 3.2.1 Arithmetic and Logic Unit (ALU)

The ALU performs the following arithmetic and logical operations.

1. Addition
2. Subtraction
3. Logical AND
4. Logical OR
5. Logical EXCLUSIVE OR
6. Complement (logical NOT)
7. Increment (add 1)
8. Decrement (subtract 1)
9. Left shift
10. Clear

The ALU is the unit that manipulates the data. ALU includes the accumulator, the temporary register, the arithmetic and logic circuits and flags. The ALU performs the actual numerical and logical operations such as 'add', 'subtract', 'AND', 'OR', etc.; uses data from memory and accumulator to perform arithmetic; always stores result of operation in accumulator.

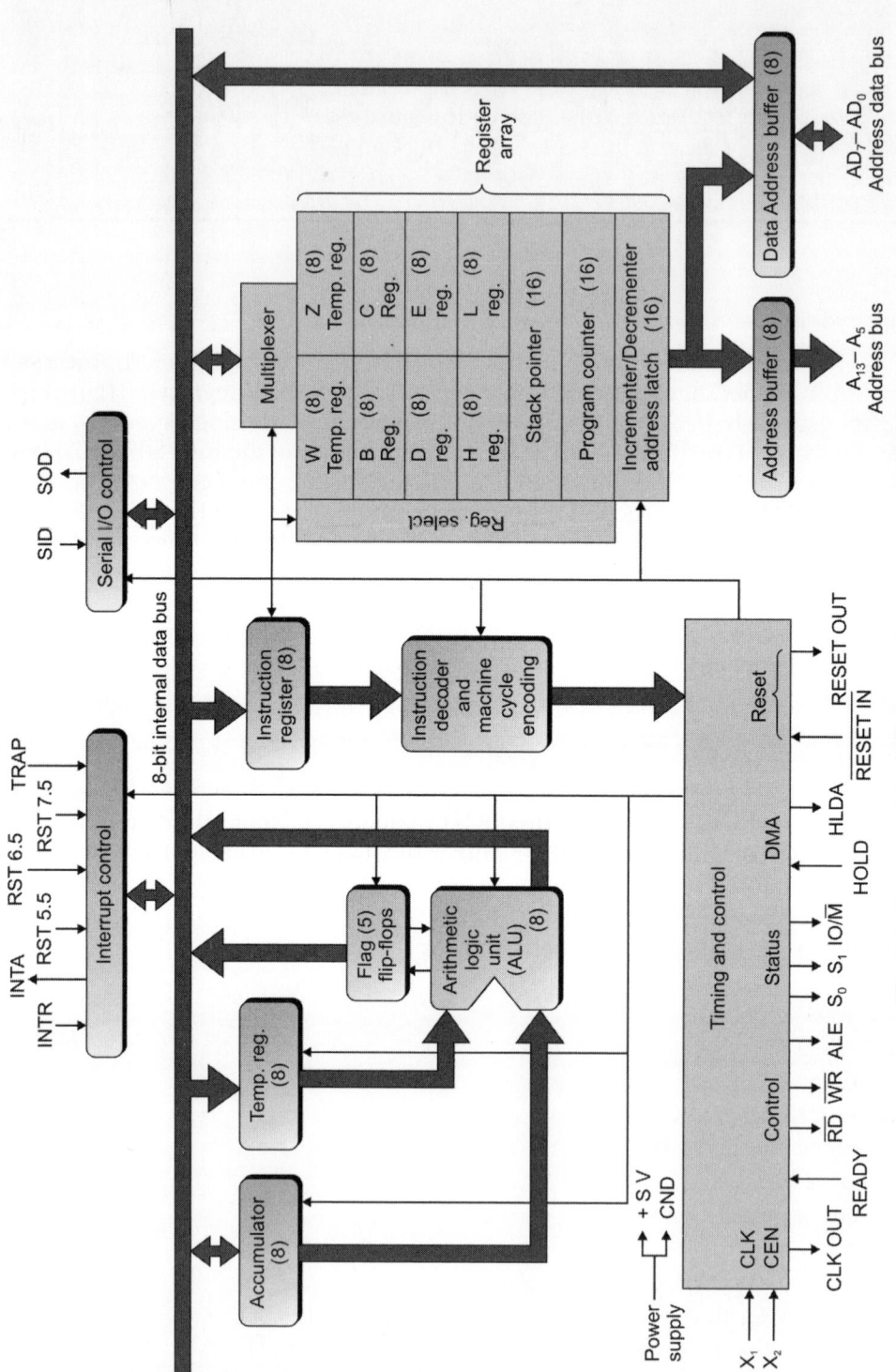

**Fig. 3.1:** Functional block diagram of microprocessor 8085

## 3.2.2 Timing and Control Unit

It has three control signals: ALE, RD (Active low) and WR (Active low) and three status signals IO/M (Active low), S0 and S1. ALE is used for providing control signals to synchronize the components of microprocessor and timing for instruction to perform the operation. RD (Active low) and WR (Active low) are used to indicate whether the operation is reading the data from memory or writing the data into memory, respectively. IO/M (Active low) is used to indicate whether the operation belongs to the memory or peripherals (Table 3.1).

**Table 3.1:** Status signals

| IO/M (Active low) | S1 | S2 | Data bus status (Output) |
|---|---|---|---|
| 0 | 0 | 0 | Halt |
| 0 | 0 | 1 | Memory write |
| 0 | 1 | 0 | Memory read |
| 1 | 0 | 1 | I/O write |
| 1 | 1 | 0 | I/O read |
| 0 | 1 | 1 | Opcode fetch |
| 1 | 1 | 1 | Interrupt acknowledge |

## 3.2.3 Serial I/O Control

It has two control signals named SID (Serial input data) and SOD (Serial output data) for serial data transmission.

SID is active high and received by processor to receive input data serially. SOD is also active high and sent by processor to transmit data serially at output port.

## 3.2.4 Registers

The 8085 can access $2^{16}$ (= 65,536) individual 8-bit memory locations, in other words, its address space is 64 KB. Unlike some other microprocessors of its era, it has a separate address space for up to $2^8$ (= 256) I/O ports.

It has the following registers:
1. One 8-bit Accumulator (ACC) A
2. Six 8-bit general purpose registers B, C, D, E, H and L
3. One 16-bit stack pointer SP
4. One 16-bit program counter PC
5. Instruction register
6. Memory address register
7. 8-bit flag register
8. Temporary register

They can be combined as register pairs—BC, DE and HL to perform some 16-bit operations. The programmer can use these registers to store or copy data into the registers by using data copy instructions.

### 3.2.4.1 Accumulator

The accumulator is an 8-bit register that is a part of arithmetic/logic unit (ALU). This register is used to store 8-bit data and to perform arithmetic and logical operations. The result of an operation is stored in the accumulator. The accumulator is also identified as register A.

### 3.2.4.2 Stack Pointer

Stack pointer is a *16-bit* register. This register is always *incremented/decremented by 2*. It always holds the address of the top most entity in the stack. The stack pointer is also used as a memory pointer. It points to a memory location in R/W memory, called the stack. The beginning of the stack is defined by loading 16-bit address in the stack pointer.

### 3.2.4.3 Program Counter

Program counter is a 16-bit register. It holds *the address of the instruction which is to be executed next*.

### 3.2.4.4 Instruction Register

It is a temporary storage for the instruction. It holds the instruction until it is decoded. This cannot be accessed by the programmer.

### 3.2.4.5 Memory Address Register

It holds the address (received from PC) of next program instruction and feeds the address bus with addresses of location of the program under execution.

### 3.2.4.6 Flag Register

The ALU includes five flip-flops that are set or reset according to the result of an operation. The microprocessor uses the flags for testing the data conditions. They are Zero (Z), Carry (CY), Sign (S), Parity (P) and Auxiliary Carry (AC) flags. The most commonly used flags are Sign, Zero and Carry. The bit position for the flags in flag register is,

| S | Z | X | AC | X | P | X | CY |
|---|---|---|----|---|---|---|----|

1. **Sign flag (S):** After execution of any arithmetic and logical operation, if D7 of the result is 1, the sign flag is set, otherwise it is reset. D7 is reserved for indicating the sign; the remaining is the magnitude of the number. If D7 is 1, the number will be viewed as negative number. If D7 is 0, the number will be viewed as positive number.
2. **Zero flag (Z):** If the result of arithmetic and logical operation is zero, then zero flag is set, otherwise it is reset.
3. **Auxiliary carry flag (AC):** If D3 generates any carry when doing any arithmetic and logical operation, this flag is set, otherwise it is reset.
4. **Parity flag (P):** If the result of arithmetic and logical operation contains even number of 1's then this flag will be set and if it is odd number of 1's, it will be reset.
5. **Carry flag (CY):** If any arithmetic and logical operation result any carry then carry flag is set, otherwise it is reset.

### 3.2.4.7 Temporary Register

It is used to hold the data during the arithmetic and logical operations.

### 3.2.5 Instruction Decoder

Current instruction sent here from instruction register prior to execution. Decoder then takes instruction and 'decodes' or interprets the instruction. Decoded instruction then passed to next stage.

### 3.2.6 Control Generator

It generates signals within µp to carry out the instruction which has been decoded. In reality, it causes certain connections between blocks of the µp to be opened or closed, so that data goes where it is required and ALU operations occur.

### 3.2.7 Register Selector

This block controls the use of the register stack in the example just as a logic circuit which switches between different registers in the set, will receive instructions from Control Unit.

### 3.2.8 8085 System Bus

Typical system uses a number of buses, collection of wires, which transmit binary numbers, one bit per wire. A typical microprocessor communicates with memory and other devices (input and output) using three buses: *Address Bus, Data Bus and Control Bus.*

#### 3.2.8.1 Address Bus

The address bus is a group of 16 lines generally identified as A0 to A15. The address bus is unidirectional: bits flow in one direction from the MPU to peripheral devices. The MPU uses the address bus to perform the first function: identifying a peripheral or a memory location.

#### 3.2.8.2 Data Bus

The data bus is a group of 8 lines used for data flow. These lines are bi-directional: data flow in both directions between the MPU and memory and peripheral devices. The MPU uses the data bus to perform the second function: transferring binary information. The 8 data lines enable the MPU to manipulate 8-bit data ranging from 00 to FF (28 = 256 numbers). The largest number that can appear on the data bus is 11111111.

#### 3.2.8.3 Control Bus

The control bus carries synchronization signals and provides timing signals. The MPU generates specific control signals for every operation it performs. These signals are used to identify a device type with which the MPU wants to communicate.

### 3.2.9 Interrupt Control Unit

It controls hardware interrupts. 8085 has five hardware interrupts: TRAP, RST7.5, RST6.5, RST5.5, INTR and an acknowledgement signal INTA (active low). It receives hardware interrupt signals and sends an acknowledgement for receiving the interrupt signal.

### 3.3 PIN CONFIGURATION OF 8085 (Fig. 3.2)

*Properties*

1. Single + 5V Supply
2. 4 Vectored Interrupts (One is non-maskable)
3. Serial In/Serial Out Port
4. Decimal, Binary, and Double Precision Arithmetic
5. Direct Addressing Capability to 64K bytes of memory

The Intel 8085A is a new generation, it completes 8-bit parallel Central Processing Unit (CPU). The 8085A uses a multiplexed data bus. The address is split between the 8-bit address bus and the 8-bit data bus.

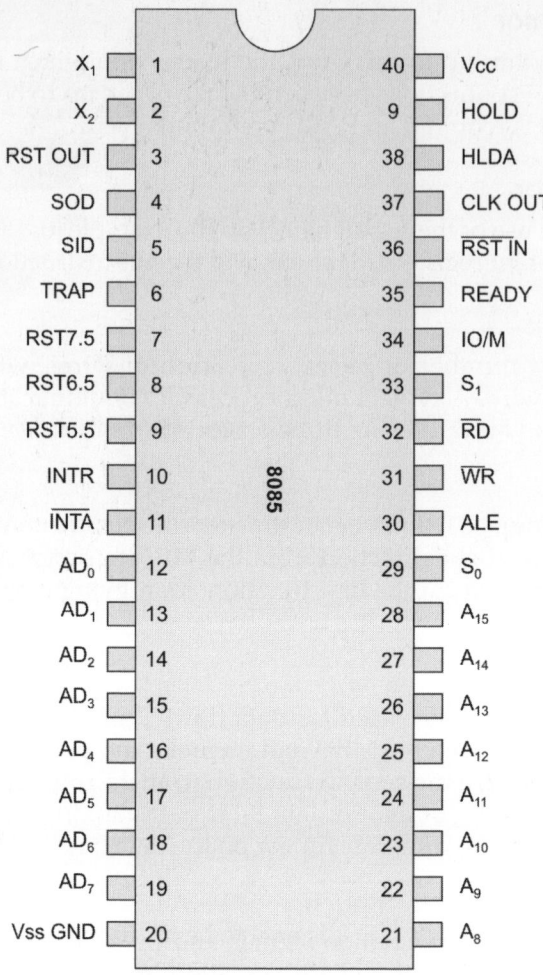

**Fig. 3.2:** Pin diagram of 8085 microprocessor

**A8 – A15 (pin-21 to 28, type-output, 3 state):** Address Bus—It is the most significant 8 bits of the memory address or the 8 bits of the I/O addresses, 3 stated during Hold and Halt modes.

**AD0 – AD7 (pin-12 to 19, type-input/output, 3 state):** Multiplexed Address/Data Bus—Lower 8 bits of the memory address (or I/O address) appear on the bus during the first clock cycle of a machine state. It then becomes the data bus during the second and third clock cycles, 3 stated during Hold and Halt modes.

**ALE (pin-30, type-output):** Address Latch Enable—It occurs during the first clock cycle of a machine state and enables the address to get latched into the on-chip latch of peripherals. The falling edge of ALE is set to guarantee setup and hold times for the address information. ALE can also be used to strobe the status information. ALE is never 3 stated.

**S0, S1 (pin-29 and 33, type-output):** Data Bus Status of the Bus Cycle. S1 can be used as an advanced R/W status (Table 3.2).

**$\overline{\text{RD}}$ (pin-32, type-output, 3 state):** READ—It indicates the selected memory or I/O device. $\overline{\text{RD}}$ is to be read and that the Data Bus is available for the data transfer.

| **Table 3.2:** Status of the Bus Cycle | | |
| --- | --- | --- |
| *S1* | *S0* | *Action* |
| 0 | 0 | Halt |
| 0 | 1 | Write |
| 1 | 0 | Read |
| 1 | 1 | Fetch |

**$\overline{WR}$ (pin-31, type-output, 3 state):** WRITE—It indicates the data on the Data Bus which is to be written into the selected memory or 1/0 location. Data is set up at the trailing edge of $\overline{WR}$, 3 stated during Hold and Halt modes.

**Ready (pin-35, type-input):** If Ready is high during a read or write cycle, it indicates that the memory or peripheral is ready to send or receive data. If Ready is low, the CPU will wait for Ready to go high before completing the read or write cycle.

**HOLD (pin-39, type-input):** HOLD—It indicates that another Master is requesting the use of the Address and Data Buses. The CPU, upon receiving the Hold request, will relinquish the use of buses as soon as the completion of the current machine cycle. Internal processing can continue. The processor can regain the buses only after the Hold is removed. When the Hold is acknowledged, the Address, Data, RD, WR and IO/M lines are 3 stated.

**HLDA (pin-38, type-output):** HOLD ACKNOWLEDGE—It indicates that the CPU has received the Hold request and that it will relinquish the buses in the next clock cycle. HLDA goes low after the Hold request is removed. The CPU takes the buses one half clock cycle after HLDA goes low.

**INTR (pin-10, type-input):** INTERRUPT REQUEST—It is used as a general purpose interrupt. It is sampled only during the next to the last clock cycle of the instruction. If it is active, the Program Counter (PC) will be inhibited from incrementing and an INTA will be issued. During this cycle, a RESTART or CALL instruction can be inserted to jump to the interrupt service routine. The INTR is enabled and disabled by software. It is disabled by Reset and immediately after an interrupt is accepted.

**$\overline{INTA}$ (Pin-11, type-output):** INTERRUPT ACKNOWLEDGE—It is used instead of (and has the same timing as) RD during the Instruction cycle after an INTR is accepted. It can be used to activate the 8259 Interrupt chip or some other Interrupt port.

### Restart Interrupts

These three inputs have the same timing as INTR except that they cause an internal RESTART to be automatically inserted.

**RST 7.5 (pin-7, type-input):** Possess Highest Priority

**RST 6.5 (pin-8, type-input)**

**RST 5.5 (pin-9, type-input):** Possess Lowest Priority

The priority of these interrupts is ordered as shown in Fig. 3.2. These interrupts have a higher priority than the INTR.

**TRAP ((pin-6, type-input):** TRAP interrupt is a non-maskable restart interrupt. It is recognized at the same time as INTR. It is unaffected by any mask or Interrupt Enable. It has the highest priority of any interrupt.

**$\overline{RST\ IN}$ (pin-36, type-input):** Reset ($\overline{RST}$ IN) sets the Program Counter to zero and resets the Interrupt Enable and HLDA flip-flops. None of the other flags or registers (except the instruction register) are affected. The CPU is held in the reset condition as long as Reset is applied.

**RST OUT (pin-3, type-output):** This pin is used as a system RESET. The signal is synchronized to the processor clock.

**X1, X2 (pin-1 and 2, type-input):** Crystal or R/C network connections to set the internal clock generator X1 can also be an external clock input instead of a crystal. The input frequency is divided by 2 to give the internal operating frequency.

**CLK OUT (pin-37, type-output):** Clock Output is used as a system clock when a crystal or R/C network is used as an input to the CPU. The period of CLK is twice the X1, X2 input period.

**IO/$\overline{\text{M}}$ (pin-34, type-output):** IO/M indicates whether the Read/Write is to memory or I/O, tri-stated during Hold and Halt modes.

**SID (pin-5, type-input):** Serial input data line, the data on this line is loaded into accumulator bit 7 whenever a RIM instruction is executed.

**SOD (pin-4, type-output):** Serial output data line, the output SOD is set or reset as specified by the SIM instruction.

**Vcc (pin-40):** +5 volt supply.

**Vss (pin-20):** Ground Reference.

## 3.4 FUNCTIONAL DESCRIPTION

The 8085A is a complete 8-bit parallel central processor. It requires a single +5 volt supply. Its basic clock speed is 3 MHz, thus, improving on the present 8080's performance with

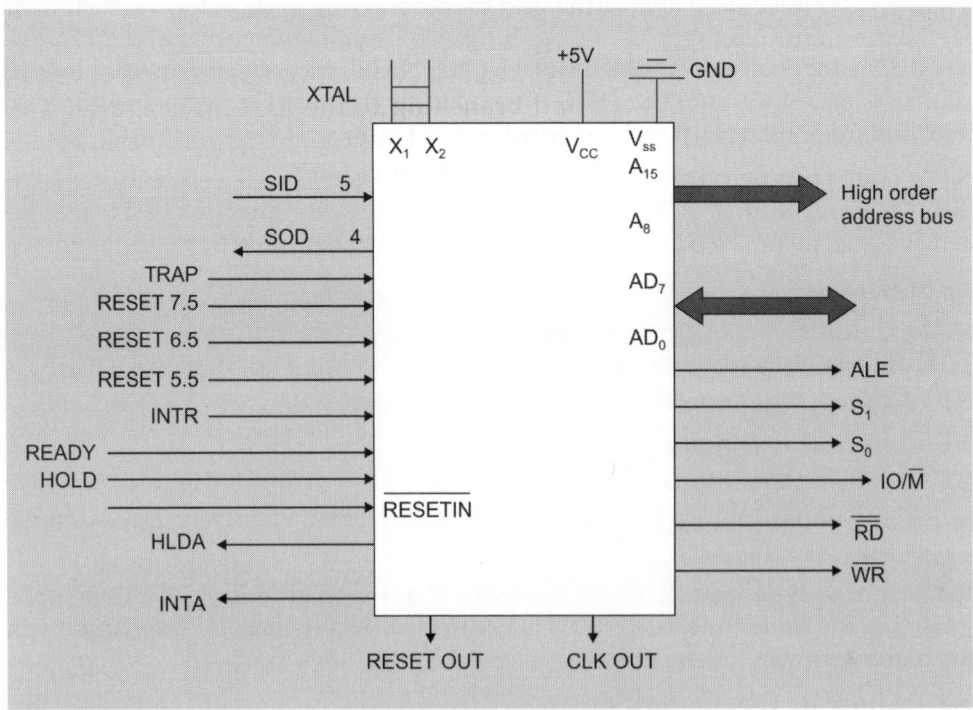

**Fig. 3.3:** Different signals of 8085

higher system speed. It is also designed to fit into a minimum system of three ICs: the CPU, a RAM/IO and a ROM or PROM/IO chip. The 8085A uses a multiplexed Data Bus. The address is split between the higher 8-bit Address Bus and the lower 8-bit Address/Data Bus. During the first cycle, the address is sent out. The lower 8-bits are latched into the peripherals by the Address Latch Enable (ALE). During the rest of the machine cycle, the Data Bus is used for memory or I/O data. The 8085A provides RD, WR and IO/Memory signals for bus control. An Interrupt Acknowledge signal (INTA) is also provided. Hold, Ready and all Interrupts are synchronized. The 8085A also provides serial input data (SID) and serial output data (SOD) lines for simple serial interface. In addition to these features, the 8085A has three maskable, restart interrupts and one non-maskable trap interrupt. The 8085A provides RD, WR and IO/M signals for Bus Control.

### 3.4.1 Status Information

Status information is directly available from the 8085A. ALE serves as a status strobe. The status is partially encoded and provides the user with advanced timing of the type of bus transfer being done. IO/M cycle status signal is provided directly also. Decoded S0, S1 carries the following status information: HALT, WRITE, READ, FETCH. S1 can be interpreted as R/W in all bus transfers. In the 8085A, the 8 LSB of address are multiplexed with the data instead of status. The ALE line is used as a strobe to enter the lower half of the address into the memory or peripheral address latch. This also frees extra pins for expanded interrupt capability.

### 3.4.2 Interrupt and Serial I/O

The 8085A has 5 interrupt inputs: INTR, RST 5.5, RST6.5, RST 7.5 and TRAP. INTR is identical in function to the 8080 INT. Each of the three RESTART inputs, 5.5, 6.5, 7.5 has a programmable mask. TRAP is also a RESTART interrupt except it is non-maskable. The three RESTART interrupts cause the internal execution of RST (saving the program counter in the stack and branching to the RESTART address), if the interrupts are enabled and the interrupt mask is not set. The non-maskable TRAP causes the internal execution of a RST independent of the state of the interrupt enable or masks. The interrupts are arranged in a fixed priority that determines which interrupt is to be recognized if more than one is pending as follows: TRAP highest priority, RST 7.5, RST 6.5, RST 5.5, INTR lowest priority. This priority scheme does not take into account the priority of a routine that was started by a higher priority interrupt. RST 5.5 can interrupt a RST 7.5 routine if the interrupts were re-enabled before the end of the RST 7.5 routine. The TRAP interrupt is useful for catastrophic errors such as power failure or bus error. The TRAP input is recognized just as any other interrupt but has the highest priority. It is not affected by any flag or mask. The TRAP input is both edge and level sensitive.

### 3.4.3 Basic System Timing

The 8085A has a multiplexed Data Bus. ALE is used as a strobe to sample the lower 8 bits of address on the Data Bus. An instruction fetch, memory read and I/O write cycle (OUT) will be discussed in Section 3.10. Note that during the I/O write and read cycle that the I/O port address is copied on both the upper and lower half of the address. As in the 8080, the READY line is used to extend the read and write pulse lengths so that the 8085A can be used with slow memory. Hold causes the CPU to relinquish the bus when it is through with it by floating the Address and Data Buses.

## 3.4.4 System Interface

8085A family includes memory components which are directly compatible to the 8085A CPU. For example, a system consisting of the three chips: 8085A, 8156 and 8355 will have the following features:

1. 2K bytes ROM
2. 256 bytes RAM
3. 1 Timer/Counter
4. 48-bit I/O Ports
5. 16-bit I/O Ports
6. 4 Interrupt Levels
7. Serial In/Serial Out Ports

In addition to standard I/O, the memory mapped I/O offers an efficient I/O addressing technique. With this technique, an area of memory address space is assigned for I/O address, thereby using the memory address for I/O manipulation. The 8085A CPU can also interface with the standard memory that does not have the multiplexed address/data bus.

## 3.5 INTERRUPTS OF 8085

Interrupt is a signal sent by an external device to the processor, to request the processor to perform a particular task or work. Mainly in the microprocessor based system, the interrupts are used for data transfer between the peripheral and the microprocessor. The processor will check the interrupts always at the 2nd T-state of last machine cycle. If there is any interrupt, it accepts the interrupt and sends the INTA (active low) signal to the peripheral. The vectored address of particular interrupt is stored in program counter. The processor executes an interrupt service routine (ISR) addressed in program counter. It returned to the main program by RET instruction.

### 3.5.1 Types of Interrupts

It supports two types of interrupts:

1. Software
2. Hardware

#### 3.5.1.1 Software Interrupts

The software interrupts are program instructions. These instructions are inserted at desired locations in a program. The 8085 has eight software interrupts from RST 0 to RST 7. The vector address for these interrupts can be calculated as follows:

- Interrupt number * 8 = Vector address
- For RST 5.5 * 8 = 40 = 28H
- Vector address for interrupt RST 5 is 0028H.

Table 3.3 shows the vector addresses of all interrupts.

#### 3.5.1.2 Hardware Interrupts

An external device initiates the hardware interrupts and places an appropriate signal at the interrupt pin of the processor. If the interrupt is accepted then the processor executes an interrupt service routine.

**Table 3.3:** Vector addresses of interrupts

| Interrupt | Vector address |
|-----------|----------------|
| RST 0 | 0000H |
| RST 1 | 0008H |
| RST 2 | 0010H |
| RST 3 | 0018H |
| RST 4 | 0020H |
| RST 5 | 0028H |
| RST 6 | 0030H |
| RST 7 | 0038H |

The 8085 has five hardware interrupts as shown in Table 3.4.

1. TRAP      2. RST 7.5      3. RST 6.5      4. RST 5.5      5. INTR

**Table 3.4:** Vector addresses of five hardware interrupts

| Interrupt | Vector address |
|-----------|----------------|
| RST 7.5 | 003CH |
| RST 6.5 | 0034H |
| RST 5.5 | 002CH |
| TRAP | 0024H |
| INTR | |

### TRAP

This interrupt is a non-maskable interrupt. It is unaffected by any mask or interrupt enable. TRAP has the highest priority and vectored interrupt. TRAP interrupt is edge and level triggered. This means that the TRAP must go high and remain high until it is acknowledged. In sudden power failure, it executes an ISR and sends the data from main memory to backup memory. The signal, which overrides the TRAP, is HOLD signal. (i.e. if the processor receives HOLD and TRAP at the same time then HOLD is recognized first and then TRAP is recognized).

There are two ways to clear TRAP interrupt:
1. By resetting microprocessor (External signal)
2. By giving a high TRAP ACKNOWLEDGE (Internal signal).

### RST 7.5

The RST 7.5 interrupt is a maskable interrupt. It has the second highest priority. It is edge sensitive, i.e. input goes to high and there is no need to maintain high state until it is recognized. Maskable interrupt is disabled by
1. DI instruction
2. System or processor reset
3. After reorganization of interrupt

It is enabled by EI instruction.

### RST 6.5 and 5.5

The RST 6.5 and RST 5.5 both are level triggered, i.e. input goes to high and stay high until it is recognized as maskable interrupt. It is disabled by

1. DI, SIM instruction
2. System or processor reset
3. After reorganization of interrupt.

It is enabled by EI instruction. The RST 6.5 has the third priority, whereas RST 5.5 has the fourth priority.

### INTR

INTR is a maskable interrupt. It is disabled by
1. DI, SIM instruction
2. System or processor reset
3. After reorganization of interrupt.

It is enabled by EI instruction. It is non-vectored interrupt. After receiving INTA (active low) signal, it has to supply the address of ISR. It has the lowest priority. It is a level sensitive interrupts, i.e. input goes to high and it is necessary to maintain high state until it is recognized.

The following sequence of events occurs when INTR signal goes high:
1. The 8085 checks the status of INTR signal during execution of each instruction.
2. If INTR signal is high, then 8085 completes its current instruction and sends active low interrupt acknowledge signal, if the interrupt is enabled.
3. In response to the acknowledge signal, external logic places an instruction OPCODE on the data bus. In the case of multi-byte instruction, additional interrupt acknowledge machine cycles are generated by the 8085 to transfer the additional bytes into the microprocessor.
4. On receiving the instruction, the 8085 saves the address of next instruction on stack and executes received instruction.

### 3.5.2 SIM and RIM

The 8085 provides additional masking facility for RST 7.5, RST 6.5 and RST 5.5 using SIM instruction. The status of these interrupts can be read by executing RIM instruction. The masking or unmasking of RST 7.5, RST 6.5 and RST 5.5 interrupts can be performed by moving an 8-bit data to accumulator and then executing SIM instruction (Table 3.5). The format of the 8-bit data is shown in Fig. 3.4.

The status of pending interrupts can be read from accumulator after executing RIM instruction. When RIM instruction is executed, an 8-bit data is loaded in accumulator, which can be interpreted as shown in Fig. 3.5.

**Table 3.5:** Status of interrupts and vector addresses

| Interrupt type | Trigger | Priority | Maskable | Vector address |
|---|---|---|---|---|
| TRAP | Edge and level | 1st | No | 0024H |
| RST 7.5 | Edge | 2nd | Yes | 003CH |
| RST 6.5 | Level | 3rd | Yes | 0034H |
| RST 5.5 | Level | 4th | Yes | 002CH |
| INTR | Level | 5th | Yes | – |

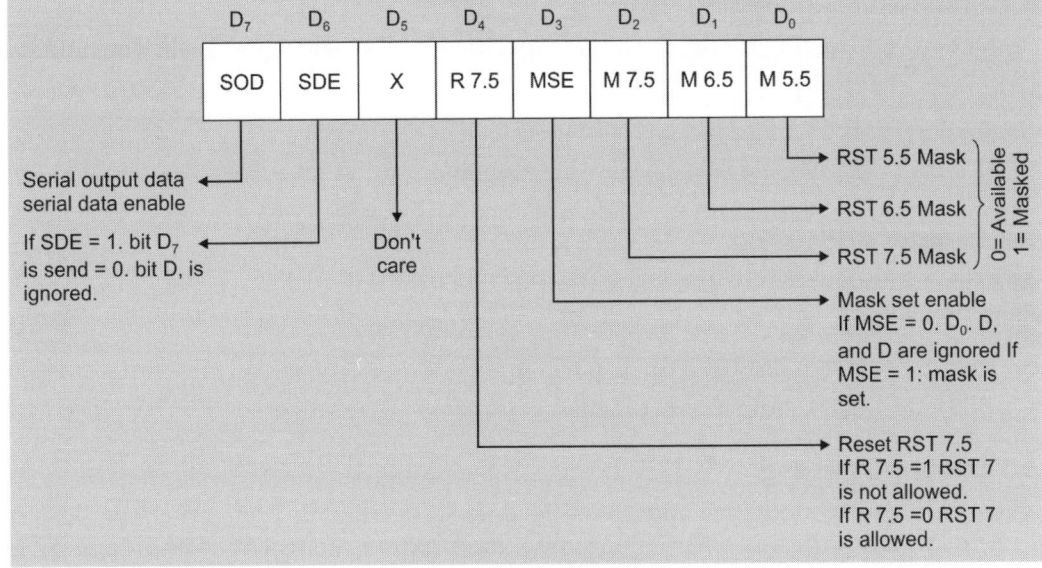

**Fig. 3.4:** Format of SIM

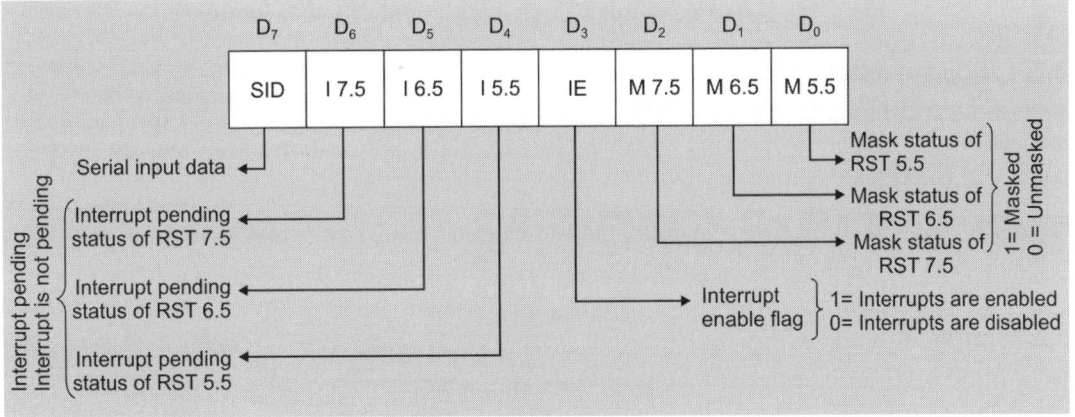

**Fig. 3.5:** Format of RIM

## 3.6 ADDRESSING MODES OF 8085

The instructions MOV B, A or MVI A, 82H are to copy data from a source into a destination. In these instructions, the source can be a register, an input port, or an 8-bit number (00H to FFH). Similarly, a destination can be a register or an output port. The sources and destination are operands. The various formats for specifying operands are called the addressing modes. For 8085, they are:

1. Immediate addressing
2. Register addressing
3. Direct addressing
4. Indirect addressing
5. Implied addressing

### 3.6.1 Immediate Addressing

In immediate addressing mode, the data is specified in the instruction itself. The data will be a part of the program instruction.

**Example:**     MVI A, 09; Move data 09H immediate in A

                   ADI 90; Add 90H data immediate with A

### 3.6.2 Register Addressing

In register addressing mode, the instruction specifies the name of the register in which the data is available.

**Example:**     MOV A, B; Move data in B into A

                   ADD C; Add data in C with A

### 3.6.3 Direct Addressing

In direct addressing mode, the address of the data is specified in the instruction. The data will be in memory. In this addressing mode, the program instructions and data can be stored in different memory.

**Example:**     STA 3000; Store data of A at memory location 3000H

                   LDA 3500; Load accumulator with data stored at 3500H location

### 3.6.4 Indirect Addressing

In register indirect addressing mode, the instruction specifies the name of the register in which the address of the data is available. Here the data will be in memory and the address will be in the register pair.

**Example:**     MOV A, M; the memory data addressed by H L pair is moved to A register

                   LDAX B; the memory data addressed by B C pair is moved to A register

### 3.6.5 Implied Addressing

In implied addressing mode, the instruction itself specifies the data to be operated.

**Example:**     CMA; Complement the content of A

                   RAL; Rotate the content of A

### 3.7 INSTRUCTION FORMAT

An **instruction** is a command to the microprocessor to perform a given task on a specified data. Each instruction has two parts: one is task to be performed, called the **operation code** (opcode), and the second is the data to be operated on, called the **operand.** The operand (or data) can be specified in various ways. It may include 8-bit (or 16-bit) data, an internal register, a memory location or 8-bit (or 16-bit) address. In some instructions, the operand is implicit. Instruction word size of 8085 instruction set is classified into the following three groups according to word size:

1. One-word or 1-byte instructions
2. Two-word or 2-byte instructions
3. Three-word or 3-byte instructions

In the 8085, "byte" and "word" are synonymous because it is an 8-bit microprocessor. However, instructions are commonly referred to in terms of bytes rather than words.

### 3.7.1 One-byte Instruction

A one-byte instruction includes the opcode and operand in the same byte. Operand(s) are internal registe and are coded into the instruction.

**For example:**

| Opcode | Operand | Binary code | Hex code |
|--------|---------|-------------|----------|
| ORA    | B       | 10110111    | B7       |
| ADC    | B       | 10001001    | 89H      |

These instructions are 1-byte instructions performing three different tasks. In the first instruction, both operand registers are specified. In the second instruction, the operand B is specified and the accumulator is assumed. Similarly, in the third instruction, the accumulator is assumed to be the implicit operand. These instructions are stored in 8-bit binary format in memory; each requires one memory location.

$$MOV\ R1,\ R2$$
$$R1 \leftarrow R2$$

Coded as 01 ddd sss where ddd is a code for one of the 7 general registers which is the destination of the data, sss is the code of the source register.

**Example:** MOV A, B

Coded as 01111000 = 78H = 170 octal (octal was used extensively in instruction design of such processors).

$$ADD\ R$$
$$A \leftarrow A+R$$

### 3.7.2 Two-byte Instruction

In a two-byte instruction, the first byte specifies the operation code and the second byte specifies the operand. Source operand is a data byte immediately following the opcode.

**For example:**

| Opcode | Operand | Binary code | Hex code | Size     |
|--------|---------|-------------|----------|----------|
| MVI    | A, 09H  | 00111110    | 3E       | 1st byte |
|        |         | 09H         | 09H      | 2nd byte |

Assume that the data is 32H. The assembly language instruction can be expressed as the following:

| Mnemonics   | Hex code |
|-------------|----------|
| MVI A, 32H  | 3E 32H   |

The instruction would require two memory locations to store in memory.

$$MVI\ R,\ Data$$
$$R \leftarrow Data$$

**Example:** MVI A, 30H coded as 3EH 30H has two contiguous bytes.

### 3.7.3 Three-byte Instruction

In a three-byte instruction, the first byte specifies the opcode and the following two bytes specify the 16-bit address. Note that the second byte is the low-order address and the third byte is the high-order address.

opcode + data byte + data byte

**For example:**

| Opcode | Operand | Binary code | Hex code | Size |
|--------|---------|-------------|----------|------|
| CALL | 2000H | 11001110 | CD | 1st byte |
| | | 00000000 | 00 | 2nd byte |
| | | 00100000 | 20 | 3rd byte |

This instruction would require three memory locations to store in memory.

Three byte instructions: opcode + data byte + data byte

LXI RP, 16-bit Data

RP is one of the pairs of registers BC, DE, HL used as 16-bit registers.

RP ← 16-bit data

P ← Lower 8 bits

R ← Upper 8 bits

**Example:**

LXI H, 2000H coded as 21H 00H 20H in three bytes. This is also an immediate addressing.

LDA 16-bit Address

A ← Data at 16-bit Address

**Example:** LDA 2000H coded as 3AH 00H 20H. This is also an example of direct addressing.

### 3.8 INSTRUCTION SET OF 8085

Instruction set of 8085 can be divided into the following categories:

1. Data transfer instructions: MOV, LDA, MVI, etc.
2. Arithmetic operation instructions: Add, subtract, increment and decrement.
3. Logical operation instructions: AND, OR, XOR and rotate.
4. Control transfer instructions: Conditional, unconditional, call subroutine, return from subroutine and restarts.
5. Input/output operation instructions, Machine control instructions.
6. Other-setting/clearing flag bits, enabling/disabling interrupts, stack operations, etc.

### 3.8.1 Data Transfer Instructions

The data transfer instructions move data between registers or between memory and registers.

**MOV** (Copy Data from Source to Destination)

**Syntax:** MOV Rd, Rs

This instruction copies the contents of the source M, Rs register into the destination register; the contents of Rd, M the source registers are not altered. If one of the operands is a memory location, its location is specified by the contents of the HL registers.

**Example:** MOV A, C or MOV A, M

**MVI** (Move Immediate 8-bit data)

**Syntax:** MVI Rd, data

The 8-bit data is transferred to the destination register or M, data memory. If the operand is a memory location, its location is specified by the contents of the HL registers.

**Example:** MVI A, 90H or MVI B, 60H

**LDA** (Load Accumulator Direct from Memory)

**Syntax:** LDA 16-bit address

The contents of a memory location, specified by a 16-bit address in the operand, are copied to the accumulator. The contents of the source are not altered.

**Example:** LDA 2000H

**LHLD** (Load H and L Registers Directly from Memory)

**Syntax:** LHLD 16-bit address

The instruction copies the contents of the memory location pointed out by the 16-bit address into register L and copies the contents of the next memory location into register H. The contents of source memory locations are not altered.

**Example:** LHLD 1200H

**LXI** (Load Register Pair with Immediate data)

**Syntax:** LXI Reg. pair, 16-bit data

The instruction loads 16-bit data in the register pair designated in the operand.

**Example:** LXI H, 2000H or LXI H, 2500H

**LDAX** (Load Accumulator from Address in Register Pair)

**Syntax:** LDAX B/D Reg. pair

The contents of the designated register pair point to a memory location. This instruction copies the contents of that memory location into the accumulator. The contents of either the register pair or the memory location are not altered.

**Example:** LDAX D

**STA** (Store Accumulator Direct in Memory)

**Syntax:** STA 16-bit address

The contents of the accumulator are copied into the memory location specified by the operand. This is a 3-byte instruction, the second byte specifies the low-order address and the third byte specifies the high-order address.

**Example:** STA 4000H

**STAX** (Store Accumulator in Address in Register Pair)

**Syntax:** STAX Reg. pair

The contents of the accumulator are copied into the memory location specified by the contents of the operand (register pair). The contents of the accumulator are not altered.

**Example:** STAX D

**SHLD (**Store H and L Registers Directly in Memory)
**Syntax:** SHLD 16-bit address

The contents of register L are stored into the memory location specified by the 16-bit address in the operand and the contents of H register are stored into the next memory location by incrementing the operand. The contents of registers HL are not altered. This is a 3-byte instruction, the second byte specifies the low-order address and the third byte specifies the high-order address.

**Example:** SHLD 3000H

**XCHG** (Exchange H and L with D and E)
**Syntax:** XCHG none

The contents of register H are exchanged with the contents of register D and the contents of register L are exchanged with the contents of register E.

**Example:** XCHG

**XTHL (**Exchange Top of Stack with H and L)
**Syntax:** XTHL none

The contents of the L register are exchanged with the stack location pointed out by the contents of the stack pointer register. The contents of the H register are exchanged with the next stack location (SP + 1); however, the contents of the stack pointer register are not altered.

**Example:** XTHL

**SPHL (**Move content of H and L to Stack Pointer)
**Syntax:** SPHL none

The instruction loads the contents of the H and L registers into the stack pointer register, the contents of the H register provide the high-order address and the contents of the L register provide the low-order address. The contents of the H and L registers are not altered.

**Example:** SPHL

### 3.8.2 Arithmetic Instructions

The arithmetic instructions are used to perform various arithmetic tasks, such as addition, subtraction, division, multiplication, increment or decrement, etc.

**ADD** (Add Register or Memory to Accumulator)
**Syntax:** ADD R

The contents of the operand (register or memory) are added to the contents of the accumulator and the result is stored in the accumulator. If the operand is a memory location, its location is specified by the contents of the HL registers. All flags are modified to reflect the result of the addition.

**Example:** ADD C

**ADI** (Add Immediate Data to Accumulator)
**Syntax:** ADI 8-bit data

The 8-bit data (operand) is added to the contents of the accumulator and the result is stored in the accumulator. All flags are modified to reflect the result of the addition.

**Example:** ADI 20H

**ADC** (Add Register to Accumulator with Carry)

**Syntax:** ADC R

The contents of the operand (register or memory) and the Carry flag are added to the contents of the accumulator and the result is stored in the accumulator. If the operand is a memory location, its location is specified by the contents of the HL registers. All flags are modified to reflect the result of the addition.

**Example:** ADC C

**ACI** (Add Immediate data to Accumulator with Carry)

**Syntax:** ACI 8-bit data

The 8-bit data (operand) and the Carry flag are added to the contents of the accumulator and the result is stored in the accumulator. All flags are modified to reflect the result of the addition.

**Example:** ACI 30H

**DAD** (Double Register Add; Add Content of Register Pair to H and L Register Pair)

**Syntax:** DAD Reg. pair

The 16-bit contents of the specified register pair are added to the contents of the HL register and the sum is stored in the HL register. The contents of the source register pair are not altered. If the result is larger than 16 bits, the CY flag is set. No other flags are affected.

**Example:** DAD H

**SUB** (Subtract Register or Memory from Accumulator)

**Syntax:** SUB R

The contents of the operand (register or memory) are subtracted from the contents of the accumulator, and the result is stored in the accumulator. If the operand is a memory location, its location is specified by the contents of the HL registers. All flags are modified to reflect the result of the subtraction.

**Example:** SUB C or SUB M

**SUI** (Subtract Immediate Data from Accumulator)

**Syntax:** SUI 8-bit data

The 8-bit data (operand) is subtracted from the contents of the accumulator and the result is stored in the accumulator. All flags are modified to reflect the result of the subtraction.

**Example:** SUI 20H

**SBB** (Subtract Source from Accumulator with Borrow)

**Syntax:** SBB R

The contents of the operand (register or memory) and the Borrow flag are subtracted from the contents of the accumulator and the result is placed in the accumulator. If the operand is a memory location, its location is specified by the contents of the HL registers. All flags are modified to reflect the result of the subtraction.

**Example:** SBB C or SBB M

**SBI** (Subtract Immediate data from Accumulator with Borrow)

**Syntax:** SBI 8-bit data

The 8-bit data (operand) and the Borrow flag are subtracted from the contents of the accumulator and the result is stored in the accumulator. All flags are modified to reflect the result of the subtraction.

**Example:** SBI 50H

**INR** (Increment Register or Memory by One)

**Syntax:** INR R

The contents of the designated register or memory are incremented by 1 and the result is stored in the same place. If the operand is a memory location, its location is specified by the contents of the HL registers.

**Example:** INR C or INR M

**DCR** (Decrement Register or Memory by One)

**Syntax:** DCR R

The contents of the designated register or memory are M decremented by 1 and the result is stored in the same place. If the operand is a memory location, its location is specified by the contents of the HL registers.

**Example:** DCR C or DCR M

**INX** (Increment Register Pair by One)

**Syntax:** INX R

The contents of the designated register pair are incremented by 1 and the result is stored in the same place.

**Example:** INX H

**DCX** (Decrement Register Pair by One)

**Syntax:** DCX R

The contents of the designated register pair are decremented by 1 and the result is stored in the same place.

**Example:** DCX H

**DAA** (Decimal Adjust after Addition)

**Syntax:** DAA none

The contents of the accumulator are changed from a binary value to two 4-bit binary coded decimal (BCD) digits. This is the only instruction that uses the auxiliary flag to perform the binary to BCD conversion, and the conversion procedure is described below. S, Z, AC, P, CY flags are altered to reflect the results of the operation. If the value of the low-order 4-bits in the accumulator is greater than 9 or if AC flag is set, the instruction adds 6 to the low-order 4-bits. If the value of the high-order 4-bits in the accumulator is greater than 9 or if the Carry flag is set, the instruction adds 6 to the high-order 4-bits.

**Example:** DAA

### 3.8.3 Logical Instructions

This group of instructions performs logical (Boolean) operations on data in registers and memory and on condition flags. The logical operations are AND, OR and Exclusive OR, etc.

**ANA** (Logical AND of Register or Memory Data with Accumulator)

**Syntax:** ANA R

The contents of the accumulator are logically ANDed with the contents of the operand (register or memory), and the result is placed in the accumulator. If the operand is a memory location, its address is specified by the contents of HL registers. S, Z, P are modified to reflect the result of the operation. CY is reset. AC is set.

**Example:** ANA B or ANA M

**ANI** (Logical AND of Immediate Data with Accumulator)

**Syntax:** ANI 8-bit data

The contents of the accumulator are logically ANDed with the 8-bit data (operand) and the result is placed in the accumulator. S, Z, P are modified to reflect the result of the operation. CY is reset. AC is set.

**Example:** ANI 40H

**ORA** (Logical OR of Register or Memory Data with Accumulator)

**Syntax:** ORA R

The contents of the accumulator are logically ORed with the contents of the operand (register or memory), and the result is placed in the accumulator. If the operand is a memory location, its address is specified by the contents of HL registers. S, Z, P are modified to reflect the result of the operation. CY and AC are reset.

**Example:** ORA B or ORA M

**ORI** (Logical OR of Immediate Data with Accumulator)

**Syntax:** ORI 8-bit data

The contents of the accumulator are logically ORed with the 8-bit data (operand) and the result is placed in the accumulator. S, Z, P are modified to reflect the result of the operation. CY and AC are reset.

**Example:** ORI 45H

**XRA** (Logical XOR of the Register or Memory Data with Accumulator)

**Syntax:** XRA R

The contents of the accumulator are Exclusive ORed with M the contents of the operand (register or memory), and the result is placed in the accumulator. If the operand is a memory location, its address is specified by the contents of HL registers. S, Z, P are modified to reflect the result of the operation. CY and AC are reset.

**Example:** XRA B or XRA M

**XRI** (Exclusive OR of Immediate Data with Accumulator)
**Syntax:** XRI 8-bit data

The contents of the accumulator are Exclusive ORed with the 8-bit data (operand) and the result is placed in the accumulator. S, Z, P are modified to reflect the result of the operation. CY and AC are reset.

**Example:** XRI 80H

**CMP** (Compare Register or Memory with Accumulator)
**Syntax:** CMP R

The contents of the operand (register or memory) are M compared with the contents of the accumulator. Both contents are preserved. The result of the comparison is shown by setting the flags of the PSW as follows:

If (A) < (reg/mem) => carry flag is set
If (A) = (reg/mem) => zero flag is set
If (A) > (reg/mem) => carry and zero flags are reset.

**Example:** CMP B or CMP M

**CPI** (Compare Immediate Data with Accumulator)
**Syntax:** CPI 8-bit data

The second byte (8-bit data) is compared with the contents of the accumulator. The values being compared remain unchanged. The result of the comparison is shown by setting the flags of the PSW as follows:

if (A) < data => carry flag is set
if (A) = data => zero flag is set
if (A) > data => carry and zero flags are reset.

**Example:** CPI 50H

**RLC** (Rotate Accumulator Left)
**Syntax:** RLC none

Each binary bit of the accumulator is rotated left by one position. Bit D7 is placed in the position of D0 as well as in the Carry flag. CY is modified according to bit D7. S, Z, P, AC are not affected.

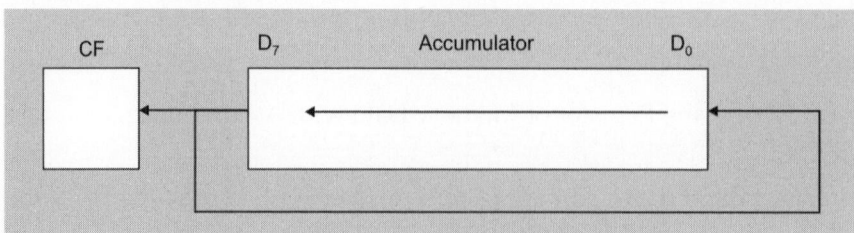

**Example:** RLC

**RRC** (Rotate Accumulator Right)

**Syntax:** RRC none

Each binary bit of the accumulator is rotated right by one position. Bit D0 is placed in the position of D7 as well as in the Carry flag. CY is modified according to bit D0. S, Z, P, AC are not affected.

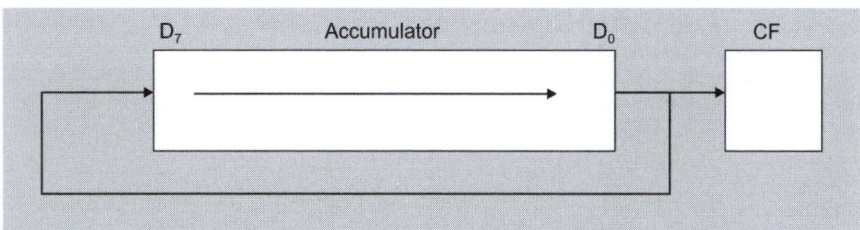

**Example:** RRC

**RAL** (Rotate Accumulator Left through Carry)

**Syntax:** RAL none

Each binary bit of the accumulator is rotated left by one position through the Carry flag. Bit D7 is placed in the Carry flag and the Carry flag is placed in the least significant position D0. CY is modified according to bit D7. S, Z, P, AC are not affected.

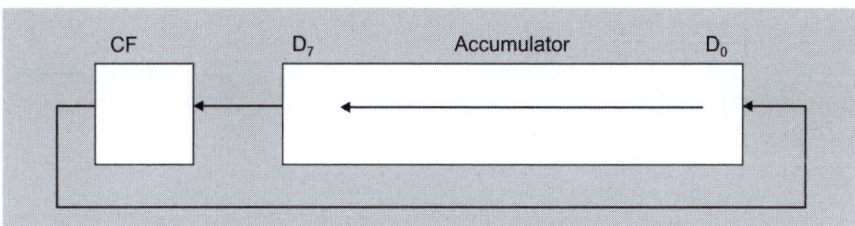

**Example:** RAL

**RAR** (Rotate Accumulator Right through Carry)

**Syntax:** RAR none

Each binary bit of the accumulator is rotated right by one position through the Carry flag. Bit D0 is placed in the Carry flag and the Carry flag is placed in the most significant position D7. CY is modified according to bit D0. S, Z, P, AC are not affected.

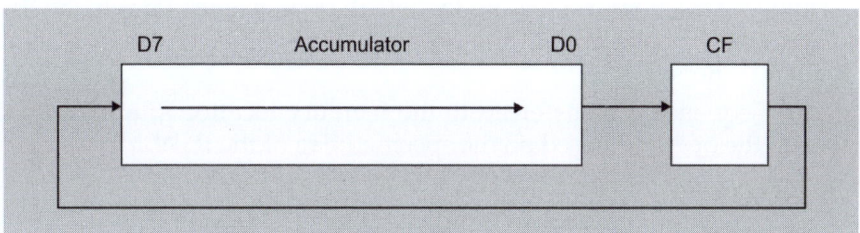

**Example:** RAR

**CMA** (Complement Accumulator)

**Syntax:** CMA none

The contents of the accumulator are complemented. No flags are affected.

Accumulator

| D7 | D6 | D5 | D4 | D3 | D2 | D1 | D0 |
|----|----|----|----|----|----|----|----|

↓

| $\overline{D7}$ | $\overline{D6}$ | $\overline{D5}$ | $\overline{D4}$ | $\overline{D3}$ | $\overline{D2}$ | $\overline{D1}$ | $\overline{D0}$ |
|----|----|----|----|----|----|----|----|

**Example:** CMA

**CMC** (Complement Carry Flag)

**Syntax:** CMC none

The Carry flag is complemented. No other flags are affected.

**Example:** CMC

**STC** (Set Carry Flag)

**Syntax:** STC none

The Carry flag is set to 1. No other flags are affected.

**Example:** STC

### 3.8.4 Branch Instructions

The branching instructions alter normal sequential program flow, either unconditionally or conditionally.

**Unconditional Branching Instruction**

**JMP** (Jump Unconditionally)

**Syntax:** JMP 16-bit address

The program sequence is transferred to the memory location specified by the 16-bit address given in the operand.

**Example:** JMP 2000H

**Conditional Branching Instructions**

This kind of jump instructions are required where the status of the flags in the flag register need to meet a specific condition in the program. When the condition is met then the program sequence alters.

The program sequence is transferred to the memory location specified by the 16-bit address given in the operand based on the specified flag of the PSW.

**Operand:** 16-bit address

Different types of conditional jump instructions are shown in Table 3.6.

**Table 3.6:** Conditional jump instructions

| Opcode | Description | Flag status |
|--------|-------------|-------------|
| JC | Jump if carry | CY = 1 |
| JNC | Jump if not carry | CY = 0 |
| JP | Jump if positive | S = 0 |
| JM | Jump if minus | S = 1 |
| JZ | Jump if zero | Z = 1 |
| JNZ | Jump if not zero | Z = 0 |
| JPE | Jump if parity even | P = 1 |
| JPO | Jump if parity odd | P = 0 |

**Example:** JZ 3000H

## Unconditional Subroutine CALL

**CALL** (Unconditional Subroutine Call)

**Syntax:** CALL 16-bit address

The program sequence is transferred to the memory location specified by the 16-bit address given in the operand. Before the transfer, the address of the next instruction after CALL (the contents of the program counter) is pushed onto the stack.

**Example:** CALL 2500H

## Conditional Subroutine CALL

The program sequence is transferred to the memory location specified by the 16-bit address given in the operand based on the specified flag of the PSW as described below. Before the transfer, the address of the next instruction after the call (the contents of the program counter) is pushed onto the stack.

**Operand:** 16-bit address

Several types of conditional CALL instructions are shown in Table 3.7.

**Table 3.7:** Conditional CALL instructions

| Opcode | Description | Flag status |
|--------|-------------|-------------|
| CC | Call if carry | CY = 1 |
| CNC | Call if not carry | CY = 0 |
| CP | Call if positive | S = 0 |
| CM | Call if minus | S = 1 |
| CZ | Call if zero | Z = 1 |
| CNZ | Call if not zero | Z = 0 |
| CPE | Call if parity even | P = 1 |
| CPO | Call if parity odd | P = 0 |

**Example:** CZ 2034H or CZ XYZ

## Unconditional Return from Subroutine

**RET** (Return from Subroutine Unconditionally)

**Syntax:** RET none

The program sequence is transferred from the subroutine to the calling program. The two bytes from the top of the stack are copied into the program counter and program execution begins at the new address.

**Example:** RET

## Conditional Return from Subroutine

The program sequence is transferred from the subroutine to the calling program based on the specified flag of the PSW as described below. The two bytes from the top of the stack are copied into the program counter and program execution begins at the new address.

**Operand:** None

Various Conditional Return instructions are shown in Table 3.8.

| **Table 3.8:** Various conditional return instructions | | |
|---|---|---|
| Opcode | Description | Flag status |
| RC | Return if carry | CY = 1 |
| RNC | Return if not carry | CY = 0 |
| RP | Return if positive | S = 0 |
| RM | Return if minus | S = 1 |
| RZ | Return if zero | Z = 1 |
| RNZ | Return if not zero | Z = 0 |
| RPE | Return if parity even | P = 1 |
| RPO | Return if parity odd | P = 0 |

**Example:** RZ

**PCHL** (Move the contents of HL to Program Counter)

**Syntax:** PCHL none

The contents of registers H and L are copied into the program counter. The contents of H are placed as the high-order byte and the contents of L as the low-order byte.

**Example:** PCHL

**RST** (Special Restart Instruction Used with Interrupts)

**RST 0–7:** The RST instruction is equivalent to a 1-byte call instruction to one of eight memory locations depending upon the number. The instructions are generally used in conjunction with interrupts and inserted using external hardware. However, these can be used as software instructions in a program to transfer program execution to one of the eight locations. The addresses are:

**Instruction Restart Address**

RST 0 0000H

RST 1 0008H

RST 2 0010H

RST 3 0018H

RST 4 0020H

RST 5 0028H

RST 6 0030H

RST 7 0038H

The 8085 has four additional interrupts and these interrupts generate RST instructions internally and thus do not require any external hardware. These instructions and their Restart addresses are:

**Interrupt Restart Address**

TRAP 0024H

RST 5.5 002CH

RST 6.5 0034H

RST 7.5 003CH

### 3.8.5 Stack I/O Instructions

The following instructions affect the Stack and/or Stack Pointer.

**PUSH** (Push Two bytes of Data onto the Stack)

**Syntax:** PUSH Reg. pair

The contents of the register pair designated in the operand are copied onto the stack in the following sequence. The stack pointer register is decremented and the contents of the high-order register (B, D, H, A) are copied into that location. The stack pointer register is decremented again and the contents of the low-order register (C, E, L, flags) are copied to that location.

**Example:** PUSH B or PUSH A

**POP** (Pop Two Bytes of Data from the Stack)

**Syntax:** POP Reg. pair

The contents of the memory location pointed out by the stack pointer register are copied to the low-order register (C, E, L, status flags) of the operand. The stack pointer is incremented by 1 and the contents of that memory location are copied to the high-order register (B, D, H, A) of the operand. The stack pointer register is again incremented by 1.

**Example:** POP H or POP A

### 3.8.6 I/O Instructions

**IN** (Input Data to Accumulator from a Port with 8-bit Address)

**Syntax:** IN 8-bit port address

The contents of the input port designated in the operand are read and loaded into the accumulator.

**Example:** IN 80H

**OUT** (Output Data from Accumulator to a Port with 8-bit Address)

**Syntax:** OUT 8-bit port address

The contents of the accumulator are copied into the I/O port specified by the operand.

**Example:** OUT 20H

## 3.8.7 Machine Control Instructions

**EI** (Enable Interrupt)

**Syntax:** EI none

The interrupt enable flip-flop is set and all interrupts are enabled. No flags are affected. After a system reset or the acknowledgement of an interrupt, the interrupt enable flip-flop is reset, thus disabling the interrupts. This instruction is necessary to re-enable the interrupts (except TRAP).

**Example:** EI

**DI** (Disable Interrupt)

**Syntax:** DI none

The interrupt enable flip-flop is reset and all the interrupts except the TRAP are disabled. No flags are affected.

**Example:** DI

**SIM** (Set Interrupt Mask)

**Syntax:** SIM

This is a multipurpose instruction, used to implement the 8085 interrupts and serial data output. When microprocessor executes this instruction then 8-bit data of the accumulator is transferred to different flip-flops. This instruction interprets the accumulator contents as follows:

| D7 | D6 | D5 | D4 | D3 | D2 | D1 | D0 |
|-----|-----|-----|------|-----|------|------|------|
| SOD | SOE | X | R7.5 | MSE | M7.5 | M6.5 | M5.5 |

Description of symbols used is given in Table 3.9.

**Table 3.9:** Description of symbols

| Symbol | Description |
|--------|-------------|
| SOD | Serial output data |
| SOE | Serial output enable |
| R7.5 | RST7.5 |
| M7.5, M6.5, M5.5 | Mask interrupts |

**Example:** SIM

**RIM** (Read Interrupt Mask)

**Syntax:** RIM

This instruction is used to read the status of interrupts 7.5, 6.5, 5.5 and to read the serial data input bits on SID pin. The instruction loads 8-bits of the accumulator with the following interpretation.

| D7 | D6 | D5 | D4 | D3 | D2 | D1 | D0 |
|----|----|----|----|----|----|----|----|
| SID | 17.5 | 16.5 | 15.5 | IE | M7.5 | M6.5 | M5.5 |

Description of symbols used is given in Table 3.10.

**Table 3.10:** Description of symbols

| Symbol | Description |
|--------|-------------|
| SID | Serial input data bit |
| 17.5, 16.5, 15.5 | Interrupt pending if bit = 1 |
| IE | Interrupt enable flip-flop is set if bit = 1 |
| M7.5, M6.5, M5.5 | Interrupt masked if bit = 1 |

**Example:** RIM

**HLT** (Halt and Enter in Wait State)

**Syntax:** HLT none

The CPU finishes executing the current instruction and halts any further execution. An interrupt or reset is necessary to exit from the halt state.

**Example:** HLT

**NOP** (No Operation)

**Syntax:** NOP none

No operation is performed. The instruction is fetched and decoded. However, no operation is executed.

**Example:** NOP

## 3.9 TIMING DIAGRAM

Timing diagram is a graphical representation. It represents the execution time taken by each instruction in a graphical format. The execution time is represented in T-states.

### 3.9.1 Instruction Cycle

The time required to execute an instruction is called instruction cycle.

### 3.9.2 Machine Cycle

The time required to access the memory or input/output devices is called machine cycle.

### 3.9.3 T-State

The machine cycle and instruction cycle takes multiple clock periods. A portion of an operation carried out in one system clock period is called as T-state.

## 3.10 MACHINE CYCLES OF 8085

The 8085 microprocessor has five basic machine cycles (Table 3.11). They are as follows:
1. Opcode fetch cycle (4T)
2. Memory read cycle (3 T)
3. Memory write cycle (3 T)
4. I/O read cycle (3 T)
5. I/O write cycle (3 T)

**Table 3.11:** Machine cycles

| Machine Cycle | Status | | | Controls | | |
|---|---|---|---|---|---|---|
| | IO/M (Active low) | S1 | S0 | RD (Active low) | WR (Active low) | INTA (Active low) |
| Opcode fetch (OF) | 0 | 1 | 1 | 0 | 1 | 1 |
| Memory read | 0 | 1 | 0 | 0 | 1 | 1 |
| Memory write | 0 | 0 | 1 | 1 | 0 | 1 |
| I/O read | 1 | 1 | 0 | 0 | 1 | 1 |
| I/O write | 1 | 0 | 1 | 1 | 0 | 1 |
| INTA | 1 | 1 | 1 | 1 | 1 | 0 |
| Bus idle (BI): DAD | 0 | 1 | 0 | 1 | 1 | 1 |
| ACK of RST, TRAP | 1 | 1 | 1 | 1 | 1 | 1 |
| HALT | Z | 0 | 0 | Z | Z | 1 |
| HOLD | Z | X | X | Z | Z | 1 |

X = Unspecified

Z = High impedance state

Each instruction of the 8085 processor consists of one to five machine cycles, i.e. when the 8085 processor executes an instruction, it will execute some of the machine cycles in a specific order. The processor takes a definite time to execute the machine cycles. The time taken by the processor to execute a machine cycle is expressed in T-states. One T-state is equal to the time period of the internal clock signal of the processor. The T-state starts at the falling edge of a clock.

### 3.10.1 Opcode Fetch Machine Cycle

Each instruction of the processor has one byte opcode. The opcodes are stored in memory. The processor executes the opcode fetch machine cycle to fetch the opcode from memory. Hence, every instruction starts with opcode fetch machine cycle. The time taken by the processor to execute the opcode fetch cycle is 4 T. During this time, the first 3 T-states are used for fetching the opcode from memory and the remaining T-states are used for internal operations by the processor (Fig. 3.6).

### 3.10.2 Memory Read Machine Cycle

The memory read machine cycle is executed by the processor to read a data byte from memory. The processor takes 3 T states to execute this cycle. The instructions which have more than one byte word size will use the machine cycle after the opcode fetch machine cycle (Fig. 3.7).

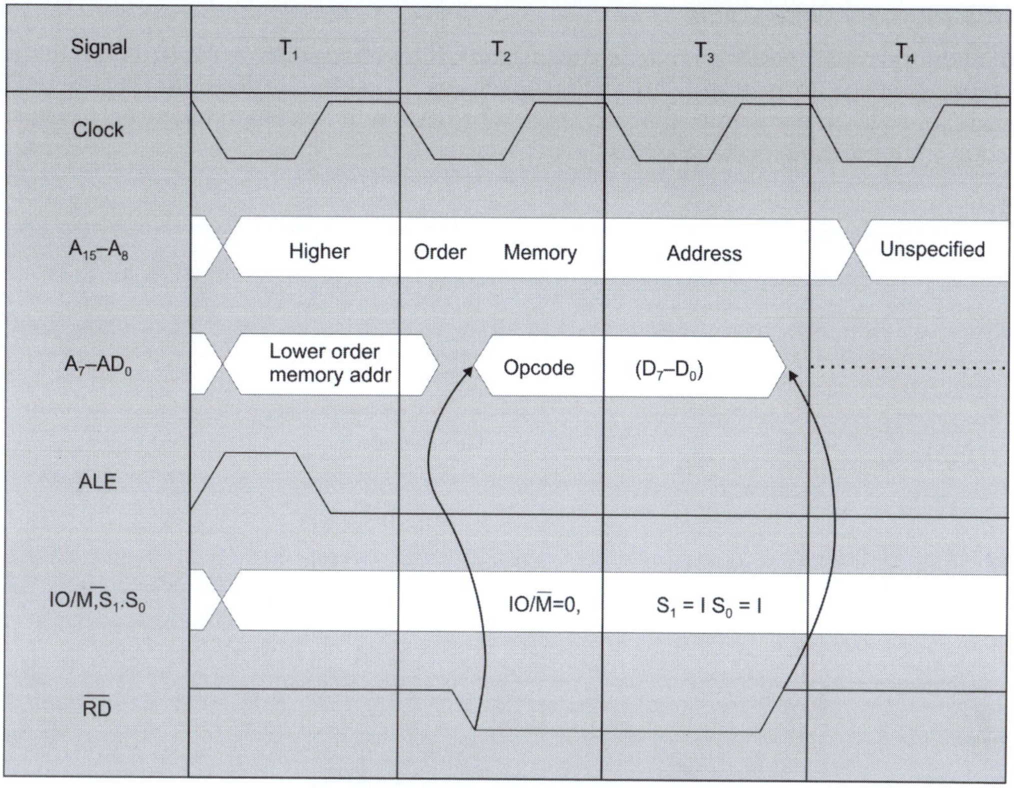

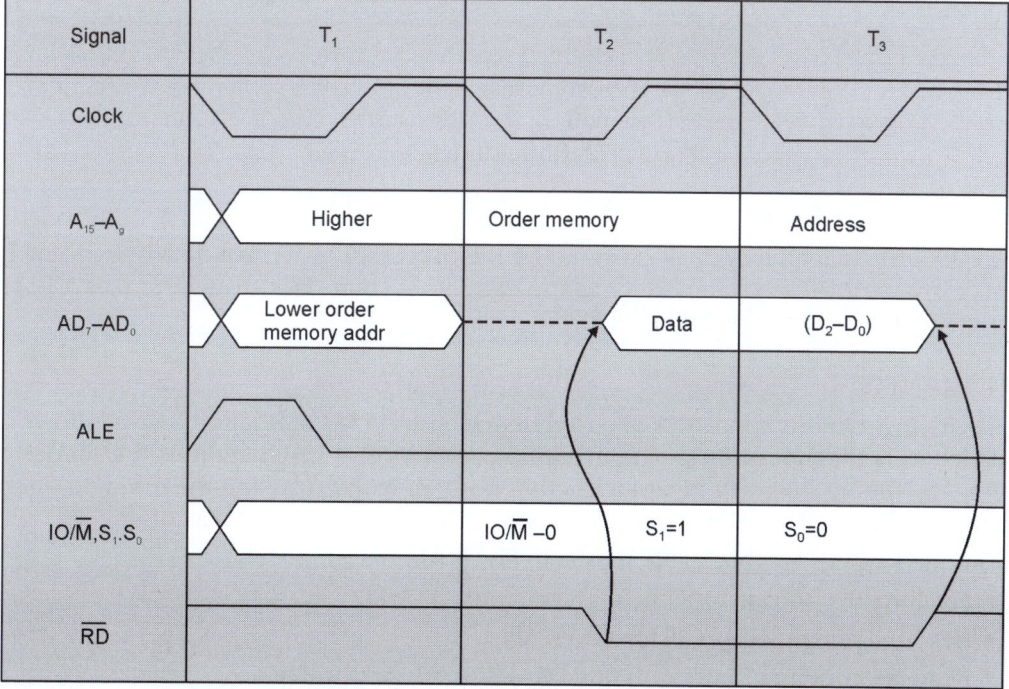

**Fig. 3.6:** Opcode fetch machine cycle

**Fig. 3.7:** Memory read machine cycle

### 3.10.3 Memory Write Cycle

The memory write machine cycle is executed by the processor to write a data byte from memory/output peripheral. The processor takes 3 T states to execute this cycle. The instructions which have more than one byte word size will use the machine cycle after the opcode fetch machine cycle (Fig. 3.8).

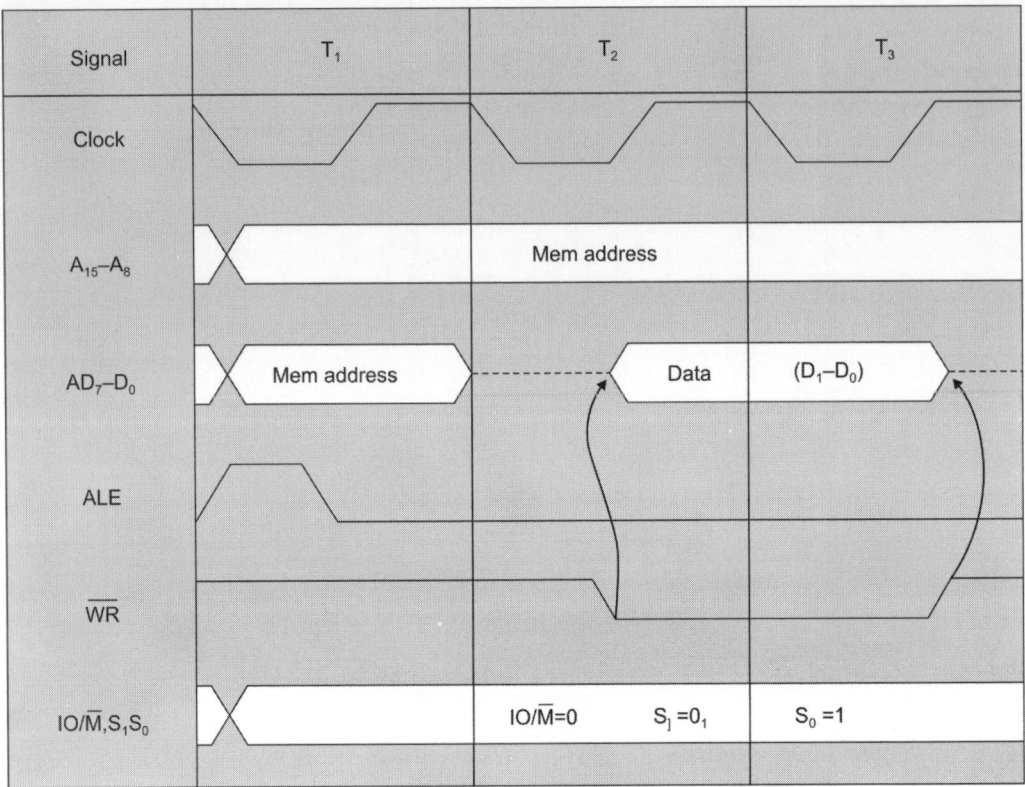

**Fig. 3.8:** Memory write machine cycle

### 3.10.4 I/O Write Cycle

The I/O write machine cycle is executed by the processor to write a data byte in the I/O port or to a peripheral, which is I/O, mapped in the system. The processor takes 3 T states to execute this machine cycle (Fig. 3.9).

### 3.11 EXAMPLES

The 8085 instructions consist of one to five machine cycles. Actually the execution of an instruction is the execution of the machine cycles of that instruction in the predefined order. The timing diagram of an instruction is obtained by drawing the timing diagrams of the machine cycles of that instruction, one-by-one in the order of execution.

Timing diagram for IN C0H is shown in Fig. 3.10.

- Fetching the opcode DBH from the memory 4125H.
- Read the port address C0H from 4126H.
- Read the content of port C0H and send it to the accumulator.
- Let the content of port is 5EH.

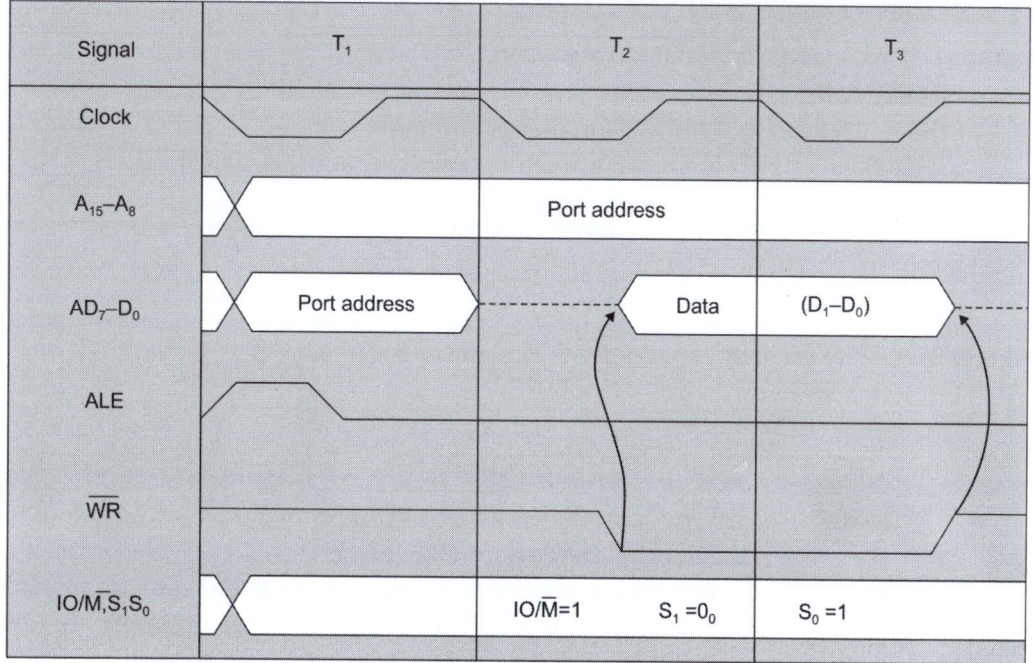

**Fig. 3.9:** I/O write cycle

| Address | Mnemonics | Opcode |
|---------|-----------|--------|
| 4125 | IN C0H | DBH |
| 4126 | | C0H |

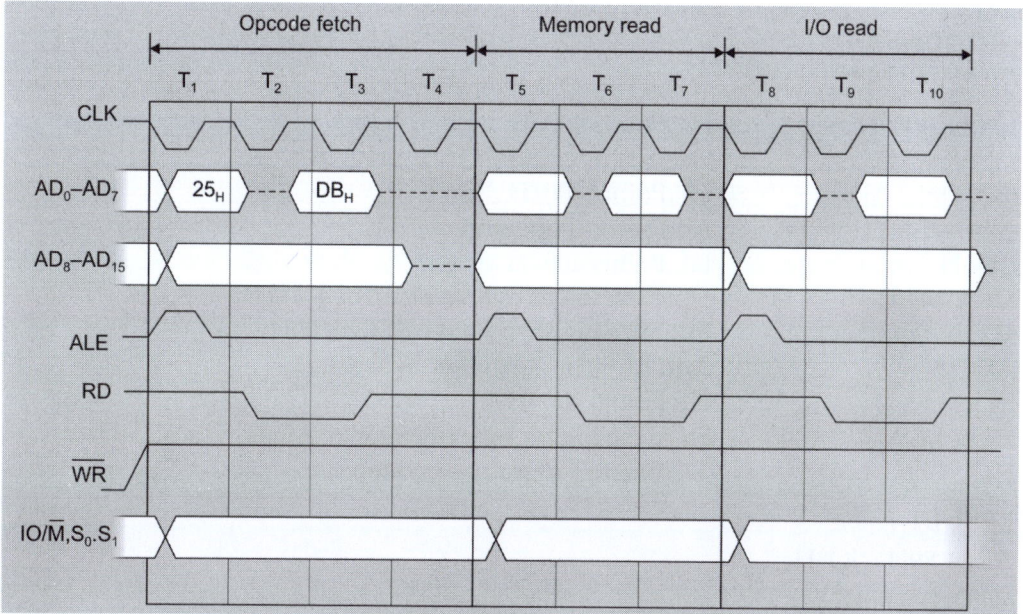

**Fig. 3.10:** Timing diagram of IN C0H

## 3.12 ASSEMBLY LANGUAGE PROGRAMMING

**Program 1:** WAP to store 8-bit data in memory.

### PROGRAM

| | | |
|---|---|---|
| MVI A, 52H | : | Store 32H in the accumulator |
| STA 4000H | : | Copy accumulator contents at address 4000H |
| HLT | : | Terminate program execution |

### RESULT

(4000H) = 52H

**Program 2:** WAP to exchange the contents of memory locations.

### PROGRAM

(2000H) = 20H

(4000H) = 30H

| | | |
|---|---|---|
| LDA 2000H | : | Get the contents of memory location 2000H into accumulator |
| MOV B, A | : | Save the contents into B register |
| LDA 4000H | : | Get the contents of memory location 4000H into accumulator |
| STA 2000H | : | Store the contents of accumulator at address 2000H |
| MOV A, B | : | Get the saved contents back into A register |
| STA 4000H | : | Store the contents of accumulator at address 4000H |

### RESULT

(2000H) = 30H

(4000H) = 20H

**Program 3:** WAP to add contents of two memory locations.

### PROGRAM

(4000H) = 7FH

(4001H) = 89H

| | | |
|---|---|---|
| LXI H, 4000H | : | HL Points 4000H |
| MOV A, M | : | Get first operand |
| INX H | : | HL Points 4001H |
| ADD M | : | Add second operand |
| INX H | : | HL Points 4002H |
| MOV M, A | : | Store the lower byte of result at 4002H |
| MVIA, 00 | : | Initialize higher byte result with 00H |
| ADC A | : | Add carry in the higher byte result |
| INX H | : | HL Points 4003H |
| MOV M, A | : | Store the higher byte of result at 4003H |
| HLT | : | Terminate program execution |

### RESULT

7FH + 89H = 1O8H

(4002H) = 08H

(4003H) = 0lH

**Program 4:** Subtract the contents of memory location 4001H from the memory location 2000H and place the result in memory location 4002H.

**PROGRAM**

(4000H) = 51H

(4001H) = 19H

| | | |
|---|---|---|
| LXI H, 4000H | : | HL points 4000H |
| MOV A, M | : | Get first operand |
| INX H | : | HL points 4001H |
| SUB M | : | Subtract second operand |
| INX H | : | HL points 4002H |
| MOV M, A | : | Store result at 4002H |
| HLT | : | Terminate program execution |

**RESULT**

(4202H) = 51H – 19H = 38H

**Program 5:** WAP to finding One's complement of a number.

**PROGRAM**

(4400H) = 55H

| | | |
|---|---|---|
| LDA 4400B | : | Get the number |
| CMA | : | Complement number |
| STA 4300H | : | Store the result |
| HLT | : | Terminate program execution |

**RESULT**

(4300B) = AAB

**Program 6:** WAP to finding Two's complement of a number.

**PROGRAM**

(4200H) = 55H

| | | |
|---|---|---|
| LDA 4200H | : | Get the number |
| CMA | : | Complement the number |
| ADI, 01 H | : | Add one in the number |
| STA 4300H | : | Store the result |
| HLT | : | Terminate program execution |

**RESULT**

(4300H) = AAH + 1 = ABH

**Program 7:** WAP to alter the contents of flag register.

**PROGRAM**

| | | |
|---|---|---|
| PUSH PSW | : | Save flags on stack |
| POP H | : | Retrieve flags in 'L' |
| MOV A, L | : | Flags in accumulator |
| CMA | : | Complement accumulator |

MOV L, A         :    Accumulator in 'L'
PUSH H           :    Save on stack
POP PSW          :    Back to flag register
HLT                :    Terminate program execution

**Program 8:** WAP to add Two 16-bit numbers and sum is 16-bit.

**PROGRAM**

1st number = 2040H

2nd number = 3020H

(8501H) = 40H

(8502H) = 20H

(8503H) = 20H

(8504H) = 30H

LHLD 8501H    :    Load the content of 8501H location in L register and H register is loaded with the content of 8502H location.

XCHG            :    DE $\leftrightarrow$ HL

LHLD 8503H    :    Load the content of 8503H location in L register and H register is loaded with the content of 8504H location.

DAD             :    HL $\leftarrow$ DE + HL

SHLD 8505H    :    (8505H) $\leftarrow$ L, (8506H) $\leftarrow$ H

HLT

**RESULT**

2040H + 3020H = 5060H

(8505H) = 60H

(8506H) = 50H

**Program 9:** WAP to pack the unpacked BCD numbers.

**PROGRAM**

Unpacked Data

(4200H) = 04

(4201H) = 09

LDA 4201H     :    Get the most significant BCD digit

RLC

RLC

RLC

RLC             :    Adjust the position of the second digit (09 is changed to 90)

ANI FOH       :    Make least significant BCD digit zero

MOV C, A      :    Store the partial result

LDA 4200H     :    Get the lower BCD digit

ADD C         :    Add lower BCD digit

STA 4300H     :    Store the result

HLT             :    Terminate program execution

**RESULT**

(4300H) = 94

**Program 10:** WAP to unpack a BCD number.

**PROGRAM**

Packed Data

(4200H) = 58

| | | |
|---|---|---|
| LDA 4200H | : | Get the packed BCD number |
| ANI FOH | : | Mask lower nibble |
| RRC | | |
| RRC | | |
| RRC | | |
| RRC | : | Adjust higher BCD digit as a lower digit |
| STA 4301H | : | Store the partial result |
| LDA 4200H | : | Get the original BCD number |
| ANI OFH | : | Mask higher nibble |
| STA 4201H | : | Store the result |
| HLT | : | Terminate program execution |

**RESULT**

(4300H) = 08

(4301H) = 05

**Program 11:** To perform addition of two 8-bit BCD numbers.

**PROGRAM**

(8200H) = 21H

(8201H) = 29H

| | | | |
|---|---|---|---|
| | LDA 8200 | : | Load data 1 into accumulator |
| | MOV B, A | : | Move accumulator contents to register B |
| | LDA 8201 | : | Load data 2 into accumulator |
| | MVI C, 00 | : | Clear register C to account for carry |
| | ADD B | : | Add data 2 to data 1 and store in accumulator |
| | DAA | : | Convert the accumulator value to BCD value |
| | JNC Ahead | : | If carry = 0, go ahead |
| | INR C | : | If carry = 1, increment register C by 1 |
| AHEAD : | STA 8203 | : | Store accumulator content (Result) to memory |
| | MOV A, C | : | Move contents of register C to accumulator |
| | STA 8204 | : | Store register C (Carry) content to memory |
| | HLT | : | End of program |

**RESULT**

(8203H) = 50H

**Program 12:** WAP to calculate the sum of series of numbers.

**PROGRAM**

(4200H) = 04H

(4201H) = 10H

(4202H) = 45H

(4203H) = 33H

(4204H) = 22H

| | | | |
|---|---|---|---|
| | LDA 4200H | | |
| | MOV C, A | : | Initialize counter |
| | SUB A | : | Sum = 0 |
| | LXI H, 4201H | : | Initialize pointer |
| BACK : | ADD M | : | SUM = Sum + data |
| | INX H | : | Increment pointer |
| | DCR C | : | Decrement counter |
| | JNZ BACK | : | If counter 0, repeat |
| | STA 4300H | : | Store sum |
| | HLT | : | Terminate program execution |

**RESULT**

10 + 41 + 30 + 12 = 83H

(4300H) = 83 H

**Program 13:** WAP to divide a 16-bit number by an 8-bit number.

**PROGRAM**

(2200H) = 60H

(2201H) = A0H

(2202H) = 12H

| | | | |
|---|---|---|---|
| | LHLD 2200H | : | Get the dividend |
| | LDA 2202H | : | Get the divisor |
| | MOV C, A | | |
| | LXI D, 0000H | : | Quotient = 0 |
| BACK : | MOV A, L | | |
| | SUB C | : | Subtract divisor |
| | MOV L, A | : | Save partial result |
| | JNC SKIP | : | If CY, 1 jump |
| | DCR H | : | Subtract borrow of previous subtraction |
| SKIP : | INX D | : | Increment quotient |
| | MOV A, H | | |
| | CPI, 00 | : | Check if dividend < divisor |
| | JNZ BACK | : | If no repeat |
| | MOV A, L | | |
| | CMP C | | |
| | JNC BACK | | |

| | | | |
|---|---|---|---|
| SHLD 2302H | : | Store the remainder |
| XCHG | | |
| SHLD 2300H | : | Store the quotient |
| HLT | : | Terminate program execution |

**Program 14:** WAP to find the square of the given numbers from memory location 6100H and store the result from memory location 7000H.

### PROGRAM

(6100H) = 04H

| | | | |
|---|---|---|---|
| | LXI H, 6200H | : | Initialize lookup table pointer |
| | LXI D, 6100H | : | Initialize source memory pointer |
| | LXI B, 7000H | : | Initialize destination memory pointer |
| BACK : | LDAX D | : | Get the number |
| | MOV L, A | : | A point to the square |
| | MOV A, M | : | Get the square |
| | STAX B | : | Store the result at destination memory location |
| | INX D | : | Increment source memory pointer |
| | INX B | : | Increment destination memory pointer |
| | MOV A, C | | |
| | CPI 05H | : | Check for last number |
| | JNZ BACK | : | If not repeat |
| | HLT | : | Terminate program execution |

### RESULT

(7000H) = 10H

**Program 15:** To find the smallest element in an array of size 'n'.

### PROGRAM

(8200H) = 05H
(8201H) = 10H
(8202H) = 13H
(8203H) = 20H
(8204H) = 40H
(8205H) = 12H

| | | | |
|---|---|---|---|
| | LXI H, 8200 | : | Setting pointer for data |
| | MOV B, M | : | Move number of data to register B from memory |
| | INX H | : | Increment the memory pointer |
| | MOV A, M | : | Move contents of memory to accumulator |
| | DCR B | : | Decrement register B content by 1 |
| Loop : | INX H | : | Increment the memory pointer |
| | CMP M | : | Compare contents of memory with Accumulator |
| | JC Ahead | : | If carry = 1, go Ahead, i.e. number is larger |
| | MOV A, M | : | Move contents of memory to accumulator |

| Ahead : | DCR B | : | Decrement register B content by 1 |
| | JNZ Loop | : | If register B is not equal to '0', go to loop |
| | STA 8300 | : | Store accumulator content (Smallest No) to memory |
| | HLT | : | End of Program |

**RESULT**

(8300H)=10H

**Program 16:** To find the largest element in an array of size 'n'.

**PROGRAM**

(8200H) = 05H

(8201H) = 10H

(8202H) = 13H

(8203H) = 20H

(8204H) = 40H

(8205H) = 12H

| | LXI H, 8200 | : | Setting pointer for data |
| | MOV B, M | : | Move number of data to register B from memory |
| | INX H | : | Increment the memory pointer |
| | MOV A, M | : | Move contents of memory to accumulator |
| | DCR B | : | Decrement register B content by 1 |
| Loop : | INX H | : | Increment the memory pointer |
| | CMP M | : | Compare contents of memory with accumulator |
| | JNC Ahead | : | If carry = 0, go ahead, i.e. number is smaller |
| | MOV A, M | : | Move contents of memory to accumulator |
| Ahead : | DCR B | : | Decrement register B content by 1 |
| | JNZ Loop | : | If register B is not equal to '0', go to loop |
| | STA 8300 | : | Store accumulator content (Sum) to memory |
| | HLT | : | End of Program |

**RESULT**

(8300H) = 40H

**Program 17:** To display 'n' elements of the Fibonacci series.

**PROGRAM**

(8200H) = 07H

| | LXI H, 8300 | : | Setting pointer for storing/displaying data |
| | LDA 8200 | : | Load contents of accumulator with no. of numbers to be generated |
| | CPI 00 | : | Check if 00 |
| | JZ End | : | If zero, end program - no display |
| | MOV C, A | : | Move contents of accumulator to register C |
| | MVI B, 01 | : | Move '01' (second number in Fibonacci series) to register B |

|         | MVI A, 00   | : | Move '00' (First number in Fibonacci series) to accumulator |
|         | MOV M, A    | : | Move contents of accumulator to memory |
|         | INX H       | : | Increment the memory pointer |
|         | DCR C       | : | Decrement register C by 1 |
|         | JZ End      | : | If zero, end program |
| Again : | MOV D, A    | : | Move contents of accumulator to register D |
|         | ADD B       | : | Add contents of register B to accumulator |
|         | MOV B, D    | : | Move contents of register D to register B |
|         | MOV M, A    | : | Move contents of accumulator to memory |
|         | INX H       | : | Increment the memory pointer |
|         | DCR C       | : | Decrement register C by 1 |
|         | JNZ Again   | : | If register C is not equal to '0', go to again |
| End :   | RST 1       | : | Reset command |

**RESULT**

(8200H) = 0
(8201H) = 1
(8202H) = 1
(8203H) = 2
(8204H) = 3
(8205H) = 5
(8206H) = 8

**Program 18:** WAP to arrange given data array in ascending order.

**PROGRAM**

(2200H) = 10H
(2201H) = 30H
(2202H) = 15H
(2203H) = 40H
(2204H) = 25H
(2205H) = 35H
(2206H) = 50H
(2207H) = 70H
(2208H) = 60H

|           | MVI B, 09    | : | Initialize counter |
| START :   | LXI H, 2200H | : | Initialize memory pointer |
|           | MVI C, 09H   | : | Initialize counter 2 |
| BACK :    | MOV A, M     | : | Get the number |
|           | INX H        | : | Increment memory pointer |
|           | CMP M        | : | Compare number with next number |
|           | JC SKIP      | : | If less, do not interchange |
|           | JZ SKIP      | : | If equal, do not interchange |

|  |  |  |
|---|---|---|
|  | MOV D, M |  |
|  | MOV M, A |  |
|  | DCX H |  |
|  | MOV M, D |  |
|  | INX H | : Interchange two numbers |
| SKIP : | DCR C | : Decrement counter 2 |
|  | JNZ BACK | : If not zero, repeat |
|  | DCR B | : Decrement counter 1 |
|  | JNZ START |  |
|  | HLT | : Terminate program execution |

**RESULT**

(2200H) = 10H
(2201H) = 15H
(2202H) = 25H
(2203H) = 30H
(2204H) = 35H
(2205H) = 40H
(2206H) = 50H
(2207H) = 60H
(2208H) = 70H

**Program 19:** WAP to arrange given data array in descending order.

**PROGRAM**

(4150H) = 08H : Size of data array
(4151H) = 30H : Start of data array
(4152H) = 15H
(4153H) = 40H
(4154H) = 25H
(4155H) = 35H
(4156H) = 50H
(4157H) = 70H
(4158H) = 60H

|  |  |  |
|---|---|---|
| START : | MVI B, 00 | : Flag = 0 |
|  | LXI H, 4150 | : Count = length of array |
|  | MOV C, M |  |
|  | DCR C | : No. of pair = count -1 |
|  | INX H | : Point to start of array |
| LOOP : | MOV A, M | : Get kth element |
|  | INX H |  |
|  | CMP M | : Compare to (K + 1)th element |
|  | JNC LOOP 1 | : No interchange if kth > = (k + 1)th |
|  | MOV D, M | : Interchange if out of order |
|  | MOV M, A |  |

```
        DCR H
        MOV M, D
        INX H
        MVI B, 01H          :  Flag = 1
LOOP 1 : DCR C              :  Count down
        JNZ LOOP
        DCR B               :  If flag = 1
        JZ START            :  Do another sort, if yes
        HLT                 :  If flag = 0, stop execution
```

**RESULT**

(4151H) = 70H
(4152H) = 60H
(4153H) = 50H
(4154H) = 40H
(4155H) = 35H
(4156H) = 30H
(4157H) = 25H
(4158H) = 15H

**Program 20:** WAP to find the square root of a given number stored at 4200H.

**PROGRAM**

```
        (4200H) = 16H
        LDA 4200H           :  Get the given data (Y) in A register
        MOV B, A            :  Save the data in B register
        MVI C, 02H          :  Call the divisor (02H) in C register
        CALL DIV            :  Call division subroutine to get initial value (X) in
                               D register
REP :   MOV E, D            :  Save the initial value in E register
        MOV A, B            :  Get the dividend (Y) in A register
        MOV C, D            :  Get the divisor (X) in C register
        CALL DIV            :  Call division subroutine to get initial value (Y/X)
                               in D register
        MOV A, D            :  Move Y/X in A register
        ADD E               :  Get the ((Y/X) + X) in A register
        MVI C, 02H          :  Get the divisor (02H) in C register
        CALL DIV            :  Call division subroutine to get ((Y/X) + X)/2 in
                               D register. This is XNEW
        MOV A, E            :  Get X in A register
        CMP D               :  Compare X and XNEW
        JNZ REP             :  If XNEW is not equal to X, then repeat
        STA 4201H           :  Save the square root in memory
        HLT                 :  Terminate program execution
```

**Division Subroutine Program:**

|          | DIV: MVI D, 00H | : | Clear D register for quotient |
|----------|-----------------|---|-------------------------------|
| NEXT :   | SUB C           | : | Subtract the divisor from dividend |
|          | INR D           | : | Increment the quotient |
|          | CMP C           | : | Repeat subtraction until the divisor is less than dividend |
|          | JNC NEXT        | : | |
|          | RET             | : | Return to the main program |

**RESULT**

(4201H) = 04H

**Program 21:** Write an assembly language program to multiply 2 BCD numbers.

**PROGRAM**

|          | Multiplier = 20H        |   | |
|----------|-------------------------|---|---|
|          | Mulplicand = 04H        |   | |
|          | MVI C, Multiplier       | : | Load BCD multiplier in C register |
|          | MVI B, 00               | : | Initialize counter B |
|          | LXI H, 0000H            | : | Result HL ← 0000 |
|          | MVI E, Multiplicand     | : | Load multiplicand in E register |
|          | MVI D, 00H              | : | D ← 00H |
| BACK :   | DAD D                   | : | Result HL ← DE + HL |
|          | MOV A, L                | : | Get the lower byte of the result L into A |
|          | ADI, 00H                |   | |
|          | DAA                     | : | Adjust the lower byte of result to BCD. |
|          | MOV L, A                | : | Store the lower byte of result in L register |
|          | MOV A, H                | : | Get the higher byte of the result H into A |
|          | ACI, 00H                |   | |
|          | DAA                     | : | Adjust the higher byte of the result to BCD |
|          | MOV H, A                | : | Store the higher byte of result in H register |
|          | MOV A, B                |   | |
|          | ADI 01H                 | : | Increment counter B through A |
|          | DAA                     | : | Adjust it to BCD and store it |
|          | MOV B,A                 |   | |
|          | CMP C                   | : | Compare if count = multiplier |
|          | JNZ BACK                | : | If not equal, repeat |
|          | SHLD 2000H              | : | Store result at location 2000H and 2001H |
|          | HLT                     | : | Stop |

**RESULT**

20H X 04H = 0080H

(2000H) = 80H

(2001H) = 00H

**Program 22:** WAP to convert two BCD numbers in memory to the equivalent HEX number.

### PROGRAM

(4150H) = 02(MSD)

(4151H) = 09(LSD)

LXI H, 4150

MOV A, M      :   Initialize memory pointer

ADD A         :   MSD X 2

MOV B, A      :   Store MSD X 2

ADD A         :   MSD X 4

ADD A         :   MSD X 8

ADD B         :   MSD X 10

INX H         :   Point to LSD

ADD M        :   Add to form HEX

INX H

MOV M, A      :   Store the result

HLT

### RESULT

(4152H) = 1D H

**Program 23:** WAP to convert given Hexadecimal number into its equivalent BCD number.

### PROGRAM

(4150H) = FFH

LXI H, 4150        :   Initialize memory pointer

MVI D, 00         :   Clear D register for most significant byte

XRA A            :   Clear accumulator

MOV C, M        :   Get HEX data

LOOP 2 :     ADI 01          :   Count the number one-by-one

DAA             :   Adjust for BCD count

JNC LOOP1

INR D

LOOP1          :   DCR C

JNZ LOOP2

STA 4151         :   Store the least significant byte

MOV A, D

STA 4152         :   Store the most significant byte

HLT

### RESULT

(4151H) = 55 (LSB)

(4152H) = 02 (MSB)

**Program 24:** WAP to convert given Hexadecimal number into its equivalent ASCII.

**PROGRAM**

(4200H) = E4H

| | | |
|---|---|---|
| LDA 4200 | : | Get Hexa data |
| MOV B, A | | |
| ANI 0F | : | Mask upper nibble |
| CALL SUB1 | : | Get ASCII code for upper nibble |
| STA 4201 | | |
| MOV A, B | | |
| ANI F0 | : | Mask lower nibble |
| RLC | | |
| RLC | | |
| RLC | | |
| RLC | | |
| CALL SUB1 | : | Get ASCII code for lower nibble |
| STA 4202 | | |
| HLT | | |

**Call Subroutine SUB 1**

| | | |
|---|---|---|
| CPI 0A | | |
| JC SKIP | | |
| ADI 07 | | |
| SKIP | : | ADI 30 |
| RET | | |

**RESULT**

(4201H) = 34 (ASCII Code for 4)

(4202H) = 45 (ASCII Code for E)

**Program 25:** WAP to convert given ASCII character into its equivalent Hexadecimal number.

**PROGRAM**

(4500H) = 31

| | | |
|---|---|---|
| LDA 4500 | | |
| SUI 30 | | |
| CPI 0A | | |
| JC SKIP | | |
| SUI 07 | | |
| SKIP | : | STA 4501 |
| HLT | | |

**RESULT**

(4501H) = 0BH

**Program 26:** WAP to insert a string of four characters from the tenth location in the given array of 50 characters.

**PROGRAM**

|  |  |  |  |
|---|---|---|---|
|  | LXI H, 2131H | : | Initialize pointer at the last location of array |
|  | LXI D, 2135H | : | Initialize another pointer to point the last location of array after insertion |
| AGAIN : | MOV A, M | : | Get the character |
|  | STAX D | : | Store at the new location |
|  | DCX D | : | Decrement destination pointer |
|  | DCX H | : | Decrement source pointer |
|  | MOV A, L | : | [Check whether desired bytes are shifted or not] |
|  | CPI 05H |  |  |
|  | JNZ AGAIN | : | If not, repeat the process |
|  | INX H | : | Adjust the memory pointer |
|  | LXI D, 2200H | : | Initialize the memory pointer to point the string to be inserted |
| REPE : | LDAX D | : | Get the character |
|  | MOV M, A | : | Store it in the array |
|  | INX D | : | Increment source pointer |
|  | INX H | : | Increment destination pointer |
|  | MOV A, E | : | [Check whether the 4 bytes are inserted] |
|  | CPI 04 |  |  |
|  | JNZ REPE | : | If not, repeat the process |
|  | HLT | : | Stop |

**Program 27:** WAP to a system (both Software and Hardware) that will cause 4 LEDs to flash 10 times when a push button switch is pressed. Use 8255. Assume persistence of vision to be 1 second.

**PROGRAM**

|  |  |  |  |
|---|---|---|---|
|  | LXI SP, 2000 H | : | Initialize stack pointe |
|  | MVI A, 90H |  |  |
|  | OUT CR | : | Initialize 8255 |
| BACK : | IN PA | : | [Read status of push button] |
|  | ANI 01 |  |  |
|  | JNZ BACK |  |  |
|  | MVI B, 0AH | : | Initialize counter |
| AGAIN : | MVI A, 00H | : | Load data to light LEDs |
|  | OUT PC | : | Send data on port C |
|  | CALL Delay | : | Call delay of 0.1 sec |
|  | MVI A, FFH | : | Load data to switch off LEDs |
|  | OUT PC | : | Send data on port C |
|  | CALL Delay | : | Call delay of 0.1 sec |
|  | DCR B | : | Decrement count |

JNZ AGAIN     : If not zero, repeat

JMP BACK     : Jump back to read status

### Delay Routine for 1 second Delay

|  |  |  |
|---|---|---|
|  | MVI B, Multiplier count | : Initialize multiplier count |
| BACK 1 : | LXI D, Initialize Count |  |
| BACK : | lDCX D | : Decrement count |
|  | MOV A, E |  |
|  | ORA D | : Logically OR D and E |
|  | JNZ BACK | : If result is not A, repeat |
|  | DCR B | : Decrement multiplier count |
|  | JNZ BACK 1 | : If not zero, repeat |
|  | RET | : Return to the main program |

**Interfacing Scheme** (Fig. 3.11)

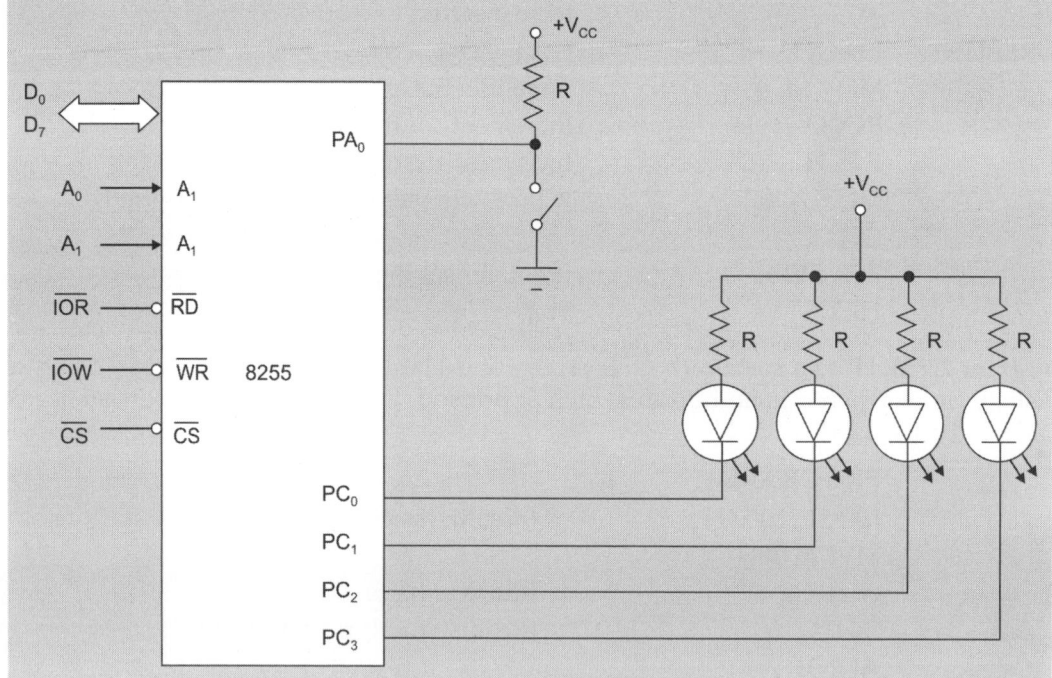

**Fig. 3.11:** Interfacing scheme

**Program 28:** WAP to design a microprocessor system to control traffic lights.

The traffic should be controlled in the following manner:

1. Allow traffic from W to E and E to W transition for 20 seconds.
2. Give transition period of 5 seconds (Yellow bulbs ON).
3. Allow traffic from N to 5 and 5 to N for 20 seconds.
4. Give transition period of 5 seconds (Yellow bulbs ON).
5. Repeat the process.

## Hardware for Traffic Light Control

Figure 3.12 shows the interfacing diagram to control 12 electric bulbs. Port A is used to control lights on N-S road and Port B is used to control lights on W-E road.

The electric bulbs are controlled by relays. The 8255 pins are used to control relay on-off action with the help of relay driver circuits. The driver circuit includes 12 transistors to drive 12 relays. Fig. 3.12 also shows the interfacing of 8255 to the system.

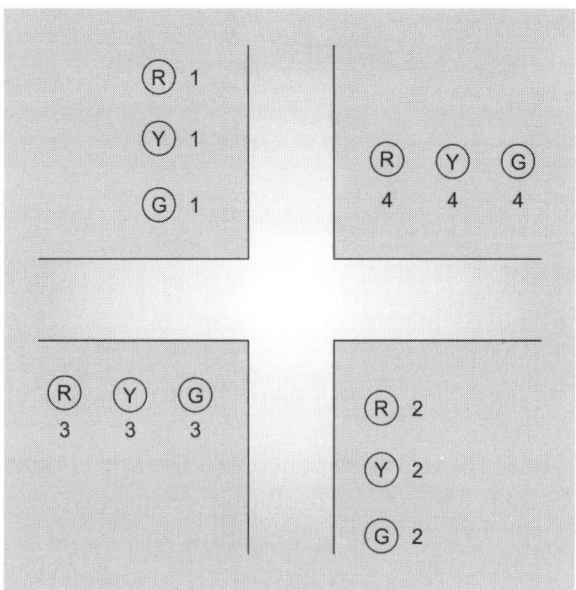

**Fig. 3.12:** Interfacing of 8255 to the system

Actual pin connections are listed in Table 3.12.

| Table 3.12: Actual pin connections | | | |
|---|---|---|---|
| *Pins* | *Light* | *Pins* | *Light* |
| $PA_0$ | $R_1$ | $PB_0$ | $R_3$ |
| $PA_1$ | $Y_1$ | $PB_1$ | $Y_3$ |
| $PA_2$ | $G_1$ | $PB_2$ | $G_3$ |
| $PA_3$ | $R_2$ | $PB_3$ | $R_4$ |
| $PA_4$ | $Y_2$ | $PB_4$ | $Y_4$ |
| $PA_5$ | $G_2$ | $PB_5$ | $G_4$ |

## Interfacing Diagram

I/O Map

| Ports/control register | Address lines | | | | | | | | Address |
|---|---|---|---|---|---|---|---|---|---|
| | *A7* | *A6* | *A5* | *A4* | *A3* | *A2* | *A1* | *A0* | |
| Port A | 1 | 0 | 0 | 0 | 0 | 0 | 0 | 0 | 80H |
| Port B | 1 | 0 | 0 | 0 | 0 | 0 | 0 | 1 | 81H |
| Port C | 1 | 0 | 0 | 0 | 0 | 0 | 1 | 0 | 82H |
| Control register | 1 | 0 | 0 | 0 | 0 | 0 | 1 | 1 | 83H |

## Software for Traffic Light Control

**Contol Word:** For initialization of 8255

| BSR/IO | MODE A | | PA | $PC_H$ | MODE B | PB | $PC_L$ |
|---|---|---|---|---|---|---|---|
| 1 | 0 | 0 | 0 | X | 0 | 0 | X |

Table shows the data bytes to be sent for specific combinations

| To glow | Port B output | Port A output |
|---|---|---|
| R1, R2, G3 and G4 | 24H | 09H |
| Y1, Y2, Y3 and Y4 | 12H | 12H |
| R3, R4, G1 and G2 | 09H | 24H |

**Program 29:**

**PROGRAM**

|  |  |  |  |
|---|---|---|---|
|  | MVI A, 80H | : | Initialize 8255, port A and port B |
|  | OUT 83H (CR) | : | in output mode |
| START : | MVI A, 09H | | |
|  | OUT 80H (PA) | : | Send data on PA to glow R1 and R2 |
|  | MVI A, 24H | | |
|  | OUT 81H (PB) | : | Send data on PB to glow G3 and G4 |
|  | MVI C, 28H | : | Load multiplier count (40ýï) for delay |
|  | CALL DELAY | : | Call delay subroutine |
|  | MVI A, 12H | | |
|  | OUT (81H) PA | : | Send data on Port A to glow Y1 and Y2 |
|  | OUT (81H) PB | : | Send data on port B to glow Y3 and Y4 |
|  | MVI C, 0AH | : | Load multiplier count (10ýï) for delay |
|  | CALL DELAY | : | Call delay subroutine |
|  | MVI A, 24H | | |
|  | OUT (80H) PA | : | Send data on port A to glow G1 and G2 |
|  | MVI A, 09H | | |
|  | OUT (81H) PB | : | Send data on port B to glow R3 and R4 |
|  | MVI C, 28H | : | Load multiplier count (40ýï) for delay |
|  | CALL DELAY | : | Call delay subroutine |
|  | MVI A, 12H | | |
|  | OUT PA | : | Send data on port A to glow Y1 and Y2 |
|  | OUT PB | : | Send data on port B to glow Y3 and Y4 |
|  | MVI C, 0AH | : | Load multiplier count (10ýï) for delay |
|  | CALL DELAY | : | Call delay subroutine |
|  | JMP START | | |

**Delay Subroutine:**

|  |  |  |  |
|---|---|---|---|
|  | LXI D, Count | : | Load count to give 0.5 sec delay |
| BACK : | DCX D | : | Decrement counter |
|  | MOV A, D | | |
|  | ORA E | : | Check whether count is 0 |

| JNZ BACK | : | If not zero, repeat |
| DCR C | : | Check if multiplier is zero, otherwise repeat |
| JNZ DELAY | | |
| RET | : | Return to the main program |

**Program 30:** WAP to control stepper motor (Fig. 3.13).

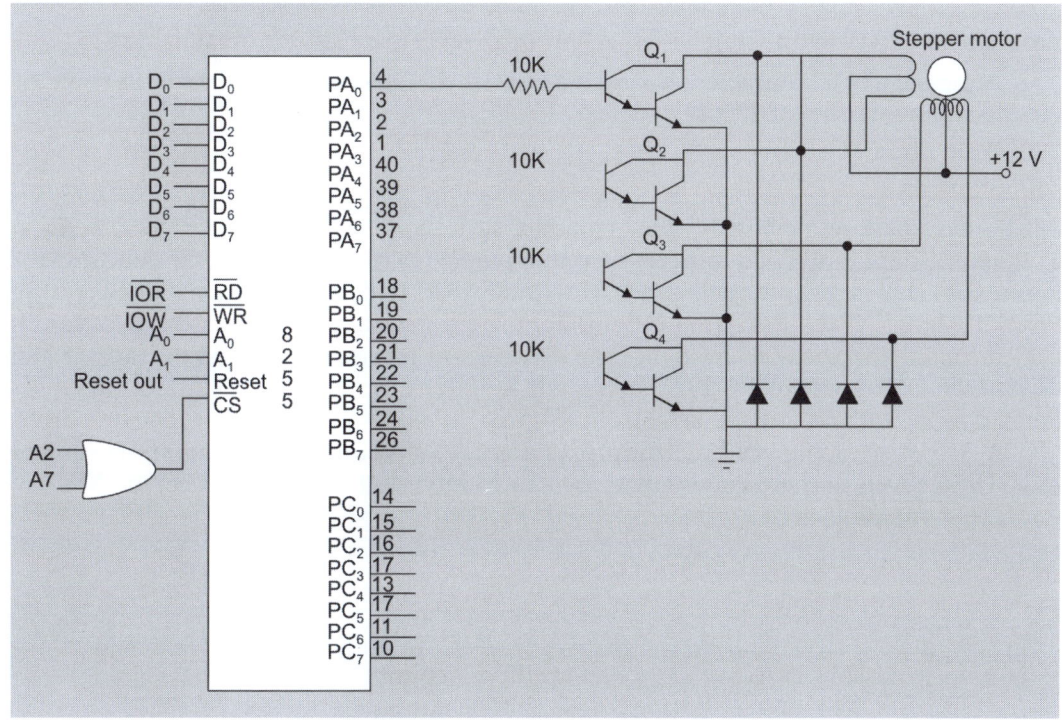

**Fig. 3.13:** WAP to control stepper motor

## Stepper Motor Control Program

6000H Excite code DB 03H, 06H, 09H, OCH: This is the code sequence for clockwise rotation. Subroutine to rotate a stepper motor clockwise by 360°—set the counts.

**Program 31:**

**PROGRAM**

| | | MVI C, 32H | : | Set repetition count to $50_{10}$ |
| START | : | MVI B, 04H | : | Counts excitation sequence |
| | | LXI H, 6000H | : | Initialize pointer |
| BACK1 | : | MOV A, M | : | Get the Excite code |
| | | OUT PORTA | : | Send Excite code |
| | | CALL DELAY | : | Wait |
| | | INX H | : | Increment pointer |
| | | DCR B | : | Repeat 4 times |
| | | JNZ BACK 1 | | |

## Delay Subroutine

```
LXI D, Count
Back        : DCX D
MOV A, D
ORA E
JNZ Back
RET
```

**Program 32:** WAP to transmit a message from an 8085 to a CRT terminal using 8251.

   i. A message of 50 characters is stored as ASCII characters (without parity) in memory locations starting at 2200H.

   ii. Baud rate x 16

   iii. Stop bits 2

CRT terminal uses normal RS 232C standard serial communication interface. Therefore, to transmit data to CRT, it is necessary to have RS 232C interface at the sending end. Figure 3.14 shows the interfacing of 8251 with RS 232C to 8085. As shown in the Fig, three RS-232C signals (TxD, RxD are Ground) are used for serial communication between the CRT terminal and the 8085 system. Line drivers and receivers are used to transfer logic levels from TTL logic to RS-232C logic. For RS-232C, the voltage level +3V to +15V is defined as logic 0 and voltage level from –3V to –15V is defined as logic 1. The line driver, MC 1488, converts logic 1 of TIL to approximately –9V and logic a of TIL to approximately +9V. These levels at the receiving end are again converted by the line receiver, MC1489, into TTL compatible logic.

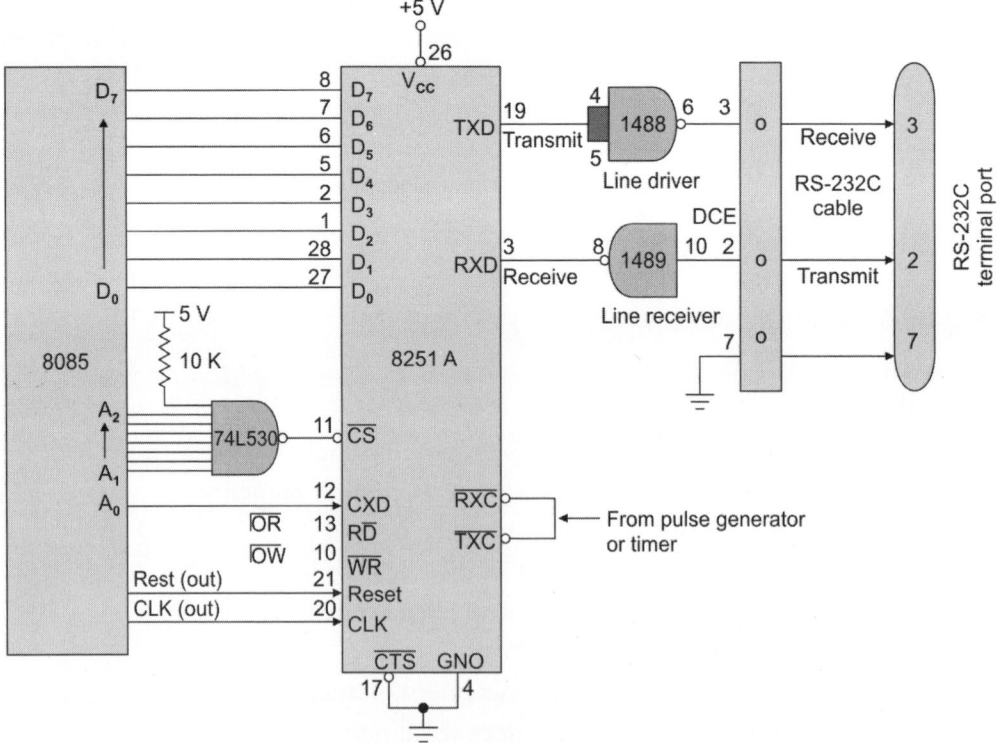

**Fig. 3.14:** Interfacing of 8251 with RS 232C

## I/O MAP

| Register | Address lines | | | | | | | | Address |
|---|---|---|---|---|---|---|---|---|---|
| | A7 | A6 | A5 | A4 | A3 | A2 | A1 | A0 | |
| Data register | 1 | 1 | 1 | 1 | 1 | 1 | 1 | 0 | FEH |
| Control register | 1 | 1 | 1 | 1 | 1 | 1 | 1 | 1 | FFH |

Mode word necessary for the given specification is as follows:

| B7 | B6 | B5 | B4 | B3 | B2 | B1 | B0 | = CAH |
|---|---|---|---|---|---|---|---|---|
| 1 | 1 | 0 | 0 | 1 | 0 | 1 | 0 | |

Command word necessary for the given specification is as follows:

| B7 | B6 | B5 | B4 | B3 | B2 | B1 | B0 | = 11H |
|---|---|---|---|---|---|---|---|---|
| X | 0 | X | 1 | X | 0 | X | 1 | |

Status word necessary for the given specification is as follows:

| B7 | B6 | B5 | B4 | B3 | B2 | B1 | B0 | = 01H |
|---|---|---|---|---|---|---|---|---|
| X | X | X | X | X | X | X | 1 | |

**Program 33:**

**PROGRAM**

```
          LXI H, 2200H      : Initialize memory pointer-to-pointer the message
          MVI C, 32H        : Initialize counter to send 50 characters
          MVI A, 00H
          OUT FFH
          OUT FFH           : Dummy mode word
          OUT FFH
          MVI A, 40H        : Reset command word
          OUT FFH           : Reset 8251A
          MVI A, CAH        : Mode word initialization
          OUT FFH
          MVI A, 11H        : Command word initialization
          OUT FFH
CHECK  :  IN FFH
          ANI 01H           : Check TxRDY
          JZ CHECK          : Is TxRDY I? If not, check again
          MOV A, M          : Get the character in accumulator
          OUT FEH           : Send character to the transmitter
          INX H             : Increment memory pointer
          DCR C             : Decrement counter
          JNZ CHECK         : If not zero, send next character
          HLT               : Stop program execution
```

**Program 34:** WAP to generate square waveform.

**PROGRAM**

```
START  :  MVI A, 00         : Move '0' into accumulator
          OUT C8            : Display 0 at port
```

|  | CALL |  |  |
| --- | --- | --- | --- |
|  | DELAY | : | Call the DELAY subroutine |
| RETURN : | MVI A, FF | : | Move 'FF' into accumulator |
|  | OUT C8 | : | Display at port |
|  | CALL DELAY | : | Call the DELAY subroutine |
|  | JMP Start | : | Start jump to start |

**Delay Subroutine**

|  | MVI B, 05 | : | Move 05 to register B to give delay |
| --- | --- | --- | --- |
| Loop 1 : | MVI C, FF | : | Move FF to register C |
| Loop 2 : | DCR C | : | Decrement register C by 1 |
|  | JNZ  Loop2 | : | If C NOT '0', decrement C again |
|  | DCR B | : | Decrement Register B by 1 |
|  | JNZ Loop1 | : | If B is NOT '0', go to Loop1 |
|  | RET | : | Return to the main program. |

## PROBLEMS

1. What are the various registers in 8085?

2. Discuss the function of ALU of 8085.

3. What are the various status flags provided in 8085? Explain in detail.

4. Discuss function of the following signals:
   INTR, HOLD, HLDA, READY, ALE, IO/M, $S_0$ and $S_1$

5. In 8085, name the 16-bit registers.

6. Discuss the different interrupts in 8085.

7. What is the difference between SIM and RIM?

8. Write a program to mask RST 7.5.

9. Write RIM format if RST 5.5 is pending.

10. Explain various types of addressing modes of 8085 with suitable examples.

11. Explain what operation will take place when the following instructions are executed:
    LXI D, 2000H; LDA 3000H; LHLD 4000H; STA 2000H; SHLD 3500H; MOV A, M; PUSH D

12. Explain the function of given instructions with examples:
    DAD, DAA, CMP, CMA, RAL, RAR, PUSH, POP, XRA

13. Discuss instruction cycle, machine cycle and state.

14. Draw the timing diagram of memory read operation.

15. Draw the timing diagram of memory write operation.

16. Draw the timing diagram of I/O read operation.

17. Draw the timing diagram of I/O write operation.

18. Draw the timing diagram for ADD B.

19. Write an assembly language program to add two BCD numbers.

20. Write an assembly language program to divide 8-bit number by another 8-bit number.

21. Write an assembly language program to find out the sum of given array of data.

22. Write an assembly language program to find out maximum number in given array of data.

# 4

# Microprocessor 8086

## 4.1 INTRODUCTION

8086 is a 16-bit microprocessor housed in a 40-pin Dual Inline Package (DIP) and capable of addressing 1 megabyte of memory. The various versions of this chip can operate with different clock frequencies:

1. 8086 (5 MHz)
2. 8086-2 (8 MHz)
3. 8086-1 (10 MHz)

It contains approximately 29,000 transistors and is fabricated using the HMOS technology. The term 16-bit means that it is an arithmetic logic unit; internal registers and most of its instructions are designed to work with 16-bit binary word. The 8086 microprocessor has a 16-bit data bus, so it can read from or write data to memory and port either 16-bits or 8-bits at a time. The 8086 microprocessor has 20-bit address bus, so it can address any one of 220 or 1,048,576 memory locations. Here 16-bit words will be stored in two consecutive memory locations. If the first byte of a word is at an even address, the 8086 can read entire word in one operation; if the first byte of the word is at an odd address; the 8086 will read the first byte with one bus operation and the byte with another bus operation.

## 4.2 DESCRIPTION OF 8086

1. It is a 16-bit microprocessor (µp). Its ALU and internal registers work with 16-bit binary word.
2. 8086 has a 20-bit address bus and can access up to $2^{20} = 1$ MB memory locations.
3. 8086 has a 16-bit data bus. It can read or write data to a memory/port either 16-bits or 8-bits at a time.
4. It can support up to 64K I/O ports.
5. It provides 14, 16-bit registers.
6. Frequency range of 8086 is 6–10 MHz.
7. It has multiplexed address and data bus AD0–AD15 and A16–A19.
8. It requires single phase clock with 33% duty cycle to provide internal timing.
9. It can prefetch up to 6 instruction bytes from memory and queues them in order to speed up instruction execution.
10. It requires +5V power supply.
11. A 40-pin dual inline package.
12. 8086 is designed to operate in two modes: minimum and maximum modes.

a. The minimum mode is selected by applying logic 1 to the MN/MX# input pin. This is a single microprocessor configuration.

b. The maximum mode is selected by applying logic 0 to the MN/MX# input pin. This is a multi-microprocessors configuration.

## 4.3 ARCHITECTURE OF 8086

The internal architecture of 8086 is shown in Fig. 4.1. 8086 has two blocks: Bus Interfacing Unit (BIU) and Execution Unit (EU). The BIU performs all bus operations, such as instruction fetching, reading and writing operands for memory and calculating the addresses of the memory operands. The instruction bytes are transferred to the instruction queue. EU executes instructions from the instruction system byte queue. Both units operate asynchronously to give the 8086 an overlapping instruction fetch and execution mechanism which is called as Pipelining. This results in efficient use of the system bus and system performance. BIU contains Instruction queue, Segment registers, Instruction pointer and Address adder. EU contains Control circuitry, Instruction decoder, ALU, Pointer and Index register, Flag register.

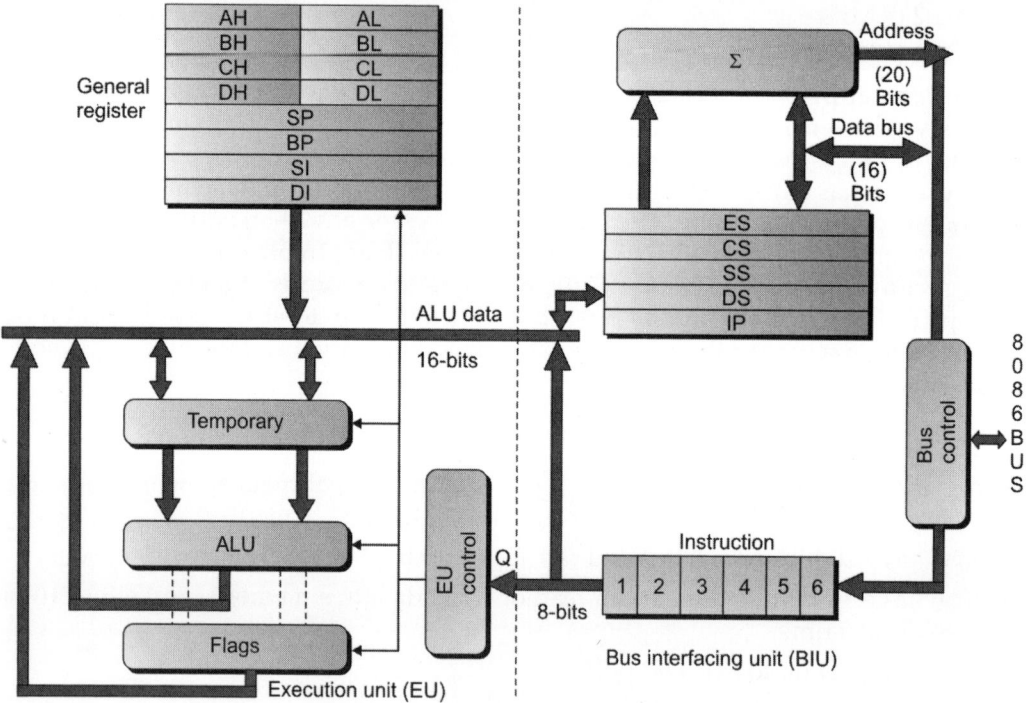

**Fig. 4.1:** Internal architecture of 8086

### 4.3.1 Bus Interface Unit (BIU)

It provides a full 16-bit bidirectional data bus and 20-bit address bus. The bus interface unit is responsible for performing all external bus operations. Specifically it has the following functions:

1. Instruction fetch
2. Instruction queuing
3. Operand fetch and storage
4. Address relocation and bus control.

The BIU uses a mechanism known as an instruction stream queue to implement a *pipeline architecture*. This queue permits prefetch of up to six bytes of instruction code. Whenever the queue of the BIU is not full, it has room for at least two more bytes and at the same time, the EU is not requesting it to read or write operands from memory, the BIU is free to look ahead in the program by prefetching the next sequential instruction. These prefetching instructions are held in its FIFO queue. With its 16-bit data bus, the BIU fetches two instruction bytes in a single memory cycle. After a byte is loaded at the input end of the queue, it automatically shifts up through the FIFO to the empty location nearest to the output. The EU accesses the queue from the output end. It reads one instruction byte after the other from the output of the queue. If the queue is full and the EU is not requesting access to operand in memory, these intervals of no bus activity, which may occur between bus cycles, are known as Idle states. If the BIU is already in the process of fetching an instruction when the EU request it to read or write operands from memory or I/O, the BIU first completes the instruction fetch bus cycle before initiating the operand read/write cycle.

The BIU also contains a dedicated adder which is used to generate the 20-bit physical address that is output on the address bus. This address is formed by adding an appended 16-bit segment address and a 16-bit offset address.

For example: The physical address of the next instruction to be fetched is formed by combining the current contents of the code segment CS register and the current contents of the instruction pointer IP register.

The BIU is also responsible for generating bus control signals such as those for memory read or write and I/O read or write. Bus interface unit has the following functional parts:

1. Segment registers

2. Instruction queue

3. Internal data bus

4. Bus control

The Bus Interface Unit consists of segment registers, adder to generate 20-bit address and instruction prefetch queue. Once this address is sent out of BIU, the instruction and data bytes are fetched from memory and they fill a First in First out 6 byte queue. The segment registers and their default offsets registers are given in Table 4.1.

**Table 4.1:** Segment registers and their default offsets registers

| Segment register | Default offset |
|:---:|:---:|
| CS | IP |
| DS | SI, DI |
| SS | SP, BP |
| ES | DI |

### 4.3.1.1 Segment Register

Most of the registers contain data/instruction offsets within 64 KB memory segment. There are four different 64 KB segments for instructions, stack, data and extra data. To specify the location of these four segments in 1 MB of processor memory, the processor uses four segment registers explained further and shown in Table 4.2.

**Table 4.2:** Segment registers and memory size

| Register | Memory size | Total memory |
|---|---|---|
| Code segment | 64 KB | 1 M |
| Data segment | 64 KB | B00000H to |
| Stack segment | 64 KB | FFFFFH |
| Extra segment | 64 KB | |

**Code segment (CS)** is a 16-bit register containing address of 64 KB segment with processor instructions. The processor uses CS segment for all accesses to instructions referenced by instruction pointer (IP) register. CS register cannot be changed directly. The CS register is automatically updated during far jump, far call and far return instructions.

**Data segment (DS)** is a 16-bit register containing address of 64 KB segment with program data. By default, the processor assumes that all data referenced by general registers (AX, BX, CX and DX) and index register (SI, DI) is located in the data segment. DS register can be changed directly using POP and LDS instructions.

**Stack segment (SS)** is a 16-bit register containing address of 64 KB segment with program stack. By default, the processor assumes that all data referenced by the stack pointer (SP) and base pointer (BP) registers is located in the stack segment. SS register can be changed directly using POP instruction.

**Extra segment (ES)** is a 16-bit register containing address of 64 KB segment, usually with program data. By default, the processor assumes that the DI register references the ES segment in string manipulation instructions. ES register can be changed directly using POP and LES instructions.

### 4.3.1.2 Instruction Pointer (IP)

IP is a 16-bit register that is used to point to or tell the MPU the instruction to execute next. Therefore, IP is used to control the sequence in which the program is executed. Each time the EU accepts an instruction to point to the next instruction in the program. It is possible to change default segments used by general and index registers by prefixing instructions with a CS, SS, DS or ES prefix.

### 4.3.1.3 Instruction Queue

The instruction queue is used as a temporary memory storage area for data instructions that are to be executed by the MPU. The BIU through the bus control unit prefetches instructions and stores them in the instruction queue. This allows the execution unit to perform its calculations at maximum efficiency. Because the BIU and EU essentially operate independently as shown in Fig. 4.2, the BIU concentrates on loading instructions

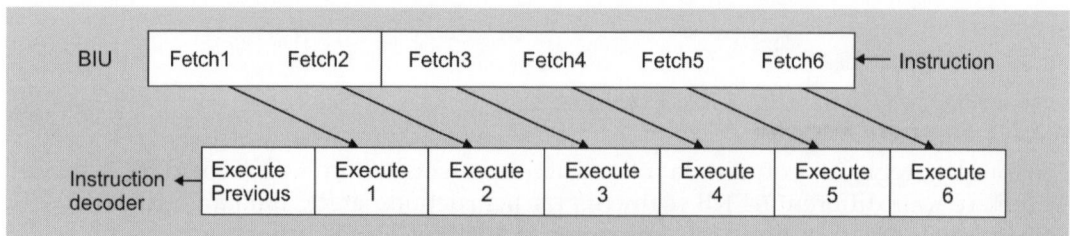

**Fig. 4.2:** Pipelining process

into the instruction queue. This usually takes more time to do than the calculations performed by the EU. In effect, the BIU and the EU works in parallel. The instruction queue is first in first out memory. This means that the first instruction loaded into the instruction queue by the bus control unit will be the first instruction to be used by the ALU.

This process of fetching the next intrusions is known as PIPELINING.

### 4.3.1.4 Bus Control Unit

Bus control unit performs the bus operations for the microprocessor. It fetches and transmits instructions, data and control signals between microprocessor and the other devices of the system.

## 4.4 EXECUTION UNIT (EU)

The Execution Unit is responsible for decoding and executing all instructions. The EU extracts instructions from the top of the queue in the BIU, decodes them, generates operands if necessary, passes them to the BIU and requests it to perform the read or write bycycles to memory or I/O and perform the operation specified by the instruction on the operands. During the execution of the instruction, the EU tests the status and control flags and updates them based on the results of executing the instruction. If the queue is empty, the EU waits for the next instruction byte to be fetched and shifted to top of the queue. When the EU executes a branch or jump instruction, it transfers control to a location corresponding to another set of sequential instructions. Whenever this happens, the BIU automatically resets the queue and then begins to fetch instructions from this new location to refill the queue. EU has the following functional parts:

1. General purpose registers
2. ALU
3. Flag register
4. EU control unit

### 4.4.1 General Purpose Registers

All general registers of the 8086 microprocessor can be used for arithmetic and logic operations. The general registers are:

| AH | | AL |
|----|----|----|
| BH | | BL |
| CH | | CL |
| DH | | DL |
| | SP | |
| | BP | |
| | SI | |
| | DI | |

### 4.4.1.1 Accumulator

Accumulator register consists of two 8-bit registers AL and AH, which can be combined together and used as a 16-bit register AX. AL in this case contains the low-order byte of the word and AH contains the high-order byte. Accumulator can be used for I/O operations and string manipulation.

### 4.4.1.2 Base Register

Base register consists of two 8-bit registers BL and BH, which can be combined together and used as a 16-bit register BX. BL in this case contains the low-order byte of the word and BH contains the high-order byte. BX register usually contains a data pointer used for based, based indexed or register indirect addressing.

### 4.4.1.3 Count Register

Count register consists of two 8-bit registers CL and CH, which can be combined together and used as a 16-bit register CX. When combined, CL register contains the low-order byte of the word and CH contains the high-order byte. Count register can be used as a counter in string manipulation and shift/rotate instructions.

### 4.4.1.4 Data Register

Data register consists of two 8-bit registers DL and DH, which can be combined together and used as a 16-bit register DX. When combined, DL register contains the low-order byte of the word and DH contains the high-order byte. Data register can be used as a port number in I/O operations. In integer 32-bit multiply and divide instruction, the DX register contains high-order word of the initial or resulting number.

### 4.4.1.5 Stack Pointer

Stack pointer (SP) is a 16-bit register pointing to program stack.

### 4.4.1.6 Base Pointer

Base pointer (BP) is a 16-bit register pointing to data in stack segment. BP register is usually used for based, based indexed or register indirect addressing.

### 4.4.1.7 Source Index

Source index (SI) is a 16-bit register. SI is used for indexed, based indexed and register indirect addressing, as well as a source data addresses in string manipulation instructions.

### 4.4.1.8 Destination Index

Destination index (DI) is a 16-bit register. DI is used for indexed, based indexed and register indirect addressing, as well as a destination data addresses in string manipulation instructions.

## 4.4.2 Arithmetic Logic Unit

It is the calculator part of the EU. It consists of electronic circuitry that performs arithmetic operations or logical operations on the binary represented electrical signals. The control system can also be considered a part of the ALU. It provides a path for the flow of instruction into the ALU, the general registers and the flag register. The results of these operations can affect the condition flags.

## 4.4.3 Flag Register

Flag register contains a group of status bits called flags that indicate the status of the CPU or the result of arithmetic operations. There are two types of flags:
1. The **status flags** which reflect the result of executing an instruction. The programmer cannot set/reset these flags directly.
2. The **control flags** enable or disable certain CPU operations. The programmer can set/reset these bits to control the CPU's operation.

Nine individual bits of the status register are used as control flags (3 of them) and status flags (6 of them). The remaining 7 are not used.

A flag can only take on the values 0 and 1. We say a flag is set if it has the value 1. The status flags are used to record specific characteristics of arithmetic and of logical instructions (Fig. 4.3).

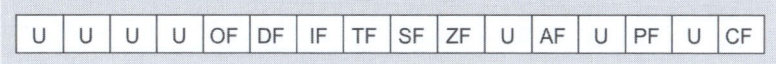

| U | U | U | U | OF | DF | IF | TF | SF | ZF | U | AF | U | PF | U | CF |

**Fig. 4.3:** Flag register

U = Undefined

**Control flags:** There are three control flags:

1. **The direction flag (D):** It affects the direction of moving data blocks by such instructions as MOVS, CMPS and SCAS. The flag values are 0 = up and 1 = down and can be set/reset by the STD (set D) and CLD (clear D) instructions.

2. **The interrupt flag (I):** It dictates whether or not system interrupts can occur. Interrupts are actions initiated by hardware block such as input devices that will interrupt the normal execution of programs. The flag values are 0 = disable interrupts or 1 = enable interrupts and can be manipulated by the CLI (clear I) and STI (set I) instructions.

3. **The trap flag (T):** It determines whether or not the CPU is halted after the execution of each instruction. When this flag is set (i.e. = 1), the programmer can single step through his program to debug any errors. When this flag = 0, this feature is off. This flag can be set by the INT 3 instruction.

**Status flags:** There are six status flags:

1. **The carry flag (C):** This flag is set when the result of an unsigned arithmetic operation is too large to fit in the destination register. This happens when there is an end carry in an addition operation or an end borrows in a subtraction operation. A value of 1 = carry and 0 = no carry.

2. **The overflow flag (O):** This flag is set when the result of a signed arithmetic operation is too large to fit in the destination register (i.e. when an overflow occurs). Overflow can occur when adding two numbers with the same sign (i.e. both positive or both negative). A value of 1 = overflow and 0 = no overflow.

3. **The sign flag (S):** This flag is set when the result of an arithmetic or logic operation is negative. This flag is a copy of the MSB of the result (i.e. the sign bit). A value of 1 means negative and 0 = positive.

4. **The zero flag (Z):** This flag is set when the result of an arithmetic or logic operation is equal to zero. A value of 1 means the result is zero and a value of 0 means the result is not zero.

5. **The auxiliary carry flag (A):** This flag is set when an operation causes a carry from bit 3 to bit 4 (or a borrow from bit 4 to bit 3) of an operand. A value of 1 = carry and 0 = no carry.

6. **The parity flag (P):** This flag reflects the number of 1s in the result of an operation. If the number of 1s is even its value = 1 and if the number of 1s is odd then its value = 0.

## 4.5 PIN DIAGRAM OF 8086

Microprocessor 8086 is available in 40-pin DIP package which is shown in Fig. 4.4. The functions of different pins are illustrated below:

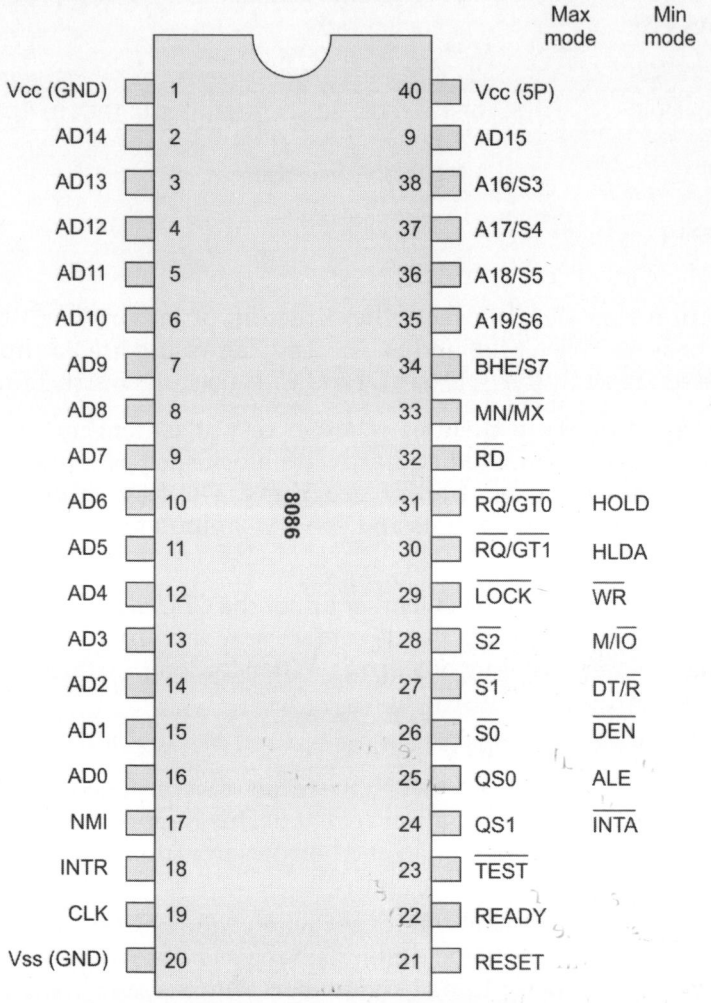

**Fig. 4.4:** Pin diagram of 8086

### AD15-AD0 (pin-2 to16 and 39, type-I/O)

**Address data bus:** These lines constitute the time multiplexed memory/IO address (T1), and data (T2, T3, TW, T4) bus. A0 is analogous to $\overline{BHE}$ for the lower byte of the data bus, pins D7-D0. It is LOW during T1 when a byte is to be transferred on the lower portion of the bus in memory or I/O operations. Eight-bit oriented devices tied to the lower half would normally use A0 to condition chip select functions. These lines are active HIGH and float to 3-state OFF during interrupt acknowledge and local bus "hold acknowledge".

### A19/S6-A16/S3 (pin-35 to 38, type-O/P)

**Address/status:** During T1, there are four most significant address lines for memory operations. During I/O operations, these lines are LOW. During memory and I/O operations, status information is available on these lines during T2, T3, TW and T4. The status of the interrupt enable FLAG bit (S5) is updated at the beginning of each CLK cycle. A17/S4 and A16/S3 are encoded as shown in Table 4.3. This information indicates which relocation register is presently being used for data accessing. These lines float to 3-state OFF during local bus "hold acknowledge."

**Table 4.3:** Encoded A17/S4 and A16/S3

| A17/S4 | A16/S3 | Characteristics |
|--------|--------|-----------------|
| 0 | 0 | Alternate data |
| 0 | 1 | Stack |
| 1 | 0 | Code or none |
| 1 | 1 | Data |

### $\overline{BHE}$ /S7 (pin-34, type-O/P)

**Bus high enable/status:** During T1, the bus high enable signal ( ) should be used to enable data onto the most significant half of the data bus, pins D15-D8. Eight-bit oriented devices tied to the upper half of the bus would normally use  to condition chip select functions. It is LOW during T1 for read, write and interrupt acknowledge cycles when a byte is to be transferred on the high portion of the bus. The S7 status information is available during T2, T3 and T4. The signal is active LOW and floats to 3-state OFF in "hold". It is LOW during T1 for the first interrupt acknowledge cycle.

**Table 4.4:** Bus high enable

| $\overline{BHE}$ | A0 | Characteristics |
|-----|-----|-----------------|
| 0 | 0 | Whole word from/to even address |
| 0 | 1 | Upper byte from/to odd address |
| 1 | 0 | Lower byte from/to even address |
| 1 | 1 | None |

### $\overline{RD}$ (Pin-32, type-O/P)

**Read:** Read strobe indicates that the processor is performing a memory or I/O read cycle, depending on the state of the S2 pin. This signal is used to read devices which reside on the 8086 local bus. $\overline{RD}$ is active LOW during T2, T3 and TW of any read cycle, and is guaranteed to remain HIGH in T2 until the 8086 local bus has floated. This signal floats to 3-state OFF in "hold acknowledge".

### READY (pin-22, type-I/P)

**Ready:** It is the acknowledgement from the addressed memory or I/O device that it will complete the data transfer. The READY signal from memory/IO is synchronized by the 8284A Clock Generator to form READY. This signal is active HIGH. The 8086 READY input is not synchronized. Correct operation is not guaranteed if the setup and hold times are not met.

### INTR (pin-18, type-I/P)

**Interrupt request:** INTR is a level triggered input which is sampled during the last clock cycle of each instruction to determine if the processor should enter into an interrupt acknowledge operation. A subroutine is vectored to via an interrupt vector lookup table located in system memory. It can be internally masked by software resetting the interrupt enable bit. INTR is internally synchronized. This signal is active HIGH.

### $\overline{\text{TEST}}$ (pin-23, type-I/P)

**Test:** $\overline{\text{TEST}}$ input is examined by the "Wait" instruction. If the $\overline{\text{TEST}}$ input is LOW, execution continues, otherwise the processor waits in an "Idle" state. This input is synchronized internally during each clock cycle on the leading edge of CLK.

### NMI (pin-17, type-I/P)

**Non-maskable interrupt:** NMI is an edge triggered input which causes a type 2 interrupt. A subroutine is vectored to via an interrupt vector lookup table located in system memory. NMI is not maskable internally by software. A transition from LOW to HIGH initiates the interrupt at the end of the current instruction. This input is internally synchronized.

### RESET (pin-21, type-I/P)

**Reset:** It causes the processor to immediately terminate its present activity. The signal must be active HIGH for at least four clock cycles. It restarts execution, as described in the Instruction Set description, when RESET returns LOW. RESET is internally synchronized.

### CLK (pin-19, type-I/P)

**Clock:** CLK provides the basic timing for the processor and bus controller. It is asymmetric with a 33% duty cycle to provide optimized internal timing.

### Vcc (pin-40)

**Vcc:** It is the pin which is used to supply a 5V power to the processor.

### GND (pin-1 and 20)

Ground

### MN/ $\overline{\text{TEST}}$ (pin-33, type- I/P)

**Minimum/Maximum:** It indicates the mode of operation of the processor; it may be minimum mode or maximum mode. These two modes will be discussed later.

### 4.6 MEMORY SEGMENTATION

In memory, data is stored as bytes. Each byte has a specific address. Intel 8086 has 20 lines address bus. With 20 address lines, the memory that can be addressed is 220 bytes. $2^{20}$ = 1,048,576 bytes (1 MB). 8086 can access memory with address ranging from 00000 H to FFFFF H.

The total memory size is divided into segments of various sizes. A segment is just an area in memory. The process of dividing memory this way is called **Segmentation**.

In 8086, memory has four different types of segments. These are:

1. Code Segment
2. Data Segment
3. Stack Segment
4. Extra Segment

Each of these segments is addressed by an address stored in corresponding segment register. These registers are 16-bit in size. Each register stores the base address (starting address) of the corresponding segment. As the segment registers cannot store 20 bits, they only store the upper 16 bits (Fig. 4.5).

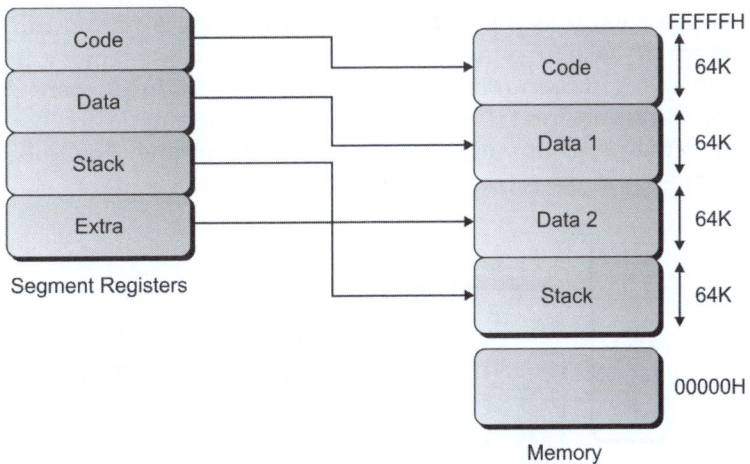

**Fig. 4.5:** Segmentation of memory

### 4.6.1 Code Segment

Program can be located anywhere in memory. Jump and call instructions can be used for short jumps within currently selected 64 KB code segment, as well as for far jumps anywhere within 1 MB of memory. All conditional jump instructions can be used to jump within approximately +127 to –127 bytes from current instruction.

### 4.6.2 Data Segment

Processor can access data in any one out of 4 available segments, which limits the size of accessible memory to 256 KB (if all four segments point to different 64 KB blocks). Accessing data from the Data, Code, Stack or Extra segments can be usually done by prefixing instructions with the DS, CS, SS or ES (some registers and instructions by default may use the ES or SS segments instead of DS segment).

Word data can be located at odd or even byte boundaries. The processor uses two memory accesses to read 16-bit word located at odd byte boundaries. Reading word data from even byte boundaries requires only one memory access.

### 4.6.3 Stack Segment

This memory can be placed anywhere in memory. The stack can be located at odd memory addresses, but it is not recommended for performance reasons (see "Data Memory" above).

### 4.7 PHYSICAL ADDRESS

The 8086 has 20 address lines to enable it to address 1M ($2^{20}$ = 1Meg) bytes of memory. However, the largest register is only 16-bits (64 K); so physical addresses have to be calculated. These calculations are done in hardware within the microprocessor.

A physical address is the actual address of a byte in memory, i.e. the value which goes out onto the address bus. The 8086 operates a segmentation technique. A segment register is used to hold the address of the start of a 'chunk' of memory called a segment. The physical address is calculated by combining the base address with an offset to form the 20-bit physical address. The segment registers CS, DS, ES and SS are used to mark the base of a segment.

A segment register value is a paragraph number where a paragraph is 16 bytes of contiguous memory, so each time the processor wants to access memory, it has to take the contents of a segment register, shift it left 4 places (multiply by 1610), then add the required offset to form the 20-bits address (Fig. 4.6). This is done automatically by hardware, so it is of no concern to the programmer. However, the programmer has to become familiar with memory addresses in the form of: **Seg: Offset.**

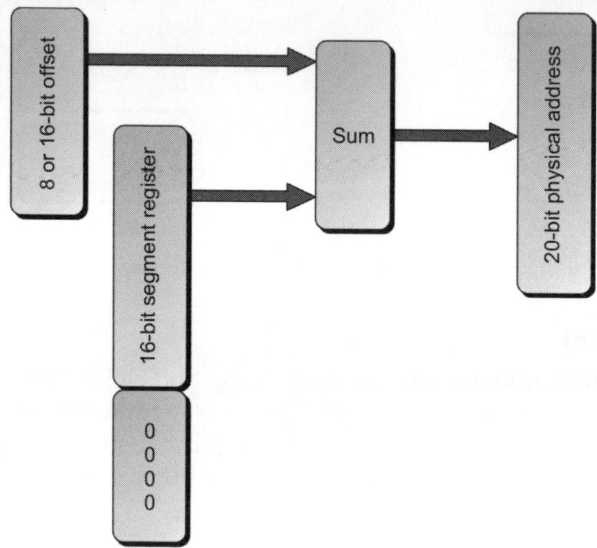

**Fig. 4.6:** 20-bit physical address generation process

The 32-bit processors support various memory management modes and use virtual memory, but as far as we are concerned, the instruction pointer is a 32-bit register that accesses linear memory. However, you do not actually need $2^{32}$ (4 GB of RAM due to virtual memory management).

BIU performs bus operation, such as instruction fetching, reading/writing of data operand for memory, inputting/outputting data for I/O peripherals. Perform other functions such as instruction queuing and data acquisitions. BIU uses instruction queue to implement a pipelined architecture (prefetch up to 6 bytes of instruction code for 8086 and then store and access the codes in FIFO order).

**Example:**

$$CS = 3000\ H$$
$$[BP] = 2300H$$
$$PA = 3000\boxed{0}$$
$$\underline{+\quad 2300\quad}$$
$$32300H$$

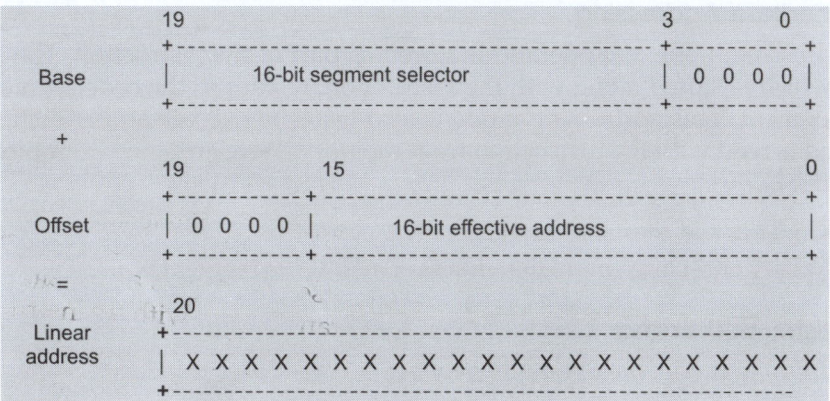

**Fig. 4.7:** Physical address

## 4.8 ADDRESSING MODES

An instruction acts on any number of operands. The way an instruction accesses its operands is called its **Addressing modes**. Operands may be of three types:

1. Implicit
2. Explicit
3. Both Implicit and Explicit

### Implicit Operands

**Implicit operands** mean that the instruction by definition has some specific operands. The programmers do NOT select these operands.

**Examples:**

XLAT: It automatically takes AL and BX as operands.

AAM: It operates on the contents of AX.

### Explicit Operands

**Explicit operands** mean the instruction operates on the operands specified by the programmer.

**Examples:**

MOV AX, BX: They take AX and BX as operands.

XCHG SI, DI: They take SI and DI as operands.

### Implicit and Explicit Operands

**Example:**

MUL BX: It automatically multiplies BX explicitly times AX.

The location of an operand value in memory space is called the **effective address (EA).**

We can classify the addressing modes of 8086 into four groups:

1. Immediate addressing
2. Register addressing
3. Memory addressing
4. I/O port addressing

### 4.8.1 Immediate Addressing

In this addressing mode, the operand is stored as part of the instruction. The immediate operand, which is stored along with the instruction, resides in the code segment—not in the data segment. This addressing mode is also faster to execute an instruction because the operand is read with the instruction from memory. Here are some examples:

MOV AL, 20: They move the constant 20 into register AL.

ADD AX, 5: They add constant 5 to register EAX.

MOV DX: They offset msg ; move the address of message to register DX.

### 4.8.2 Register addressing

In this addressing mode, the operands may be:

- **reg16:** 16-bit general registers: AX, BX, CX, DX, SI, DI, SP or BP.
- **reg8:** 8-bit general registers: AH, BH, CH, DH, AL, BL, CL or DL.
- **Sreg:** Segment registers: CS, DS, ES or SS. There is an exception: CS cannot be a destination.

For register addressing modes, there is no need to compute the effective address. The operand is in a register and to get the operand, there is no memory access involved.

MOV AX, BX: They mov reg16, reg16.

ADD AX, SI: They add reg16, reg16.

MOV DS, AX: They mov Sreg, reg16.

### 4.8.3 Memory Addressing Modes

Memory (RAM) is the main component of a computer to store temporary data and machine instructions. In a program, programmers often need to read from and write into memory locations.

There are different forms of memory addressing modes:

1. Direct addressing
2. Register indirect addressing
3. Based addressing
4. Indexed addressing
5. Based indexed addressing
6. Based indexed with displacement

#### *Direct Addressing*

The instruction mov al, ds:[8088h] loads the AL register with a copy of the byte at memory location 8088h. Likewise, the instruction mov ds:[1234h], dl stores the value in the dl register to memory location 1234h. By default, all displacement-only values provide offsets into the data segment. If you want to provide an offset into a different segment, you must use a segment override prefix before your address. For example, to access location 1234h in the extra segment (es), you would use an instruction of the form mov ax, es:[1234h]. Likewise, to access this location in the code segment, you would use the instruction mov ax, cs:[1234h]. The ds: prefix in the previous examples is not a segment override.

The instruction mov al, ds:[8088h] is same as mov al, [8088h]. If not mentioned, DS register is taken by default.

### Register Indirect Addressing

The 80x86 CPUs let you access memory indirectly through a register using the register indirect addressing modes. There are four forms of this addressing mode on the 8086, best demonstrated by the following instructions:

mov al, [bx]

mov al, [bp]

mov al, [si]

mov al, [di]

*Code example*

MOV BX, 100H

MOV AL, [BX]

The [bx], [si] and [di] modes use the ds segment by default. The [bp] addressing mode uses the stack segment (ss) by default. You can use the segment override prefix symbols if you wish to access data in different segments. The following instructions demonstrate the use of these overrides:

mov al, cs:[bx]

mov al, ds:[bp]

mov al, ss:[si]

mov al, es:[di]

Intel refers to [bx] and [bp] as base addressing modes and bx and bp as base registers (in fact, bp stands for base pointer). Intel refers to the [si] and [di] addressing modes as indexed addressing modes (si stands for source index, di stands for destination index). However, these addressing modes are functionally equivalent. This text will call these forms register indirect modes to be consistent.

### Based Addressing

8-bit or 16-bit instruction operand is added to the contents of a base register (BX or BP), the resulting value is a pointer to location where data resides.

Mov al, [bx],[si]

Mov bl, [bp],[di]

Mov cl, [bp],[di]

*Code example*

If bx = 1000h

si = 0880h

Mov AL, [1000 + 880]

Mov AL, [1880]

### Indexed Addressing

The indexed addressing modes use the following syntax:

mov al, [bx + disp]

mov al, [bp + disp]

mov al, [si + disp]

mov al, [di + disp]

*Code example*

MOV BX, 100H

MOV AL, [BX + 15]

MOV AL, [BX + 16]

If bx contains 1000h, then the instruction mov cl, [bx + 20h] will load cl from memory location ds:1020h. Likewise, if bp contains 2020h, mov dh, [bp + 1000h] will load dh from location ss:3020. The offsets generated by these addressing modes are the sum of the constant and the specified register. The addressing modes involving bx si and di all use the data segment, the [bp + disp] addressing mode uses the stack segment by default. As with the register indirect addressing modes, you can use the segment override prefixes to specify a different segment:

mov al, ss:[bx + disp]

mov al, es:[bp + disp]

mov al, cs:[si + disp]

mov al, ss:[di + disp]

**Example:** MOV AX, [DI + 100]

### Based Indexed Addressing

The based indexed addressing modes are simply combinations of the register indirect addressing modes. These addressing modes form the offset by adding together a base register (bx or bp) and an index register (si or di). The allowable forms for these addressing modes are:

mov al, [bx+si]

mov al, [bx+di]

mov al, [bp+si]

mov al, [bp+di]

*Code example*

MOV BX, 100H

MOV SI, 200H

MOV AL, [BX + SI]

INC BX

INC SI

Suppose that bx contains 1000h and si contains 880h. Then the instruction mov al, [bx][si] would load al from location DS:1880h. Likewise, if bp contains 1598h and di contains 1004, mov ax, [bp+di] will load the 16 bits in ax from locations SS:259C and SS:259D. The addressing modes that do not involve bp use the data segment by default. Those that have bp as an operand use the stack segment by default.

### Based Indexed Plus Displacement Addressing

These addressing modes are a slight modification of the based/indexed addressing modes with the addition of an eight bit or sixteen bit constant. The following are some examples of these addressing modes:

mov al, disp[bx][si]

mov al, disp[bx+di]

mov al, [bp+si+disp]

mov al, [bp][di][disp]

*Code example*

MOV BX, 100H

MOV SI, 200H

MOV AL, [BX + SI +100H]

INC BX

INC SI

## 4.9 BUS OPERATION

The 8086 has a combined address and data bus commonly referred as a time multiplexed address and data bus. The main reason behind multiplexing address and data over the same pins is the maximum utilisation of processor pins and it facilitates the use of 40-pin standard DIP package. The bus can be demultiplexed using a few latches and transreceivers, whenever required. Basically, all the processor bus cycles consist of at least four clock cycles. These are referred to as T1, T2, T3, T4 (Fig. 4.8). The address is transmitted by the processor during T1. It is present on the bus only for one cycle. The negative edge of this ALE pulse is used to separate the address and the data or status information. In maximum mode, the status lines S0, S1 and S2 are used to indicate the type of operation. Status bits S3 to S7 are multiplexed with higher order address bits and the BHE signal. Address is valid during T1 while status bits S3 to S7 are valid during T2 through T4.

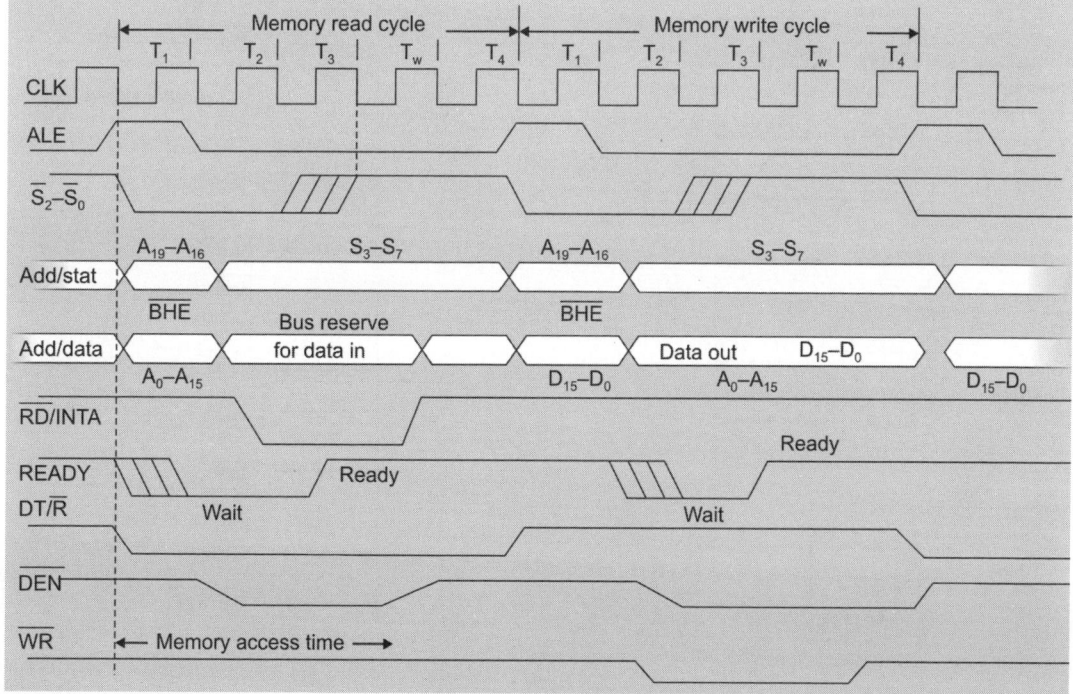

**Fig. 4.8**: Bus operation

## 4.10 MINIMUM AND MAXIMUM MODES

The requirements for supporting minimum and maximum 8086 systems are sufficiently different that they cannot be done efficiently with 40 uniquely defined pins. Consequently, the 8086 is equipped with a strap pin MN/ $\overline{MX}$ which defines the system configuration. The definition of a certain subset of the pins changes dependent on the condition of the strap pin. When MN/ $\overline{MX}$ pin is strapped to GND, the 8086 treats pins 24 through 31 in maximum mode. An 8288 bus controller interprets status information coded into S0, S2, S2 to generate bus timing and control signals compatible with the MULTIBUS architecture. When the MN/ $\overline{MX}$ pin is strapped to Vcc, the 8086 generates bus control signals itself on pins 24 through 31.

### 4.10.1 Minimum Mode

Minimum mode is one of two different hardware modes of the Intel 8086 processor (CPU). The other is maximum mode. Mode selection is accomplished by how the chip is hard-wired in the circuit. Specifically, pin no 33 (MN/) is used to select the mode, depending on whether it is wired to voltage or to ground.

Changing the state of pin no 33 changes the function of certain other pins. If pin 33 asserted high then the 8086 will work in MINIMUM mode. Mode cannot be changed by software. Figure 4.9 shows the minimum mode signals.

**M/ $\overline{IO}$ :** It is the 28th pin of 8086 microprocessor and it is output type pin.

**Status line:** It is logically equivalent to S2 in the maximum mode. It is used to distinguish a memory access from an I/O access. M/ $\overline{IO}$ becomes valid in the T4

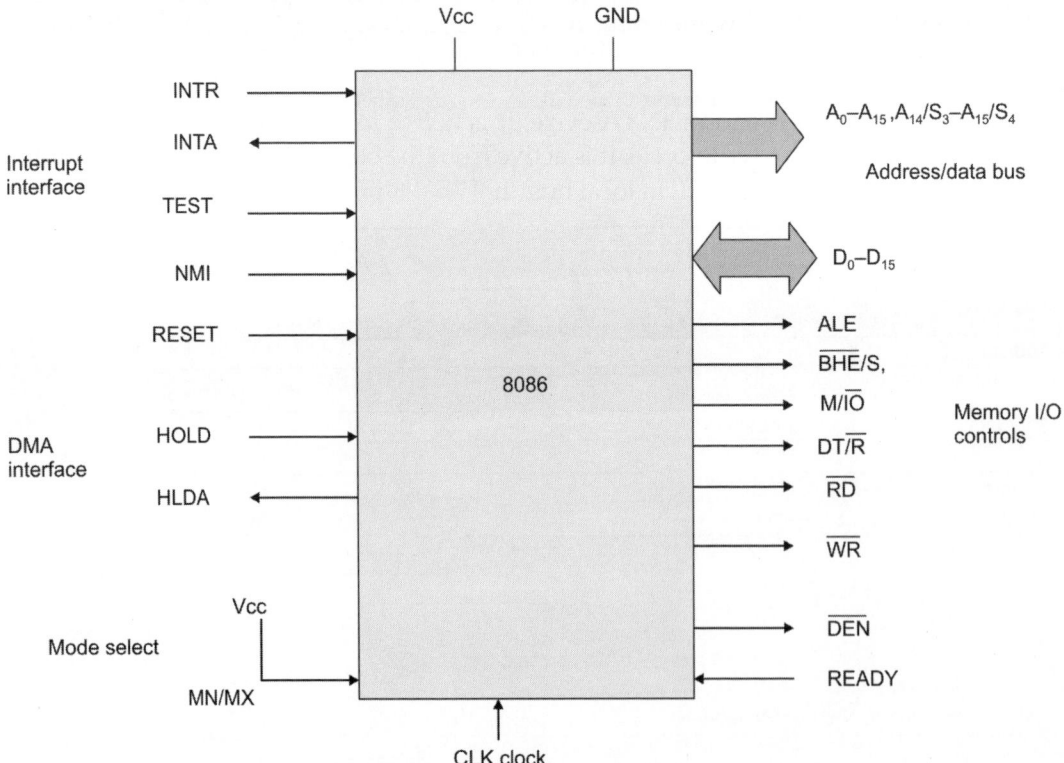

**Fig. 4.9:** Minimum mode signals

preceding a bus cycle and remains valid until the final T4 of the cycle (M = HIGH, IO = LOW). M/$\overline{\text{IO}}$ floats to 3-state OFF in local bus "hold acknowledge".

$\overline{\text{WR}}$ : It is the 29th pin of 8086 microprocessor and it is output type pin.

**Write:** Indicates that the processor is performing a write memory or write I/O cycle, depending on the state of the M/$\overline{\text{IO}}$ signal. $\overline{\text{WR}}$ is active for T2, T3 and $T_W$ of any write cycle. It is active LOW and floats to 3-state OFF in local bus "hold acknowledge".

$\overline{\text{INTA}}$ : It is the 24th pin of 8086 microprocessor and it is output type pin.

**INTA:** It is used as a read strobe for interrupt acknowledge cycles. It is active LOW during T2, T3 and TW of each interrupt acknowledge cycle.

**ALE:** It is the 25th pin of 8086 microprocessor and it is output type pin.

**Address latch enable:** It is provided by the processor to latch the address into the 8282/ 8283 address latch. It is a HIGH pulse active during T1 of any bus cycle. Note that ALE is never floated.

DT/$\overline{R}$ : It is the 27th pin of 8086 microprocessor and it is output type pin.

**Data transmit/receive:** It is needed in minimum system that desires to use an 8286/8287 data bus transceiver. It is used to control the direction of data flow through the transceiver. Logically, DT/$\overline{R}$ is equivalent to S1 in the maximum mode and its timing is the same as for M/$\overline{\text{IO}}$ (T = HIGH, R = LOW). This signal floats to 3-state OFF in local bus "hold acknowledge".

$\overline{\text{DEN}}$ : It is the 26th pin of 8086 microprocessor and it is output type pin.

**Data enable:** It is provided as an output enable for the 8286/8287 in a minimum system which uses the transceiver. $\overline{\text{DEN}}$ is active LOW during each memory and I/O access and for $\overline{\text{INTA}}$ cycles. For a read or INTA cycle, it is active from the middle of T2 until the middle of T4, while for a write cycle, it is active from the beginning of T2 until the middle of T4. $\overline{\text{DEN}}$ floats to 3-state OFF in local bus "hold acknowledge".

### Unique Minimum Mode Signals of 8086 *(Table 4.5)*

**Table 4.5:** Mode signals

| Name | Function | Type |
| --- | --- | --- |
| HOLD | Hold request | Input |
| HLDA | Hold acknowledge | Output |
| $\overline{\text{WR}}$ | Write control | Output, 3-state |
| M/$\overline{\text{IO}}$ | Memory/IO control | Output, 3-state |
| DT/$\overline{R}$ | Data transmit/receive | Output, 3-state |
| $\overline{\text{DEN}}$ | Data enable | Output, 3-state |
| ALE | Address latch enable | Output |
| $\overline{\text{INTA}}$ | Interrupt acknowledge | Output |

**HOLD, HLDA:** These are the 31st and 30th pin of 8086 microprocessor and these are input and output type pins.

**Hold:** It indicates that another master is requesting a local bus "hold." To be acknowledged, HOLD must be active HIGH. The processor receiving the "hold" request will issue HLDA (HIGH) as an acknowledgement in the middle of a T4 or $T_i$ clock cycle.

Simultaneous with the issuance of HLDA, the processor will float the local bus and control lines. After HOLD is detected as being LOW, the processor will lower the HLDA, and when the processor needs to run another cycle, it will again drive the local bus and control lines. Holds acknowledge (HLDA) and HOLD have internal pull-up resistors. The same rules as for $\overline{RQ}/\overline{GT}$ apply regarding when the local bus will be released. HOLD is not an asynchronous input. External synchronization should be provided if the system cannot otherwise guarantee the setup time.

## 4.10.2 Maximum Mode

The 8086 is said to be in maximum mode if its $MN/\overline{MX}$ pin is grounded. This mode requires additional circuitry to translate control signals. The device used is 8288 bus controller which converts status signals into the I/O and memory transfer signals. Figure 4.10 shows the signals of 8086 in maximum mode.

### 4.10.2.1 Pin Function Description in Maximum Mode

$\overline{S2}\ \overline{S1}\ \overline{S0}$ : These are the 26, 27 and 28 pins of 8086 and these are output type pins.

**Status:** Active during T4, T1 and T2 and is returned to the passive state (1, 1, 1) during T3 or during $T_W$ when READY is HIGH. This status is used by the 8288 Bus Controller to generate all memory and I/O access control signals. Any change by $\overline{S2}\ \overline{S1}$ or $\overline{S0}$ during T4 is used to indicate the beginning of a bus cycle, and the return to the passive state in T3 or $T_W$ is used to indicate the end of a bus cycle. These signals float to 3-state OFF in "hold acknowledge". These status (Continued) lines are encoded as shown in Table 4.6.

$\overline{RQ}/\overline{GT0}, \overline{RQ}/\overline{GT1}$ : These are the 30th and 31st pins of 8086 microprocessor and these are I/O type pins.

**Request/grant:** Pins are used by other local bus masters to force the processor to release the local bus at the end of the processor's current bus cycle. Each pin is bidirectional with $\overline{RQ}/\overline{GT0}$ having higher priority than $\overline{RQ}/\overline{GT0}, \overline{RQ}/\overline{GT1}$ pins have internal pull-up resistors and may be left unconnected.

1. A pulse of 1 CLK wide from another local bus master indicates a local bus request ("hold") to the 8086 (pulse 1).

2. During a T4 or T1 clock cycle, a pulse 1 CLK wide from the 8086 to the requesting master (pulse 2), indicates that the 8086 has allowed the local bus to float and that it will enter the "hold acknowledge" state at the next CLK. The CPU's bus interface unit is disconnected logically from the local bus during "hold acknowledge".

3. A pulse 1 CLK wide from the requesting master indicates to the 8086 (pulse 3) that the "hold" request is about to end and that the 8086 can reclaim the local bus at the next CLK.

Each master-master exchange of the local bus is a sequence of 3 pulses. There must be one dead CLK cycle after each bus exchange. Pulses are active LOW. If the request is made

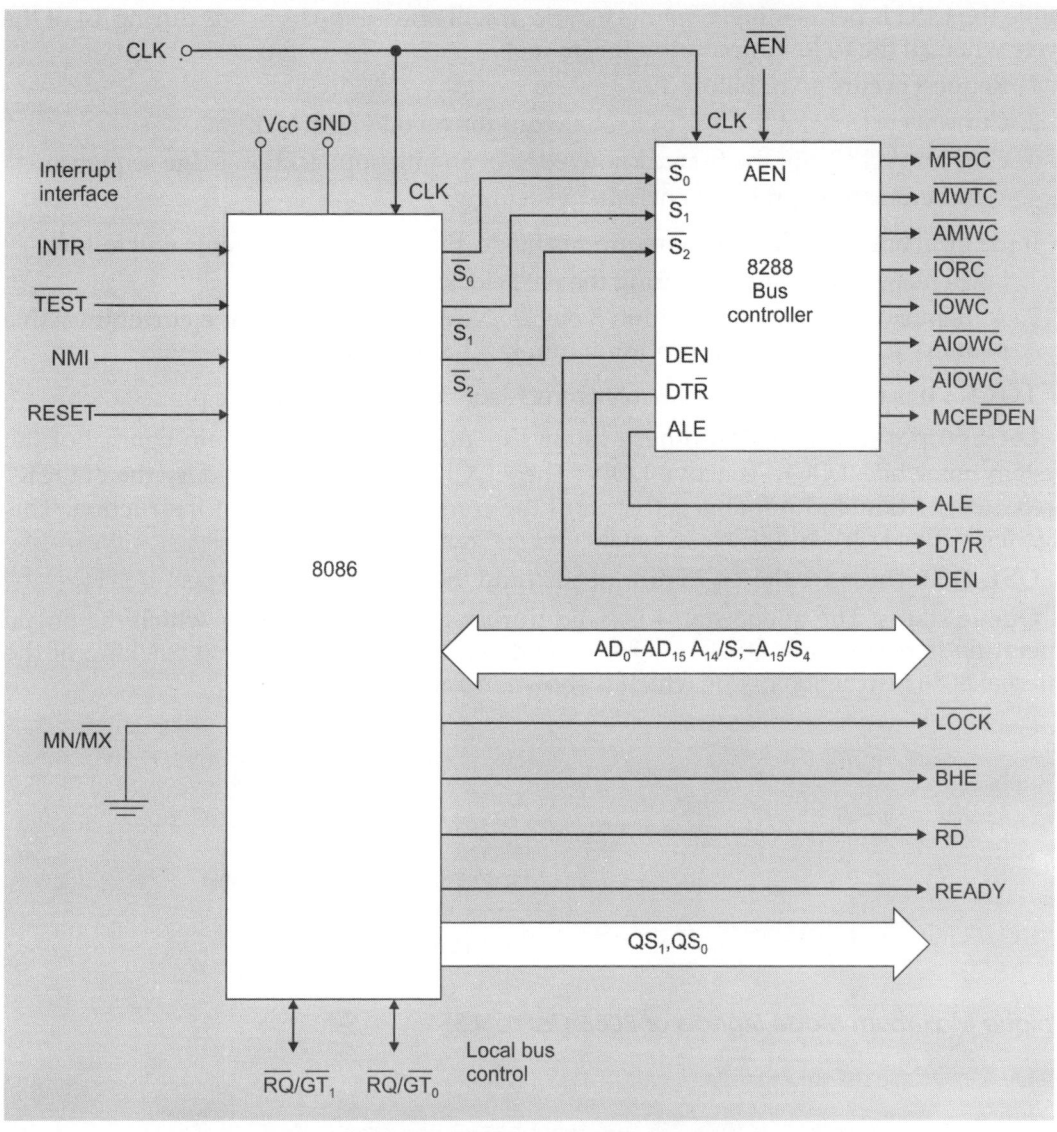

**Fig. 4.10:** Maximum mode signals

**Table 4.6:** Status lines

| $\overline{S2}$ | $\overline{S1}$ | $\overline{S0}$ | Action |
|---|---|---|---|
| 0 | 0 | 0 | Interrupt ACK |
| 0 | 0 | 1 | Read I/O port |
| 0 | 1 | 0 | Write I/O port |
| 0 | 1 | 1 | Halt |
| 1 | 0 | 0 | Code access |
| 1 | 0 | 1 | Read memory |
| 1 | 1 | 0 | Write memory |
| 1 | 1 | 1 | Passive |

while the CPU is performing a memory cycle, it will release the local bus during T4 of the cycle when all the following conditions are met:

1. Request occurs on or before T2.
2. Current cycle is not the low byte of a word (on an odd address).
3. Current cycle is not the first acknowledge of an interrupt acknowledge sequence.
4. A locked instruction is not currently executing.

If the local bus is idle when the request is made, the two possible events will follow:

1. Local bus will be released during the next clock.
2. A memory cycle will start within 3 clocks. Now the four rules for a currently active memory cycle apply with condition number 1 already satisfied.

$\overline{\text{LOCK}}$ : It is the 29th pin of 8086 microprocessor and it is output type pin.

**Lock** output indicates that other system bus masters are not to gain control of the system bus while $\overline{\text{LOCK}}$ is active LOW. The $\overline{\text{LOCK}}$ signal is activated by the "LOCK" prefix instruction and remains active until the completion of the next instruction. This signal is active LOW and floats to 3-state OFF in "hold acknowledge".

**QS1, QS0:** These are the 24, 25 pins of 8086 and these are of output type.

**Queue status:** The queue status is valid during the CLK cycle after which the queue operation is performed. QS1 and QS0 provide status to allow external tracking of the internal 8086 instruction queue which is shown in Table 4.7.

**Table 4.7:** Queue status

| QS1 | QS0 | Action |
|-----|-----|--------|
| 0 | 0 | NOP |
| 0 | 1 | First byte of opcode from queue |
| 1 | 0 | Empty the queue |
| 1 | 1 | Subsequent byte from queue |

### Unique Maximum Mode Signals of 8086 (Table 4.8)

**Table 4.8:** Unique maximum mode signals

| Name | Function | type |
|------|----------|------|
| $\overline{\text{RQ}}/\overline{\text{GT0}}$, $\overline{\text{RQ}}/\overline{\text{GT1}}$ | Request/grant bus access control | Bi-directional |
| $\overline{\text{LOCK}}$ | Bus priority lock control | Output 3-state |
| $\overline{\text{S2}}$ $\overline{\text{S1}}$ $\overline{\text{S0}}$ | Bus cycle status | Output 3-state |
| QS1, QS0 | Instruction queue status | Output |

### Typical Configuration of 8086 System

Figure 4.11 shows a typical 8086 system in the minimum mode configuration. The 8286 is a transceiver which provide buffer for data bus. The 8286 and 8287 are 8-bit bipolar transceivers with 3-state outputs. The 8287 inverts the input data at its output terminals whereas the 8286 does not.

Figure 4.12 shows a typical 8086 system in the maximum mode configuration. In addition to latches and bus transceivers, a bus controller is also employed for this

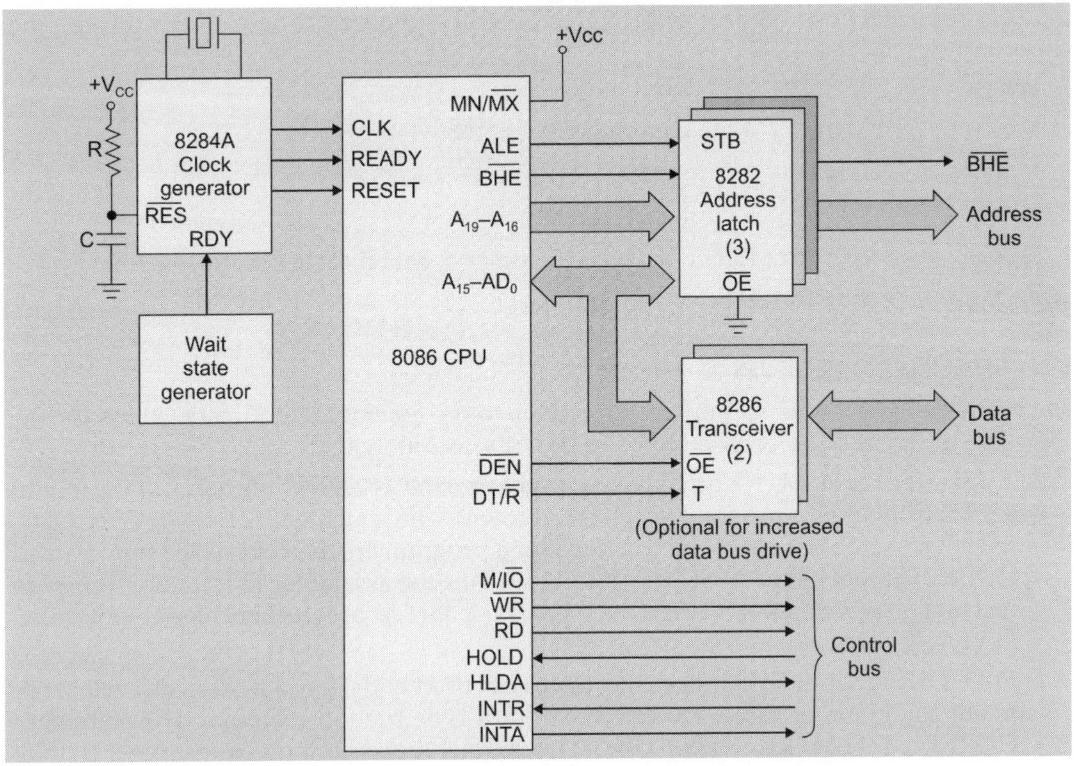

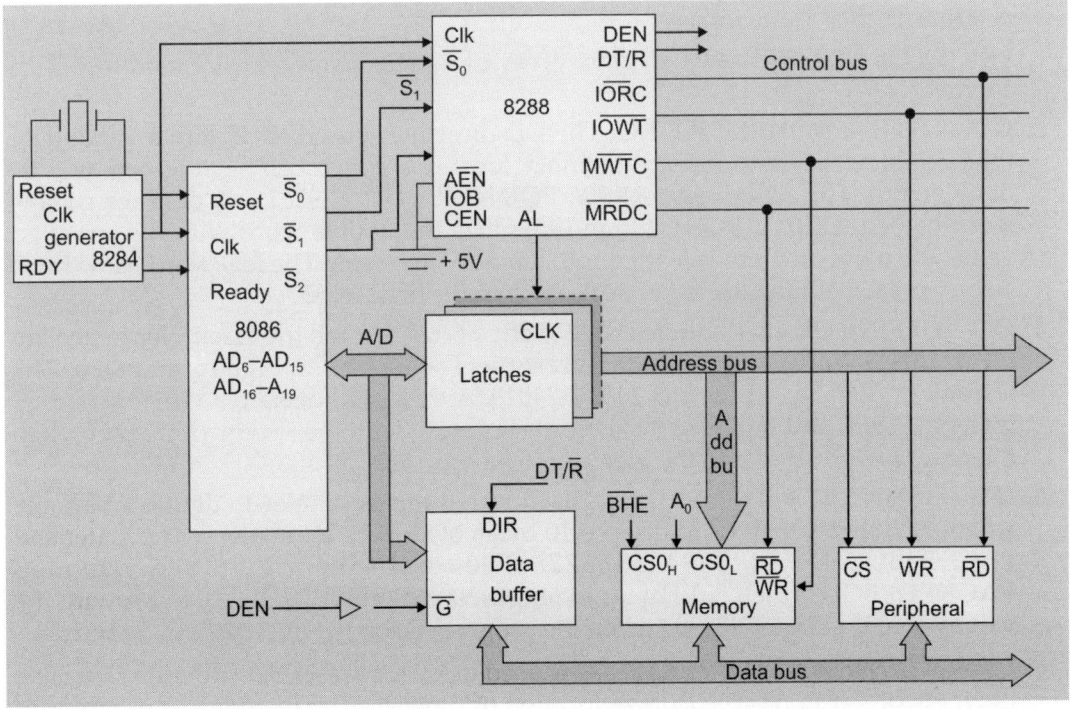

**Fig. 4.11:** 8086 system in the minimum mode of configuration

**Fig. 4.12:** 8086 system in the maximum mode of configuration

configuration. The bus controller provides control signals as shown in the figure. The important signals are:

| | |
|---|---|
| $\overline{\text{WRDC}}$ | Memory read command |
| $\overline{\text{MWTC}}$ | Memory write command |
| $\overline{\text{IORC}}$ | I/O read command |
| $\overline{\text{IOWC}}$ I/O | I/O write command |
| $\overline{\text{AMWC}}$ | Advanced memory write command issued earlier in the machine cycle |
| $\overline{\text{AIOWC}}$ | Advanced I/O write command |

## 4.11 ASSEMBLER DIRECTIVES

The word defined in this section are directions to the assembler, not instructions for the 8086. The different types of assembler are describe as following:

1. **ASSUME:** The ASSUME directives is used to tell the assembler the name of the logical segments it should use for a specified segment. The statement ASSUME CS: CODE, tells the assembler that the instructions for a program are in a logical segment named CODE. The statement ASSUME DS: DATA tells the assembler that for any program instruction which refers to the data segment; it should use the logical segment called DATA.

2. **DB—DIRECTIVE BYTE:** The DB directive is used to declare a byte–type variable or to set aside one or more storage location of type byte in memory. The statement CURRENT_TEMPERATURE DB 42H, foretells the assembler to reserve 1 byte of memory for a variable named CURRENT_TEMPERATURE and to put the value 42H in that memory location when the program is loaded into RAM to be run.

   A few examples are as follows:

   PRICES DB 49H, 87H, 29H; declare array of 3 bytes named PRICE and initialize 3 bytes.

3. **DD—DEFINE DOUBLE WORD:** The DD directive is used to declare a variable of type doubleword or to reserve memory locations which can be accessed as type doubleword. The statement ARRAY_POINTER DD 25629261 will define a double word named ARRAY_POINTER and initialize the double word with the specified value when the program is loaded into memory to be run. The low word 9261H will be put in the memory at a lower address than the high word.

4. **DQ—DEFINE QUADWORD:** This directive is used to tell the assembler to declare a variable 4 words in length or to reserve 4 words of storage in memory. The statement BIG_NUMBER DQ 243598740192A92BH will declare a variable named BIG_NUMBER and initialize the 4 words set aside with the specified number when the program is loaded into the memory to be run.

5. **DT—DEFINE TEB BYTES:** This is used to tell the assembler to define a variable which is 10 bytes long or to reserve 10 bytes of storage in memory. The statement PACED_BCD DT 11223344556677889900 will declare an array named PACKED_BCD which is 10 bytes in length. It will initialize the 10 bytes with the values 11223344556677889900 when the program is loaded into memory to be run.

6. **DW—DEFINE WORD:** DW directive is used to tell the assembler to define a variable of type word or to reserve storage location of type word in memory. The statement MULTIPLIER DW 437A declares a variable of type word named MULTIPLIER. The

statement also tells the assembler that the variable MULTIPLIER should be initialized with the value 437A when the program is loaded into memory to be run.

7. **END—END PROGRAM:** This directive is put after the last statement of a program to tell the assembler that this is the end of the program module. The assembler will ignore any statement after an END directive.

8. **ENDP—END PROCEDURE:** This directive is used along with the name of the procedure to indicate the end of a procedure to the assembler.

| | | |
|---|---|---|
| SQUARE_ROOT PROC | ; | Start procedure |
| | ; | Procedure instructions |
| | ; | Statements |
| SQUARE_ROOT ENDP | ; | End of procedure |

9. **ENDS—END STATEMENT:** This directive is used with the name of a segment to indicate the end of that logical segment. ENDS is used with the SEGMENT directive to "bracket" a logical segment continuing instructions or data.

| | | |
|---|---|---|
| CODE SEGMENT | ; | Start logical segment |
| | ; | Containing code |
| | ; | Instructions Statements |
| CODE ENDS | ; | End of segment named |
| | ; | CODE |

10. **EQU—EQUATE:** EQU is used to give a name to some value or symbol. Each time the assembler finds the given name in the program, it will replace the name with the value or symbol you equate with that name. Let CORRECTION_FACTOR EQU 03H at the start of the program and later in the program you write the instruction statement ADD AL, CORRECTION_FACTOR. When it codes this instruction statement, the assembler will code it as if you had written ADD AL, 03H. The advantage of using EQU in this manner is that if CORRECTION_FACTOR is used 27 times in a program, and you want to change the value, all you have to do is change the EQU statement and reassemble the program. The assembler will automatically put in the new value each time it finds the name CORRECTION_FACTOR.

11. **EVEN—ALIGN ON EVEN MEMORY ADDRESS:** The even directive tells the assembler to increment the location counter to the next even address. The 8086 can read a word from memory in one bus cycle if the word is at even addresses. When even is used in data segment, the location counter will simply be incremented to the next even address if necessary. When even is used in code segment, the location counter will simply be incremented to the next even address if necessary.

12. **EXTRN:** The EXTRN directive is used to tell the assembler that the name or levels following the directive are in some other assembly module. For example, if you want to call a procedure which is in a program module assembled at a different time from that which contains the CALL instruction, you must tell the assembler that the procedure is external. For a reference to an external named variable, you must specify the type of the variable such as EXTRN DIVISOR: WORD.

13. **GLOBAL:** The global directive can be used in place of a PUBLIC directive. For a name or symbol defined in the current assembly module, the GLOBAL directive is used to make the symbol available to the other module. The statement GLOBAL DIVISOR makes the variable DIVISOR public so that it can be accessed from other assembly module.

14. **GROUP—Group Related Segments:** The GROUP directive is used to tell the assembler to group the logical segments named after the directive into one logical group segment. This allows the contents of all the segments to be accessed from the same group segment base, such that GROUP CODE, DATA, STACK_SEG.

15. **INCLUDE—INCLUDE SOURCE CODE FROM FILE:** This directive is used to tell the assembler to insert a block of source code from the named file into the current source module. This shortens the source code.

16. **LABEL:** The label directive is used to give a name to the current value in the location counter. The LABEL directive must be followed by a term which specifies the type you want associated with that name. If the label is going to be used as the destination for a jump or a call, then the label must be specified as type near or type far. If the label is going to be used to reference a data item, then the label must be specified as type byte, type word or type doubleword.

17. **LENGTH:** LENGTH is an operator which tells the assembler to determine the number of elements in some named data item, such as a string or an array. When the assembler reads the statement MOV CX, LENGTH STRING1, it will determine the number of elements in STRING1 and codes this number in as part of the instruction. When the instruction executes, then the length of the string will be loaded into CX. If the string was declared as a string of bytes, LENGTH will produce the number of bytes in the string. If the string was declared as a word string, LENGTH will produce the number of words in the string.

18. **NAME:** The name directive is used to give a specific name to each assembly module when programs consisting of several modules are written. The statement NAME PC_BOARD, for example, might be used to name an assembly module which contains the instructions for controlling printed-circuit-board-making machines.

19. **OFFSET:** OFFSET is an operator which tells the assembler to determine the offset or displacement of a named data item (variable) or procedure from the start of the segment which contains it. This operator is usually used to load the offset of a variable into a register so that the variable can be accessed with one of the indexed addressing modes. When the assembler reads the statement MOV BX, OFFSET PRICES, for example, it will determine the offset of the variable PRICES from the start of the segment in which PRICES are defined and code this displacement in as part of the instruction.

20. **ORG—ORIGINATE:** As the assembler assembles a section of data declarations or instruction statements, it uses a location counter to keep track of how many bytes it is from the start of a segment at any time. The location counter is automatically set to 0000 when the assembler starts reading a segment. The ORG directive allows you to set the location counter to a desired value at any point in the program. The statement ORG 2000H tells the assembler to set the location counter to 2000H, for example.

A "$" is often used to symbolically represent the current value of the location counter. The "$" actually represents the next available byte location where the assembler can put a data or code byte. The $ is often used in ORG statement to tell the assembler to make some changes in the location counter relative to its current value. The statement ORG $+100 tells the assembler to increment the value of the location counter by 100 from its current value.

21. **PROC—PROCEDURE:** The PROC directive is used to identify the start of a procedure. The PROC directive follows a name you give the procedure. After the PROC directive, the term near or the term far is used to specify the type of the procedure.

The statement SMART_DIVIDE PROC FAR, for example, identifies the start of a procedure named SMART_DIVIDE and tells the assembler that the procedure is far (in a segment with a different name from the one that contains the instruction which calls the procedure).

22. **PTR—POINTER:** The PTR operator is used to assign a specific type to a variable or to a label. It is necessary to do this in any instruction where the type of the operand is not clear. When the assembler reads the instruction INC [BX], for example, it will not know whether to increment the byte pointed to by BX or to increment the word pointed to by BX. We use PTR operator to clarify how we want the assembler to code the instruction. The statement IN BYTE PTR [BX] tells the assembler that we want to increment the byte pointed to by BX. The statement INC WORD PTR [BX] tells the assembler that we want to increment the word pointed to by BX.

The PTR operator can be used to override the declared type of a variable. Suppose, for example, that we have declared an array of words with the statements WORDS DW437AH, 0B972H, 7C41H.

23. **PUBLIC:** Large programs are usually written as several separate modules. Each module is individually assembled, tested and debugged. When all the modules are working correctly, their object code files are linked together to form the complete program. In order for the modules to link together correctly, any variable name or label referred to in other modules must be declared public in the module in which it is defined. The PUBLIC directive is used to tell the assembler that a specified name or label will be accessed from other modules. Example is the statement PUBLIC DIVISOR, DIVIDEND, which makes the two variables DIVISOR and DIVIDEND available to other assembly modules.

24. **SEGMENT:** The SEGMENT directive is used to indicate the start of a logical segment. Preceding the SEGMENT directive is the name you want to give the segment. The statement CODE SEGMENT, for example, indicates to the assembler the start of a logical segment called CODE. The SEGMENT and ENDS directives are used to "bracket" a logical segment containing code or data. Refer to the ENDS directive for an example of how this is done.

Additional terms are often added to a SEGMENT directive statement to indicate some special way in which we want the assembler to treat the segment. The CODE SEGMENT WORD tells the assembler that we want the contents of this segment located on the next available word (even) address when segments are combined and given absolute addresses.

25. **SHORT:** The SHORT operator is used to tell the assembler that only a 1-byte displacement is needed to code a jump instruction. If jump destination is after the jump instruction in the program, the assembler will automatically reserve two bytes for the displacement. Using short operator saves one byte of memory by telling the assembler that it needs to reserve only one byte for this particular jump. The statement JMP SHORT NEARBY_LABEL is an example of the use of SHORT.

26. **TYPE:** The type operator tells the assembler to determine the type of a specified variable. The assembler actually determines the number of bytes in the type of the variable. For a byte type variable, the assembler will give a value of one. For a word type variable, the assembler will give a value of two. For a doubleword type variable, the assembler will give a value of four.

**For example:** ADD BX, TYPE WORD_ARRAY.

## 4.12 INTERRUPT STRUCTURE

The meaning of 'interrupts' is to break the sequence of operation. While the CPU is executing a program, 'interrupt' breaks the normal sequence of execution of instructions, diverts its execution to some other program called Interrupt Service Routine (ISR). After executing ISR, the control is transferred back again to the main program. Interrupt processing is an alternative to polling.

### 4.12.1 Need for Interrupt

Interrupts are particularly useful when interfacing I/O devices provide or require data at relatively low data transfer rate.

#### Types of Interrupt

There are two types of Interrupts in 8086. They are:

1. Hardware interrupts
2. Software interrupts

#### Hardware Interrupts

The Intel microprocessors support hardware interrupts through:

1. Two pins that allow interrupt requests, INTR and NMI
2. One pin that acknowledges, INTA, the interrupt requested on INTR.

#### INTR and NMI

1. INTR is a maskable hardware interrupt. The interrupt can be enabled/disabled using STI/CLI instructions or using more complicated method of updating the FLAGS register with the help of the POPF instruction.
2. When an interrupt occurs, the processor stores FLAGS register into stack, disables further interrupts, fetches from the bus one byte representing interrupt type and jumps to interrupt processing routine address of which is stored in location 4* <interrupt type>. Interrupt processing routine should return with the IRET instruction.
3. NMI is a non-maskable interrupt. Interrupt is processed in the same way as the INTR interrupt. Interrupt type of the NMI is 2, i.e. the address of the NMI processing routine is stored in location 0008h. This interrupt has higher priority than the maskable interrupt.

   **Examples:** NMI, INTR.

#### Software Interrupt

Software interrupts can be caused by:

1. INT instruction—breakpoint interrupt. This is a type 3 interrupt.
2. INT <interrupt number> instruction—any one interrupt from available 256 interrupts.
3. INTO instruction—interrupt on overflow
4. Single-step interrupt—generated if the TF flag is set. This is a type 1 interrupt. When the CPU processes this interrupt, it clears TF flag before calling the interrupt processing routine.
5. Processor exceptions: Divide Error (Type 0), Unused Opcode (type 6) and Escape Opcode (type 7).

6. Software interrupt processing is the same as for the hardware interrupts.

   **Examples:** INT n (Software Instructions)

7. Control is provided through:

   i. IF and TF flag bits

   ii. IRET and IRETD

Performance of Software Interrupt (Fig. 4.13)

1. It decrements SP by 2 and pushes the flag register on the stack.
2. Disables INTR by clearing the IF.
3. It resets the TF in the flag register.
4. It decrements SP by 2 and pushes CS on the stack.
5. It decrements SP by 2 and pushes IP on the stack.
6. Fetch the ISR address from the interrupt vector (Fig. 4.14).

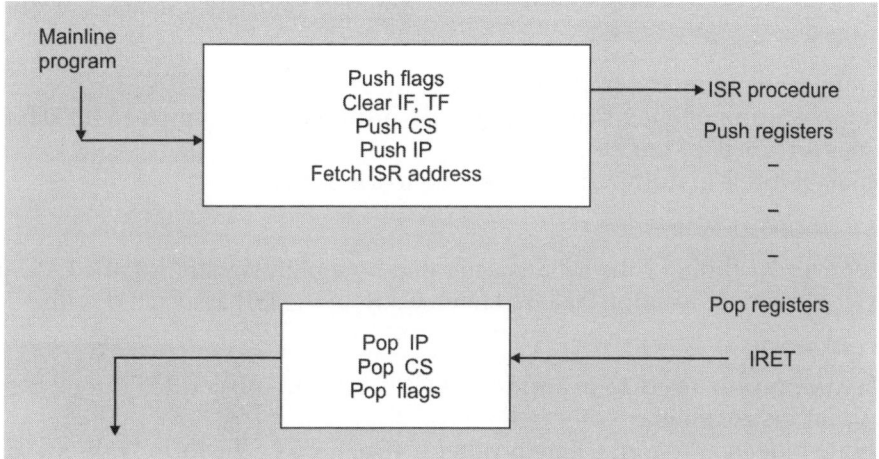

**Fig. 4.13:** Performance of software interrupt

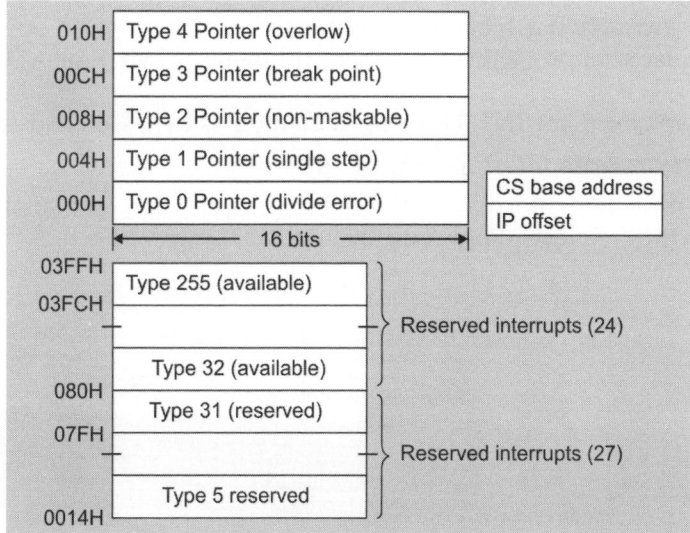

**Fig. 4.14:** Interrupt vector

**Table 4.9:** INT number and physical address

| INT number | Physical address |
|------------|------------------|
| INT 00 | 00000 |
| INT 01 | 00004 |
| INT 02 | 00008 |
| ⋮ | ⋮ |
| ⋮ | ⋮ |
| INT FF | 003FC |

### Functions Associated with INT 00 to INT 04

*INT 00 (Divide error)*

- INT 00 is invoked by the microprocessor whenever there is an attempt to divide a number by zero.
- ISR is responsible for displaying the message "Divide Error" on the screen.

*INT 01*

- For single stepping, the trap flag must be 1.
- After execution of each instruction, 8086 automatically jumps to 00004H to fetch 4 bytes for CS: IP of the ISR.
- The job of ISR is to dump the registers on to the screen.

*INT 02 (Non-maskable interrupt)*

- Whenever NMI pin of the 8086 is activated by a high signal (5v), the CPU jumps to physical memory location 00008 to fetch CS:IP of the ISR associated with NMI.

*INT 03 (Break point)*

- A break point is used to examine the CPU and memory after the execution of a group of instructions.
- It is one-byte instruction, whereas other instructions of the form "INT nn" are 2 byte instructions.

*INT 04 (Signed number overflow)*

- There is an instruction associated with this INT 0 (interrupt on overflow).
- If INT 0 is placed after a signed number arithmetic as IMUL or ADD, the CPU will activate INT 04 if 0F = 1.
- In case where 0F = 0, the INT 0 is not executed but is bypassed and acts as a NOP.

### Performance of Hardware Interrupt (Fig. 4.15)

1. **NMI:** Non-maskable interrupts—TYPE 2 interrupt
2. **INTR:** Interrupt request—Between 20H and FFH

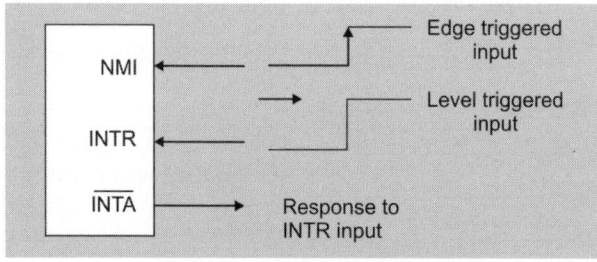

**Fig. 4.15:** Hardware interrupt

## Interrupt Priority Structure

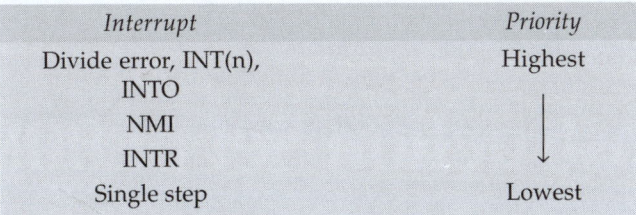

| Interrupt | Priority |
|---|---|
| Divide error, INT(n), INTO | Highest |
| NMI | |
| INTR | |
| Single step | Lowest |

## PROBLEMS

1. Differentiate between 8085 and 8086.

2. Explain the internal architecture of 8086 with neat and clean diagram.

3. What are the important signals of Intel 8086? Discuss in brief.

4. Describe the pin configuration of 8086 microprocessor.

5. Discuss various addressing modes used in 8086 with suitable examples.

6. Discuss conditional and control flags of 8086.

7. What are LOCK and LOCK (active low)? Discuss their role.

8. Discuss concept of segmentation. Explain different segment registers used in 8086.

9. Explain the minimum and maximum modes of operation of 8086 with details of each pin.

10. What do you understand by assembler directives?

11. Discuss the following assembler directives with examples:
    DT, DD, EXTRN, NAME

12. Discuss 8086 based system configured in maximum mode with a neat block diagram.

13. Discuss 8086 based system configured in minimum mode with a neat block diagram.

14. Draw the register organization of 8086 and explain typical applications of each register.

15. What is bus cycle? Draw memory-read bus cycle in minimum mode of operation of 8086.

16. Discuss various interrupts of 8086 in detail.

17. Discuss how physical address is generated.

# Instruction Set of 8086

## 5.1 INSTRUCTION FORMAT

The instruction format of 8086 has one or more number of fields associated with it. The first field is called operation code field or opcode field, which indicates the type of operation. The instruction format also contains other fields known as operand fields.

There are six general formats of instructions in 8086 instruction set. The length of an instruction may vary from one byte to six bytes.

### 5.1.1 One-byte Instruction

This format is only one byte long and may have the implied data or register operands. The least significant 3 bits of the opcode are used for specifying the register operand, if any. Otherwise, all the eight bits form an opcode and the operands are implied.

**For example:**

$$1 1 1 1 1 0 0 0 \quad F8_H \quad CLC: \text{Clear Carry}$$

This is an operation without any operand, which clears the carry flag bit.

Exchange register with accumulator | 10010 | reg |

Depending on the register (reg = $RRR$), the contents of the specified register will be exchanged with the accumulator. This operation is having one operand which is specified in a register.

*ASCII* Adjust for addition AAA   00110111  $37_H$

Here, the operand to this instruction is implicit and it takes the contents of register *AL*.

### 5.1.2 Register to Register

This format is two bytes long. The first byte of the code specifies the operation code and the width of the operand specifies by $w$ bit. The second byte of the opcode shows the register operands and *RIM* field.

| $D_7$ | | $D_1$ | $D_0$ | | $D_7$ | $D_6$ | $D_5$ | $D_4$ | $D_3$ | $D_2$ | $D_1$ | $D_0$ |
|---|---|---|---|---|---|---|---|---|---|---|---|---|
| *Opcode* | | *d* | *w* | | 11 | | REG | | | R/M | | |

The register represented by the *REG* field is one of the operands. The *RIM* field specifies another register or memory location, i.e. the other operand. The register specified by *REG* is a source operand if $D = 0$, else it is a destination operand.

**For example:**

*MOV*: Data Transfer Operation from Register to Register.

Opcode is

$$\boxed{1\,0\,0\,0\,1\,0\,dw} \qquad\qquad \boxed{1\,1\,REG}$$

10001000          11 000  001          $88_H\,Cl_H$

$REG = 0\,0\,0$      indicates register $AL$

$REG = 0\,0\,1$      indicates register $CL$

$w = 0$ indicates it is a byte operation (8 bit).

$d = 0$ indicates $AL$ is a source register.

This instruction indicates MOV $CL$, $AL$, i.e. $CL \leftarrow AL$.

## 5.1.3  Register to/from Memory with no Displacement

This format is also two bytes long and similar to the register format except for the $MOD$ field.

| $D_7$ | | $D_1$ | $D_0$ | $D_7$ | $D_6$ | $D_5$ | $D_4$ | $D_3$ | $D_2$ | $D_1$ | $D_0$ |
|---|---|---|---|---|---|---|---|---|---|---|---|
| Opcode | | d | w | MOD | | REG | | | R/M | | |

The $MOD$ field shows the $MOD$ of addressing. In case of no displacement, $MOD = 00$.

**For example:**

*MOV:* Data Transfer Register/Memory to/from Register.

$$\boxed{1\,0\,0\,0\,1\,0\,dw} \qquad\qquad \boxed{mod \quad reg \quad rim}$$

This format is similar to register to register transfer. The difference is in mod field.

For register to register, mod = 11.

For register to/from memory with no displacement, $mod = 00$.

When $mod = 0\,0$, the r/m fields indicate the address to memory location.

As for example $r/m = 1\,1\,1$ indicates $(Bx)$.

The instruction

$1\,0\,0\,0\,1\,0\,1\,0\,0\,0\,0\,0\,0\,0\,1\,1\,1$  indicates the instruction MOV $AX$, $[BX]$. In hexadecimal, the instruction is $8A_H\,O7_H$.

Here, the data is present in a memory location in $DS$ whose offset address is in $BX$. The effective address of the data is given as $10H \times DS + [BX]$.

There $d = 1$ indicates $AX$ is a destination register, so it moves the data from memory to register.

## 5.1.4  Register to/from Memory with Displacement

This type of instruction format contains one or two additional bytes for displacement along with 2-bytes of the format of the register to/from memory without displacement.

| $D_7$ | $D_7$ | $D_6$ | $D_5$ | $D_4$ | $D_3$ | $D_2$ | $D_1$ | $D_0$ | $D_1$ | $D_0$ |
|---|---|---|---|---|---|---|---|---|---|---|---|
| Opcode | MOD | | REG | | | R/M | | | Low byte of displacement | |

| $D_1$ | $D_0$ |
|---|---|
| High byte of displacement | |

$MOD = 0\,1$ indicates displacement of 8 bytes (Instruction is of size 3 bytes).

$MOD = 1\,0$ indicates displacement of 16 bytes (Instruction is of size 4 bytes).

We have already seen the other two options of $MOD$.

$MOD = 1\,1$ indicates register to register transfer.

$MOD = 0\,0$ indicates memory without displacement

In this case, $R/M$ fields indicate a memory when $MOD$ is not $1\,1$.

$R/M = 1\,1\,1$ indicates $(BX)$.

When MOD = $0\,1$, the offset address is $(BX) + D8$.

When MOD = $1\,0$, the offset address is $(BX) + D16$.

### 5.1.5 Immediate Operand to Register

In this format, the first byte as well as the 3 bytes from the second byte which are used for $REG$ field in case of register-to-register format are used for opcode. It also contains one or two bytes of immediate data.

| $D_7$ | | $D_0$ |
|---|---|---|
| Opcode | | $w$ |

| $D_7$ | $D_6$ | $D_5$ | $D_4$ | $D_3$ | $D_2$ | $D_1$ | $D_0$ |
|---|---|---|---|---|---|---|---|
| 1 1 | | opcode | | | R/M | | |

| $D_7$ | | $D_0$ |
|---|---|---|
| Lower byte data | | |

| $D_7$ | | $D_0$ |
|---|---|---|
| Higher byte data | | |

When $w = 0$, the size of immediate data is 8 bits and the size of instruction is 3 bytes.

When $w = 1$, the size of immediate data is 16 bits and the size of instruction is 4 bytes.

### 5.1.6 Immediate Operand to Memory with 16-Bit Displacement

This type of instruction format requires 5 to 6 bytes for coding. The first two bytes contain the information regarding $OPCODE$, $MOD$ and $R/M$ fields. The remaining 4 bytes contain 2 bytes of displacement and 2 bytes of data.

| $D_7$ | | $D_0$ |
|---|---|---|
| Opcode | | $w$ |

| $D_7$ | $D_6$ | $D_5$ | $D_4$ | $D_3$ | $D_2$ | $D_1$ | $D_0$ |
|---|---|---|---|---|---|---|---|
| MOD | | opcode | | | R/M | | |

| $D_7$ | | $D_0$ |
|---|---|---|
| Higher byte of displacement | | |

| $D_7$ | | $D_0$ |
|---|---|---|
| Lower byte of data | | |

| $D_7$ | | $D_0$ |
|---|---|---|
| Lower byte of data | | |

The $REG$ code of the different registers (either a source or destination operands) in the opcode byte are assigned with binary code.

Table 5.1 shows coding of different registers.

### 5.2 ADDRESSING MODES OF 8086

*Addressing mode indicates a way of locating data or operands.* The different addressing modes of the 8086 instructions along with corresponding $MOD$, $REG$ and $R/M$ fields are given in Table 5.2.

**Table 5.1:** Coding of different registers

| w-bit | Register code (3 bit) | Registers | Segment register code (2 bit) | Segment register |
|---|---|---|---|---|
| 0 | 0 0 0 | AL | 0 0 | ES |
| 0 | 0 0 1 | CL | 0 1 | CS |
| 0 | 0 1 0 | DL | 1 0 | SS |
| 0 | 0 1 1 | BL | 1 1 | DS |
| 0 | 1 0 0 | AH | | |
| 0 | 1 0 1 | CH | | |
| 0 | 1 1 0 | DH | | |
| 0 | 1 1 1 | BH | | |
| 1 | 0 0 0 | AX | | |
| 1 | 0 0 1 | CX | | |
| 1 | 0 1 0 | DX | | |
| 1 | 0 1 1 | BX | | |
| 1 | 1 0 0 | SP | | |
| 1 | 1 0 1 | BP | | |
| 1 | 1 1 0 | SI | | |
| 1 | 1 1 1 | DI | | |

**Table 5.2:** Different addressing modes

| Operands | Memory operands | | | Register operands | |
|---|---|---|---|---|---|
| | No displacement | displacement 8 bits | displacement 16 bits | 11 | |
| MOD R/M | 00 | 01 | 10 | w = 0 | w = 1 |
| 0 0 0 | (BX) + (SI) | (BX) + (SI) + D8 | (BX) + (SI) + D16 | AL | AX |
| 0 0 1 | (BX) + (DI) | (BX) + (DI) + D8 | (BX) + (DI) + D16 | CL | CX |
| 0 1 0 | (BP) + (SI) | (BP) + (SI) + D8 | (BP) + (SI) + D16 | DL | DX |
| 0 1 1 | (BP) + (DI) | (BP) + (DI) + D8 | (BP) + (DI) + D16 | BL | BX |
| 1 0 0 | (SI) | (SI) + D8 | (SI) + D16 | AH | SP |
| 1 0 1 | (DI) | (DI) + D8 | (DI) + D16 | CH | BP |
| 1 1 0 | D16 | (BP) + D8 | (BP) D16 | DH | SS |
| 1 1 1 | (BX) | (BX) + D8 | (BX) + D16 | BH | DI |

D8 and D16 represent 8 and 16-bits displacement, respectively.

The default segment for the addressing modes using BP and SP is SS. For all other addressing modes, the default segments are DS or ES.

## 5.2.1 Different Addressing Modes of 8086

### 5.2.1.1 Immediate Addressing

In this addressing mode, immediate data is a part of instruction and appears in the form of successive byte or bytes.

**Example:** *MOV  AX, 005 OH*

Here, *005OH* is the immediate data and it is moved to register *AX*. The immediate data may be 8-bit or 16-bit in size.

### 5.2.1.2 Direct Addressing

In the direct addressing mode, a 16-bit address (offset) is directly specified in the instruction as a part of it.

**Example:** *MOV AX [1 0 0 0 H]*

Here, data resides in a memory location in the data segment, whose effective address is $10H \times DS + 1000\ H$.

### 5.2.1.3 Register Addressing

In register addressing mode, the data is stored in a register and it is referred using the particular register. All the registers except IP may be used in this mode.

**Example:** *MOV AX, BX*

**Register indirect:** In this addressing mode, the address of the memory location which contains data or operand is determined in an indirect way using offset registers. The offset address of data is in either *BX* or *SI* or *DI* register. The default segment register is either *DS* or *ES*.

**Example:** *MOV AX [BX]*

The data is present in a memory location in *DS* whose offset is in *BX*. The effective address is $10H \times DS + [BX]$.

### 5.2.1.4 Indexed Addressing

In this addressing mode, offset of the operand is stored in one of the index registers. *DS* and *ES* are the default segments for index registers *SI* and *DI*, respectively.

**Example:** *MOV AX, [SI]*

The effective address of the data is $10H \times DS + [SI]$.

**Register relative:** In this addressing mode the data is available at an effective address formed by adding an 8-bit or 16-bit displacement with the content of any one of the registers: *BX, BP, SI* and *DI* in the default either *DS* or *ES* segment.

**Example:** *MOV AX, 50H [BX]*

The effective address of the data is $10H \times DS + 50H + [BX]$.

### 5.2.1.5 Based Indexed Addressing

In this addressing mode, the effective address of the data is formed by adding the content of a base register (any one of *BX* or *BP*) to the content of an index register (any one of *SI* or *DI*). The default segment register may be *ES* or *DS*.

**Example:** *MOV*

The effective address is $10H \times DS + [BX] + [SI]$.

### 5.2.1.6 Relative Based Indexed Addressing

The effective address is formed by adding an 8-bit or 16-bit displacement with the sum of contents of any one of the base registers (*BX* or *BP*) and any one of the index registers in a default segment.

**Example:** *MOV AX, 50H*

Here, 50H is an immediate displacement. The effective address is $10H \times DS + [HX] + [SI] + 50H$.

### 5.2.1.7 Intrasegment Direct Mode

In this mode, the address to which the control is to be transferred lies in the segment in which the control transfer instruction lies and appears directly in the instruction as an immediate displacement value. The displacement is computed relative to the content of the instruction pointer *IP*.

### 5.2.1.8 Intrasegment Indirect Mode

This mode is similar to intrasegment direct mode except the displacement to which control is to be transferred is passed to the instruction indirectly. Here the branch address is found as the content of a register or a memory location.

### 5.2.1.9 Intersegment Direct Mode

In this mode, the address to which the control is to be transferred is in a different segment. This addressing mode provides a means of branching from one code segment to another code segment. Here, the *CS* and *IP* of the destination address are specified directly in the instruction.

### 5.2.1.10 Intersegment Indirect Mode

This mode is similar to intersegment direct mode except the address to which the control is to be transferred is passed to the instruction indirectly. This information is kept in a memory block of 4 bytes: *IP(LSB)*, *IP(MSB)*, *LS(LSR)* and *CS(MSB)* sequentially. The starting address of the memory block may be referred using any of the addressing modes except immediate mode.

### 5.2.1.11 Instruction Set 8086

The instruction of 8086 can be divided into the following categories according to their functions.

1. Data transfer instructions or data movement instructions.
2. Arithmetic instructions
3. Logical instructions
4. Shift instructions
5. Rotate instructions
6. Process or machine control instructions
7. Flow control or branch instructions
8. Interrupt instructions
9. Loop instructions
10. String operation instructions

## 5.3 DATA TRANSFER INSTRUCTIONS

These types of instructions are used to transfer data from source to destination. These include MOV, XLAT, IN, OUT, PUSH, POP, XCHG, MOVS, LODS, STOS, LDS, etc. The explanation of these instructions is given further.

1. **MOV:** MOV (Move) transfers a byte, word or double word from the source operand to the destination operand.

| Mnemonic | Description | Syntax | Operation | Flag affected |
|----------|-------------|--------|-----------|---------------|
| MOV | Move | MOV D, S | (D) ← (S) | None |

The MOV instruction is useful for transferring data along any of the following paths. There are also variants of MOV that operate on segment registers.

| Destination | Source |
|-------------|--------|
| Memory | Accumulator |
| Accumulator | Memory |
| Register | Register |
| Register | Memory |
| Memory | Register |
| Register | Immediate |
| Memory | Immediate |
| Segment register | Register 16 |
| Segment register | Memory 16 |
| Register 16 | Segment register |
| Memory | Segment register |

The MOV instruction cannot move data from memory-to-memory or from segment register to segment register. Memory-to-memory moves can be performed, however, by the string move instruction MOVS.

**Example:** MOV AX, 2000H    AX ← 2000H

MOV AX, BX    AX ← BX

MOV [3000], CX    [3000] ← CX

2. **XCHG:** This instruction exchanges the content of a register with the content of another register or the content of a register with the content of a memory location. The instruction cannot exchange the content of two memory locations. The segment register cannot be used in this instruction. No flag is affected.

| Mnemonic | Description | Syntax | Operation | Flag affected |
|----------|-------------|--------|-----------|---------------|
| XCHG | Exchange | XCHG D, S | (D) ↔ (S) | None |

| Destination | Source |
|-------------|--------|
| Accumulator | Register 16 |
| Memory | Register |
| Register | Register |
| Register | Memory |

**Example:** XCHG    CX, BX    CX ↔ BX

XCHG [2000], AX    [2000] ↔ AX

3. **IN:** This instruction is used to input a byte or word from port. The instruction copies data from a port to the accumulator. No flag is affected.

| Mnemonic | Description | Syntax | Operation | Flag affected |
|----------|-------------|--------|-----------|---------------|
| IN | Input (read) from port | IN AL, Port 8 | Read byte from 8-bit port (direct) | None |
| | | IN AX, Port 16 | Read word from 16-bit port (indirect) | None |

This instruction can be executed in two different addressing modes.

**Direct:** In direct addressing mode, 8-bit address of the port is a part of the instruction itself.

**Example:**  IN AL, 08H            AL ← Port at address 08

IN AX, 2000H          AX ← Port at address 2000

**Indirect:** In indirect addressing mode, the address of the port is referred from DX register. Since DX is a 16-bit register, the port address can be any number between 0000H and FFFFH.

**Example:**  MOV CX, 2000H

IN AL, CX             Port at address 2000 ← AL

IN AH, CX             Port at address 2000 ← AH

4. **OUT:** It sends a byte or word to the port. The out instruction copies a byte from AL or a word from AX to the specified port.

| Mnemonic | Description | Syntax | Operation | Flag affected |
|----------|-------------|--------|-----------|---------------|
| OUT | Output (write) to port | OUT Port 8, AL | Write byte in AL to 8-bit port | None |
| | | OUT Port 16, AX | Write word in AX to 16-bit port | None |

This instruction can be executed in two different modes as the IN instruction.

**Example:**  OUT 08, AL          → Direct Mode

MOV BX, 2000H       → Indirect Mode

OUT BX, AL

OUT BX, AX

5. **XLAT:** XLAT exchanges the byte in AL register from the user table index to the table entry, addressed by BX. It transfers 16-bit information at a time. The no-operands form (XLATB) provides a "short form" of the XLAT instructions.

| Mnemonic | Description | Syntax | Operation | Flag affected |
|----------|-------------|--------|-----------|---------------|
| XLAT | Translate byte in AL | XLAT | $((AL) + (BX) + (DS)\boxed{0}) \rightarrow AL$ | None |

**Example:**  DS = 2000H

BX = 1000H

AL = 08H

PA = DS0 + BX + AL

= 20000 + 1000 + 08 = 21008

6. **LEA (Load Effective Address):** This instruction indicates the offset of the variable or memory location named as the source and put this offset in the indicated 16-bit register.

**Syntax:** LEA R, S

| Mnemonic | Description | Syntax | Operation | Flag affected |
|---|---|---|---|---|
| LEA | Load effective address | LEA R16, EA | EA $\rightarrow$ R16 | None |
| LDS | Load register and DS | LDS R16, MEM32 | MEM32 $\rightarrow$ R16<br>MEM32+2 $\rightarrow$ DS | None |
| LES | Load register and ES | LES R16, MEM32 | MEM32 $\rightarrow$ R16<br>MEM32 +2 $\rightarrow$ DS | None |

**Example:** LEA CX, [BX] [DI]    CX $\rightarrow$ EA = BX + DI

7. **LDS:** This instruction loads a far pointer from the memory address specified by op2 into the DS segment register and the op1 to the register.
   **Syntax:** LDS register, memory address of first word or LDS op1, op2

   **Example:** LDS BX, [4326] ;   Copy the contents of the memory at displacement 4326H in DS to BL, contents of the 4327H to BH. Copy contents of 4328H and 4329H in DS to DS register.

8. **LES:** This instruction loads a 32-bit pointer from the memory address specified to destination register and Extra Segment. The offset is placed in the destination register and the segment is placed in Extra Segment. Using this instruction, the loading of far pointers may be simplified.
   **Syntax:** LES register, memory address of first word

   **Example**: LES SI, [200H] ; This instruction loads SI from memory location at physical location PA = 12000H + 200H = 12200H and ES is loaded with the contents of the following two bytes at address 12202H.

9. **PUSH:** PUSH instruction decrements the stack pointer by 2 and copies a word from a specified source to the location in the stack segment where the stack pointer points.

   **Syntax:**    PUSH source

   **Example:**   PUSH DX
   SP (Stack Pointer) = 1000H
   SS (Stack Segment) = 2000H
   [SS : SP] = 21000H ; Memory location pointed by pointer in stack segment
   [SS : SP]–1 $\leftarrow$ DH
   [SS : SP]–2 $\leftarrow$ DL
   SP $\leftarrow$ SP–2

| Mnemonic | Description | Syntax | Operation | Flag affected |
|---|---|---|---|---|
| PUSH | PUSH word onto the stack | PUSH R16 | SP = SP–2 then<br>[SS:SP] = R16 | None |
| POP | POP word from the stack | POP R16 | R16 = [SS:SP] then<br>SP = SP + 2 | None |

10. **POP:** POP instruction copies the word at the current top of the stack to the operand specified by op then increments the stack pointer to point to the next stack.

   **Example:** POP DX

   SP (Stack pointer) = 1000H

   SS (Stack segment) = 2000H

   [SS: SP] = 21000H ; Memory location pointed by pointer in
                                         stack segment

   [SS:SP] → DL

   [SS:SP]+1 → DH

   SP ← SP+2

11. **LAHF:** This instruction loads the lower byte of the flag register in AH, then copies the content of lower byte of 8086 flag register to AH register.

12. **SAHF:** This instruction stores the content of AH to the low byte of the flags, then copies the content of the AH register into the lower byte of the 8086 flag register.

## 5.4 ARITHMETIC INSTRUCTIONS

These instructions are used to perform arithmetic operations, such as addition, subtraction, multiplication, division, comparison, etc.

1. **ADD:** This instruction is used to add a no. from source to a no. from destination.

| Mnemonic | Description | Syntax | Operation | Flag affected |
|----------|-------------|--------|-----------|---------------|
| ADD | Addition | ADD D, S | $D \leftarrow D + S$ | CF, SF, ZF, AF, OF, PF |

| Source | Destination |
|--------|-------------|
| Memory | Memory |
| Register | Register |
| Immediate data | |

  **Example:**  ADD AL, BH      $AL \leftarrow AL + BH$

                 ADD AL, 20H     $AL \leftarrow AL + 20H$

                 ADD BX, [SI]     $BX \leftarrow BX + [SI]$

2. **ADC:** This instruction adds the status of carry flag into the result.

| Mnemonic | Description | Syntax | Operation | Flag affected |
|----------|-------------|--------|-----------|---------------|
| ADC | Addition with carry | ADC D, S | $D \leftarrow D + S + CY$ | CF, SF, ZF, AF, OF, PF |

| Source may be | Destination may be |
|---------------|--------------------|
| Memory | Memory |
| Register | Register |
| Immediate data | |

  **Example:**  ADC BL, BH       $BL \leftarrow BL + BH + CY$

                 ADC AX, 2000H    $AX \leftarrow AX + 2000 + CY$

                 ADC [SI], 20H       $[SI] \leftarrow [SI] + 20 + CY$

**3. SUB:** This instruction is used to subtract the number in the indicated source from a no. in the indicated destination and put the result in the indicated destination.

| Mnemonic | Description | Syntax | Operation | Flag affected |
|---|---|---|---|---|
| SUB | Subtraction | SUB D, S | DßD-S | CF, SF, ZF, AF, OF, PF |

| Source may be | Destination may be |
|---|---|
| Memory | Memory |
| Register | Register |
| Immediate data | |

**Example:** SUB AL, BH      AL ← AL-BH

SUB AL, 20H      AL ← AL-20H

SUB BX, [SI]      BX ← BX - [SI]

**4. SBB:** This instruction is used to subtract source operand and borrow (carry) bit from the destination operand.

| Mnemonic | Description | Syntax | Operation | Flag affected |
|---|---|---|---|---|
| SBB | Subtraction with borrow | SBB D, S | D ← D-S-CY | CF, SF, ZF, AF, OF, PF |

| Source may be | Destination may be |
|---|---|
| Memory | Memory |
| Register | Register |
| Immediate data | |

**Example:** SBB BL, BH      BL ← BL-BH-CY

SBB AX, 2000H      AX ← AX-2000-CY

SBB [SI], 20H      [SI] ← [SI]-20-CY

**5. MUL:** This instruction multiplies an unsigned multiplication of the accumulator by the operand specified by source. The op may be a register or memory operand.

**Syntax:** MUL OP

**Example:** MUL BH      AX ← ALXBL

MUL BX      RESULT 32 BIT = AXXBX

AX ← LOWER 16 BITS

DX ← UPPER 16 BITS

MUL 20H      AX ← ALX20

**6. IMUL:** This instruction performs a signed multiplication. There are two types of syntax for this instruction. They are:

**Syntax:**      IMUL S      ;      In this form, the accumulator is the multiplicand and op is the multiplier. OP may be a register or a memory operand.

IMUL OP1, OP2   ;      In this form, OP1 is always be a register operand and OP2 may be a register or a memory operand.

**Example:** $-28 \times 59 = -1652$

AL = 11100100 (2's Complement of $-28$)

BL = 00111011

IMUL BL                            AX ← ALXBL = F98C

                                        MSB = 1, so negative result in 2's complement

IMUL CX, BX                    CX ← CXXBX (Lower 16-bits of result)

| Mnemonic | Description | Syntax | Operation | Flag affected |
|---|---|---|---|---|
| MUL | Unsigned multiply | MUL S | AX ← ALXS8 | CF, SF, ZF, AF, OF, PF |
| | | | (DX)(AX) ← AXXS16 | |
| IMUL | Integer (Signed) multiply | IMUL S | AX ← ALXS8 | CF, SF, ZF, AF, OF, PF |
| | | | (DX)(AX) ← AXXS16 | |

| Source |
|---|
| Memory 8 |
| Register 8 |
| Register 16 |
| Memory 16 |

7. **DIV:** When a double word is divided by a word, the most significant word of the double word must be in DX and the least significant word of the double word must be in AX. After the division, AX will contain the 16-bit result (quotient) and DX will contain a 16-bit remainder. Again, if an attempt is made to divide by zero or quotient is too large to fit in AX (greater than FFFFH), the 8086 will do a type of 0 interrupt.

| Mnemonic | Description | Syntax | Operation | Flag affected |
|---|---|---|---|---|
| DIV | Division unsigned | DIV S | AL ← Q{(AX)/(S8)} | CF, SF, ZF, |
| | | | AH ← R{(AX)/(S8)} | AF, OF, PF |
| | | | AX ← Q{(DX, AX)/(S16)} | |
| | | | DX ← R{(DX, AX)/(S16)} | |
| IDIV | Integer (Signed) divide | IDIV S | AL ← Q{(AX)/(S8)} | CF, SF, ZF, |
| | | | AH ← R{(AX)/(S8)} | AF, PF, OF |
| | | | AX ← Q{(DX, AX)/(S16)} | |
| | | | DX ← R{(DX, AX)/(S16)} | |

| Source |
|---|
| Memory 8 |
| Register 8 |
| Register 16 |
| Memory 16 |

For DIV, the dividend must always be in AX or DX and AX, but the source of the divisor can be a register or a memory location specified by one of the 24 addressing modes.

If you want to divide a byte by a byte, you must first put the dividend byte in AL and fill AH with all 0's. The SUB AH, AH instruction is a quick way to do. If you want to divide a word by a word, put the dividend word in AX and fill DX with all 0's. The SUB DX, DX instruction does this quickly.

**Example:**  DIV BH ; AX = 37D7H = 14, 295 decimal and BH = 97H = 151 decimal
AX/BH
AX = Quotient = 5EH = 94 decimal and AH = Remainder = 65H = 101 decimal.

8. **IDIV:** This instruction is used to divide a signed word by a signed byte or to divide a signed double word by a signed word. If source is a byte value, AX is divided by register and the quotient is stored in AL and the remainder in AH. If source is a word value, DX:AX is divided by register, and the quotient is stored in AL and the remainder in DX.

**Example:** IDIV BL ; Signed word in AX is divided by signed byte in BL

9. **INC:** This instruction increments destination by 1. Destination may be register or memory.

| Mnemonic | Description | Syntax | Operation | Flag affected |
|---|---|---|---|---|
| INC | Increment by one | INC D | $D \leftarrow D + 1$ | SF, ZF, AF, OF, PF |

**Example:**  INC DL DL $\leftarrow$ DL + 1
INC [SI] [SI] $\leftarrow$ SI + 1

10. **DEC:** This instruction decreases the destination by 1. Destination may be register or memory.

| Mnemonic | Description | Syntax | Operation | Flag affected |
|---|---|---|---|---|
| DEC | Decrement by one | DEC D | $D-1 => D$ | SF, ZF, AF, OF, PF |

**Example:**  DEC CL          CL $\leftarrow$ CL-1
DEC [SI]         [SI] $\leftarrow$ SI-1

11. **CMP:** The CMP instruction compares the destination and source, i.e. it subtracts the source from destination. The result is not stored anywhere. It neglects the results but sets the flags accordingly. This instruction is usually used before a conditional jump instruction.

| Mnemonic | Description | Syntax | Operation | Flag affected |
|---|---|---|---|---|
| CMP | Compare two operands | CMP D, S | D-S, discarding result but setting the flag | CF, SF, ZF, AF, OF, PF |

**Example:**  CMP AL, BL       ; AL $\leftarrow$ AL-BL; if CY = 0, AL>BL
If CY = 1, AL<BL
If Z = 1, AL = BL

CMP [2000], 20H ; [2000] $\leftarrow$ [2000]-20; if CY = 0, [2000] > 20
If CY = 1, [2000] < 20
If Z = 1, [2000] = 20

Destination and source may be register, memory and immediate data.

12. **NEG:** NEG performs the two's complement subtraction of the operand from zero and sets the flags according to the result.

| Mnemonic | Description | Syntax | Operation | Flag affected |
|----------|-------------|--------|-----------|---------------|
| NEG | Two's complement negate | NEG D | D = 0-D | CF, SF, ZF, AF, OF, PF |

**Example:** AX = 2CBh

NEG AX ; After executing NEG result, AX = FD35H

13. **AAA (ASCII Adjust After Addition):** AAA converts the result of the addition of two valid unpacked BCD digits to a valid 2-digit BCD number and takes the AL register as its implicit operand.

Two operands of the addition must have its lower 4-bits that contain a number in the range from 0 to 9. The AAA instruction then adjust AL so that it contains a correct BCD digit. If the addition produce carry (AF = 1), the AH register is incremented and the carry CF and auxiliary carry AF flags are set to 1. If the addition did not produce a decimal carry, CF and AF are cleared to 0 and AH is not altered. In both cases, the higher 4 bits of AL are cleared to 0. AAA will adjust the result of the two ASCII characters that were in the range from 30h ("0") to 39h("9"). This is because the lower 4 bits of those character fall in the range of 0–9. The result of addition is not a ASCII character but it is a BCD digit.

| Mnemonic | Description | Syntax | Operation | Flag affected |
|----------|-------------|--------|-----------|---------------|
| AAA | ASCII adjust after addition | AAA | Corrects result in AH and AL after addition when working with BCD values | CF, AF |

**Algorithm for AAA:** It works according to the following Algorithm:

If low nibble of AL > 9 or AF = 1 then:

$$AL = AL + 6$$
$$AH = AH + 1$$
$$AF = 1$$
$$CF = 1$$

Else

$$AF = 0$$
$$CF = 0$$

**In both cases:** Clear the high nibble of AL.

**Example:**   MOV AL, 05H

MOV BL, 06H

ADD AL, BL ; AL ← AL + BL = 0BH

; 0BH > 09

AAA      ; 0B + 06 = 11

; AH = 01, AL = 01

14. **AAS (ASCII Adjust After Subtraction):** AAS converts the result of the subtraction of two valid unpacked BCD digits to a single valid BCD number and takes the AL register as an implicit operand. The two operands of the subtraction must have its lower 4 bit contain number in the range from 0 to 9. The AAS instruction then adjust AL so that it contain a correct BCD digit.

| Mnemonic | Description | Syntax | Operation | Flag affected |
|----------|-------------|--------|-----------|---------------|
| AAS | ASCII adjust after subtraction | AAS | Corrects result in AH and AL after subtraction when working with BCD values | CF, AF |

**Algorithm for AAS:** If low nibble of AL > 9 or AF = 1 then:

$$AL = AL - 6$$
$$AH = AH - 1$$
$$AF = 1$$
$$CF = 1$$

Else

$$AF = 0$$
$$CF = 0$$

**In both cases:** Clear the high nibble of AL.

**Example:**
```
                    ; AL = 00111001 = ASCII 9
                    ; BL = 00110101 = ASCII 5
        SUB AL, BL  ; (9 – 5) Result: AL = 00000100 = BCD
                      04, CF = 0
        AAS         ; Result: AL = 00000100 = BCD 04
                      CF = 0 NO Borrow required
```

**Example:**
```
                    ; AL = 0011 0101 = ASCII 5
                    ; BL = 0011 1001 = ASCII 9
        SUB AL, BL  ; ( 5–9) Result: AL = 1111 1100 = – 4 in 2's complement CF = 1
        AAS         ; Result: AL = 0000 0100 = BCD 04
                      CF = 1 borrow needed
```

15. **AAM (ASCII Adjust After Multiplication):** AAM converts the result of the multiplication of two valid unpacked BCD digits into a valid 2-digit unpacked BCD number and takes AX as an implicit operand. To give a valid result, the digits that have been multiplied must be in the range of 0 – 9 and the result should have been placed in the AX register. Since both operands of multiply are required to be 9 or less, the result must be less than 81 and thus is completely contained in AL. AAM unpacks the result by dividing AX by 10, placing the quotient (MSD) in AH and the remainder (LSD) in AL.

| Mnemonic | Description | Syntax | Operation | Flag affected |
|----------|-------------|--------|-----------|---------------|
| AAM | ASCII adjust after multiplication | AAM | Corrects the result of multiplication of two BCD values | PF, ZF, SF |

**Algorithm for AAM:**

AH = AL/10

AL = remainder

**Example:**
```
        MOV AL, 8
        MOV BL, 6
        MUL BL      ; AL ← ALXBL = 30H
        AAM         ; AH ← AL/10 = 04
                    ; AL = 08
```

16. **AAD (ASCII Adjust Before Division):** This instruction appears before division and it prepares two BCD values for division. AAD performs the inverse operation to AAM. It multiplies AH by ten, adds it to AL and sets AH to zero. Again, the multiplier 10 can be changed. SF, ZF and PF flags are affected by this instruction; the OF, AF and CF flags are left in an indeterminate condition.

| Mnemonic | Description | Syntax | Operation | Flag affected |
|----------|-------------|--------|-----------|---------------|
| AAD | ASCII adjust before division | AAD | Prepares two BCD values for division | PF, ZF, SF |

**Algorithm for AAD:**

AL = (AH * 10) + AL
AH = 0

**Example:**  MOV AX, 0206h ; AH = 02, AL = 06
                AAD    ; AH = 00, AL = 1AH (26)

17. **DAA (Decimal Adjust After Addition):** The contents after addition are changed from a binary value to two 4-bit binary coded decimal (BCD) digits. S, Z, AC, P, CY flags are altered to reflect the results of the operation. If the value of the low-order 4-bits in the accumulator is greater than 9 or if AC flag is set, the instruction adds 6 to the low-order 4-bits. If the value of the high-order 4-bits in the accumulator is greater than 9 or if the Carry flag is set, the instruction adds 6 to the high-order 4-bits.

| Mnemonic | Description | Syntax | Operation | Flag affected |
|----------|-------------|--------|-----------|---------------|
| DAA | Decimal adjust after addition | DAA | Corrects the result of addition of two packed BCD values | CF, AF, PF, ZF, SF |

**Algorithm for DAA:**

If low nibble of AL > 9 or AF = 1 then:

$$AL = AL + 6$$
$$AF = 1$$

If AL > 9Fh or CF = 1 then:

$$AL = AL + 60h$$
$$CF = 1$$

**Example:** MOV AL, 0Fh ; AL = 0FH (15); 0F >09
                    ; 0F + 06 = 15H
              DAA ; AL = 15H

18. **DAS (Decimal Adjust After Subtraction):** This instruction corrects the result (in AL) of subtraction of two packed BCD values. The flags which modify are AF, CF, PF, SF, and ZF.

| Mnemonic | Description | Syntax | Operation | Flag affected |
|----------|-------------|--------|-----------|---------------|
| DAS | Decimal adjust after subtraction | DAS | Corrects the result of subtraction of two packed BCD values | CF, AF, PF, ZF, SF |

**Algorithm for DAS:**

If low nibble of AL > 9 or  AF = 1 then:

$$AL = AL - 6$$
$$AF = 1$$

If AL > 9Fh or CF = 1 then:

$$AL = AL - 60h$$
$$CF = 1$$

**Example:** MOV AL, 0FFh ; AL = 0FFh (–1)

DAS ; AL = 99h, CF = 1

## 5.5 LOGICAL INSTRUCTIONS

These instructions are used to perform logical operations, such as AND, OR, XOR, NOT, etc.

1. **AND:** This performs a bitwise Logical AND of two operands. The result of the operation is stored in the op1 and used to set the flags. AND op1, op2 perform a bitwise AND of the two operands, each bit of the result is set to 1, if and only if the corresponding bit in both of the operands is 1, otherwise the bit in the result I cleared to 0.

| Mnemonic | Description | Syntax | Operation | Flag affected |
|---|---|---|---|---|
| AND | Logical AND operation | AND OP1, OP2 | Logical AND between all bits of two operands. Result is stored in operand | (CF, OF both reset to 0) PF, SF, ZF |

OP1 and OP2 may be register, memory and immediate data.

**Example:**   ; AL = 00000101 = 05H

; BL = 11111111 = FFH

AND AL, BL ; AL ← AL AND BL = 05H

2. **OR:** OR instruction perfor the bitwise logical OR of two operands. Each bit of the result is cleared to 0, if and only if both corresponding bits in each operand are 0, otherwise the bit in the result is set to 1.1 OR 1 = 1.

| Mnemonic | Description | Syntax | Operation | Flag affected |
|---|---|---|---|---|
| OR | Logical OR operation | OR OP1, OP2 | Logical OR between all bits of two operands. Result is stored in first operand | (CF, OF both reset to 0) PF, SF, ZF |

OP1 and OP2 may be register, memory and immediate data.

**Example:** ; AL = 00000101 = 05H

; BL = 11111111 = FFH

OR AL, BL ; AL ← AL OR BL = FFH

3. **XOR:** XOR performs a bitwise logical XOR of the operands specified by op1 and op2. The result of the operand is stored in op1 and is used to set the flag.

| Mnemonic | Description | Syntax | Operation | Flag affected |
|---|---|---|---|---|
| XOR | Logical XOR operation | XOR OP1, OP2 | Logical XOR (Exclusive OR) between all bits of two operands. Result is stored in first operand | (CF, OF both reset to 0) PF, SF, ZF |

OP1 and OP2 may be register, memory and immediate data.

**Example:** ; AL = 00000101 = 05H
        ; BL = 11111111 = FFH
        XOR AL, BL ; ALAL XOR BL = 11111010 = FAH

4. **NOT and NEG Instruction:** NOT perfor the bitwise complement of operand and stores the result back into operand itself.

| Mnemonic | Description | Syntax | Operation | Flag affected |
|---|---|---|---|---|
| NOT | Logical NOT operation | NOT OP | Invert each bit of the operand | None |

OP may be register or memory.

**Examples:** ; AL = 00000101 = 05H
        NOT AL ; AL ← 11111010 = FAH

## 5.6 SHIFT INSTRUCTIONS

Shift operation moves the numbers to the left or right within a register or memory location. There are two types of shift operations: logical shift and arithmetic shift. The logical shift operation functions with unsigned numbers and arithmetic shift operation functions with signed numbers. All flags are affected when shift instruction executes.

1. **SAL/SHL (shift operand bits left):** SAL instruction shifts the bits in the operand specified by op1 to its left by the count specified in op2. As a bit is shifted out of LSB position, a 0 is kept in LSB position. CF will contain MSB bit.

| Mnemonic | Description | Syntax | Operation | Flag affected |
|---|---|---|---|---|
| SAL/SHL | Shift operand bits left, put zero in LSB(s) | SAL/SHL OP1, OP2 | Shift operand1 left. The number of shifts is set by operand2. Put zero in LSB(s) | CF, SF, ZF, PF |

CF ← MSB ← LSB 0

**Algorithm for SAL/SHL:**

- Shift all bits left, the bit that goes off is set to CF.
- Zero bit is inserted to the right-most position.

| OP1 | OP2 |
|---|---|
| Register | Immediate data |
| Register | CL |
| Memory | Immediate data |
| Memory | CL |

**Example:** ; CF = 0, CX = 11100101 11010011

SAL CX, 1; Shift BX register contents by 1 bit position towards left

; CF = 1, BX = 11001011 1010011

; AL = F4H = 11110100

MOV CL, 04

SAL AL, CL ; AL = 40H = 01000000

2. **SHR (Shift operand bits right):** SHR instruction shifts the bits in op1 to right by the number of times specified by op2.

| Mnemonic | Description | Syntax | Operation | Flag affected |
|----------|-------------|--------|-----------|---------------|
| SHR | Shift operand bits right, put zero in MSB | SHR OP1, OP2 | Shift operand1 right. The number of shifts is set by operand2. Put zero in MSB | CF, SF, ZF, PF |

$$0 \rightarrow MSB \rightarrow LSB \rightarrow CF$$

**Algorithm for SHR:**

- Shift all bits right, the bit that goes off is set to CF.
- Zero bit is inserted to the left-most position.

| OP1 | OP2 |
|-----|-----|
| Register | Immediate data |
| Register | CL |
| Memory | Immediate data |
| Memory | CL |

**Example:** ; CF = 0, CX = 11100101 11010011

SHR CX, 1 ; Shift BX register contents by 1 bit position towards left

; CF = 1, BX = 011100101 1101001

; AL = F4H = 11110100

MOV CL, 04

SHR AL, CL ; AL = 0FH = 00001111

3. **SAR (Shift operand bits right, New MSB = OLD MSB):** SAR instruction shifts the bits in the operand specified by op1 towards right by count specified in op2. As bit is shifted out, a copy of old MSB is taken in MSB. MSB position and LSB is shifted to CF.

| Mnemonic | Description | Syntax | Operation | Flag affected |
|----------|-------------|--------|-----------|---------------|
| SAR | Shift operand bits right, put zero in MSB | SHR OP1, OP2 | Shift arithmetic operand1 right. The number of shifts is set by operand2 | CF, SF, ZF, PF, OF |

$$MSB \rightarrow MSB \rightarrow LSB \rightarrow CF$$

**Algorithm for SAR:**

- Shift all bits right, the bit that goes off is set to CF.
- The sign bit that is inserted to the left-most position has the same value as before shift.

**Example:** ; CF = 0, CX = 11100101 11010011

SAR CX, 1 ; Shift BX register contents by 1 bit position towards left

; CF = 1, BX = 111100101 1101001a

; AL = F4H = 11110100

MOV CL, 04

SAR AL, CL ; AL = FFH = 11111111

## 5.7 ROTATE INSTRUCTIONS

Rotate instruction position binary data by rotating the information in a register or memory location, either from one end to another or through the carry flag. Number rotates through a register or memory location, through the C flag or through a register or memory location only. A rotate count can be immediate or located in the register CL.

1. **ROL ( Rotate all bits of operand left, MSB to LSB):** ROL instruction rotates the bits in the operand specified by op1 towards left by the count specified in oper2. ROL moves each bit in the operand to next higher bit position. The higher order bit is moved to lower order position. Last bit rotated is copied into carry flag.

| Mnemonic | Description | Syntax | Operation | Flag affected |
|----------|-------------|--------|-----------|---------------|
| ROL | Rotate all bits of operand left, MSB to LSB | ROL OP1, OP2 | Rotate operand1 left. The number of rotates is set by operand2 | CF, OF |

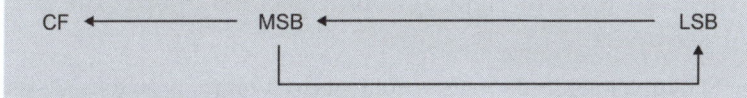

**Algorithm for ROL:**
- Shift all bits left, the bit that goes off is set to CF and the same bit is inserted to the right-most position.

| OP1 | OP2 |
|-----|-----|
| Register | Immediate data |
| Register | CL |
| Memory | Immediate data |
| Memory | CL |

**Example:** ; CF = 1, [SI] = 20H = 00100000

ROL [SI], 1 ; Rotate content at [SI] location left by 1

; CF = 0, [SI] = 01000000

2. **RCL (Rotate operand around to the left through CF):** RCL instruction rotates the bits in the operand specified by op1 towards left by the count specified in oper2. The operation is circular, the MSB of operand is rotated into a carry flag and the bit in the CF is rotated around into the LSB of operand.

| Mnemonic | Description | Syntax | Operation | Flag affected |
|----------|-------------|--------|-----------|---------------|
| RCL | Rotate operand around to the left through carry flag | RCL OP1, OP2 | Rotate operand1 left through carry flag. The number of rotates is set by operand2. | CF, OF |

### Algorithm for RCL:

- Shift all bits left, the bit that goes off is set to CF and previous value of CF is inserted to the right-most position.

| OP1 | OP2 |
|-----|-----|
| Register | Immediate data |
| Register | CL |
| Memory | Immediate data |
| Memory | CL |

**Example:** ; CF = 1, [SI]= 20H=00100000

ROL [SI], 1 ; Rotate content at [SI] location left by 1 through carry flag

; CF = 0, [SI]=01000001

3. **ROR (Rotate all bits of operand right LSB to MSB):** ROR instruction rotates the bits in the operand op1 towards right by count specified in op2. The last bit rotated is copied into CF.

| Mnemonic | Description | Syntax | Operation | Flag affected |
|----------|-------------|--------|-----------|---------------|
| ROR | Rotate all bits of operand right LSB to MSB | ROR OP1, OP2 | Rotate operand1 right. The number of rotates is set by operand2 | CF, OF |

### Algorithm for ROR:

- Shift all bits right, the bit that goes off is set to CF and the same bit is inserted to the left-most position.

**Example:** ; CF = 1, [SI] = 20H = 00100000

ROR [SI], 1 ; Rotate content at [SI] location right by 1 LSB to MSB

; CF = 0, [SI]= 00010000

4. **RCR (Rotate operand around to the right through carry):** RCR instruction rotates the bits in the operand specified by op1 towards right by the count specified in op2.

| Mnemonic | Description | Syntax | Operation | Flag affected |
|----------|-------------|--------|-----------|---------------|
| RCR | Rotate operand around to the right through carry flag | RCR OP1, OP2 | Rotate operand1 right through carry flag. The number of rotates is set by operand2 | CF, OF |

**Algorithm for RCR:**

• Shift all bits right, the bit that goes off is set to CF and previous value of CF is inserted to the left-most position.

| OP1 | OP2 |
|-----|-----|
| Register | Immediate data |
| Register | CL |
| Memory | Immediate data |
| Memory | CL |

**Example:**  ; CF = 1, [SI] = 20H = 00100000
ROR [SI], 1 ; Rotate content at [SI] location right by 1 LSB to MSB
; CF = 0, [SI] = 10010000

## 5.8 FLAG MANIPULATION INSTRUCTIONS

### 1. CLC

| Mnemonic | Description | Syntax | Operation | Flag affected |
|----------|-------------|--------|-----------|---------------|
| CLC | Clear carry flag | CLC | CF = 0 | CF |

### 2. CMC

| Mnemonic | Description | Syntax | Operation | Flag affected |
|----------|-------------|--------|-----------|---------------|
| CMC | Complement carry flag | CMC | Inverts value of CF | CF |

### 3. STC

| Mnemonic | Description | Syntax | Operation | Flag affected |
|----------|-------------|--------|-----------|---------------|
| STC | Set carry flag | STC | CF = 1 | CF |

### 4. CLD

| Mnemonic | Description | Syntax | Operation | Flag affected |
|----------|-------------|--------|-----------|---------------|
| CLD | Clear direction flag | CLD | DF = 0 | DF |

### 5. STD

| Mnemonic | Description | Syntax | Operation | Flag affected |
|----------|-------------|--------|-----------|---------------|
| STD | Set direction flag | STD | DF = 1 | DF |

## 6. CLI

| Mnemonic | Description | Syntax | Operation | Flag affected |
|----------|-------------|--------|-----------|---------------|
| CLI | Clear interrupt enable flag | CLI | IF = 0, this disables hardware interrupts | IF |

## 7. STI

| Mnemonic | Description | Syntax | Operation | Flag affected |
|----------|-------------|--------|-----------|---------------|
| STI | Set interrupt enable flag | STI | IF = 1, this enables hardware interrupts | IF |

## 5.9 PROCESS CONTROL INSTRUCTIONS

### 1. HLT

| Mnemonic | Description | Syntax | Operation | Flag affected |
|----------|-------------|--------|-----------|---------------|
| HLT | Halt the system | HLT | Stop μP operation | None |

### 2. NOP

| Mnemonic | Description | Syntax | Operation | Flag affected |
|----------|-------------|--------|-----------|---------------|
| NOP | No operation | NOP | Do nothing | None |

## 5.10 FLOW CONTROL INSTRUCTIONS

1. **JMP (JUMP):** This instruction transfers program control to a different point in the instruction stream without recording return information. The destination (target) operand specifies the address of the instruction being jumped to. This destination can be an 8 or 16-bit signed immediate, a general-purpose register or a memory location. JUMP instruction is referred to as unconditional jump instruction.

   If the destination is in the same code segment then only the instruction pointer will be changed to get to the destination location. This is referred to as near jump. If the destination is in a segment with a name different from that of the segment containing the jump instruction, then both the instruction pointer and code segment register contents will be changed to get the destination location.

| Mnemonic | Description | Syntax | Operation | Flag affected |
|----------|-------------|--------|-----------|---------------|
| JMP | Unconditional jump | JMP D | Transfers control to another part of the program | None |

This instruction can be used to execute four different types of jumps:

**Near jump**   A jump to an instruction within the current code segment (the segment currently pointed to by the CS register), sometimes referred to as an intrasegment jump.

**Short jump**   A near jump where the jump range is limited to −128 to +127 from the current EIP value.

**Far jump**   A jump to an instruction located in a different segment than the current code segment but at the same privilege level, sometimes referred to as an intersegment jump.

**Example:**  JMP 2000H      IP ← 2000H

While executing a program, it is desired to jump the program execution from one point to the other when certain condition is satisfied. Such jump instruction is called conditional jump.

| Mnemonic | Description | Syntax | Operation | Flag affected |
|---|---|---|---|---|
| JCC | Conditional jump | JCC operand | If the specified condition is true, jump to the specified address in the operand field otherwise the instruction is executed | None |
| JA | Jump if above | JA D | Short jump if first operand is above second operand (as set by CMP instruction). Unsigned if (CF = 0) and (ZF = 0), then jump | None |
| JAE | Jump if above or equal | JAE D | Short jump if first operand is above or equal to second operand (as set by CMP instruction). Unsigned if CF = 0, then jump | None |
| JB | Jump if below | JB D | Short jump if first operand is below second operand (as set by CMP instruction). Unsigned if CF = 1, then jump | None |
| JBE | Jump if below or equal | JBE D | Short jump if first operand is below or equal to second operand (as set by CMP instruction). Unsigned if CF = 1 or ZF = 1, then jump | None |
| JC | Jump if carry | JC D | Short jump if carry flag is set to 1. If CF = 1, then jump | None |
| JE | Jump if equal | JE D | Short jump if first operand is greater than second operand (as set by CMP instruction). Signed if (ZF = 0) and (SF = OF), then jump | None |
| JZ | Jump if zero | JZ D | Short jump if zero (equal) set by CMP, SUB, ADD, TEST, AND, OR, XOR instructions. If ZF = 1, then jump | None |
| JG | Jump if greater than | JG D | Short jump if first operand is greater than second operand (as set by CMP instruction). Signed if (ZF = 0) and (SF = OF), then jump | None |

*(Contd...)*

(Contd...)

| Mnemonic | Description | Syntax | Operation | Flag affected |
|---|---|---|---|---|
| JGE | Jump if greater than or equal | JGE D | Short jump if first operand is greater or equal to second operand (as set by CMP instruction). Signed if SF = OF, then jump | None |
| JL | Jump if less than | JL D | Short jump if first operand is less than second operand (as set by CMP instruction). Signed if SF <> OF, then jump | None |
| JLE | Jump if less than or equal | JLE D | Short jump if first operand is less or equal to second operand (as set by CMP instruction). Signed if SF <> OF or ZF = 1, then jump | None |
| JNA | Jump if above | JNA D | Short jump if first operand is not above second operand (as set by CMP instruction). unsigned if CF = 1 or ZF = 1, then jump | None |
| JNAE | Jump if neither above nor equal | JNAE D | Short jump if first operand is neither above nor equal to second operand (as set by CMP instruction). Unsigned if CF = 1, then jump | None |
| JNB | Jump if not below | JNB D | Short jump if first operand is not below second operand (as set by CMP instruction). Unsigned if CF = 0, then jump | None |
| JNBE | Jump if neither below nor equal | JNBE D | Short jump if first operand is neither below nor equal to second operand (as set by CMP instruction). Unsigned if (CF = 0) and (ZF = 0), then jump | None |
| JNC | Jump if not carry | JNC D | Short jump if carry flag is set to 0. If CF = 0, then jump | None |
| JNE | Jump if not equal | JNE D | Short jump if first operand is not equal to second operand (as set by CMP instruction). Signed/ unsigned if ZF = 0, then jump | None |
| JNG | Jump if not greater than | JNG D | Short jump if first operand is not greater than second operand (as set by CMP instruction). Signed if (ZF = 1) and (SF <> OF), then jump | None |
| JNGE | Jump if neither greater than nor equal | JNGE D | Short jump if first operand is not greater than second operand (as set by CMP instruction). Signed if (ZF = 1) and (SF <> OF), then jump | None |

(Contd...)

*(Contd...)*

| Mnemonic | Description | Syntax | Operation | Flag affected |
|---|---|---|---|---|
| JNL | Jump if not less than | JNL D | Short jump if first operand is not less than second operand (as set by CMP instruction). Signed if SF = OF, then jump | None |
| JNLE | Jump if neither less than nor equal | JNLE D | Short jump if first operand is neither less nor equal to second operand (as set by instruction). Signed if (SF = OF) and (ZF = 0), then jump | None / CMP |
| JNO | Jump if not overflow | JNO D | Short jump if not overflow. If OF = 0, then jump | None |
| JNP | Jump if not parity | JNP D | Short jump if no parity (odd). Only 8 low bits of results are checked set by CMP, SUB, ADD, TEST, AND, OR, XOR instructions. If PF = 0, then jump | None |
| JNS | Jump if not sign | JNS D | Short jump if not signed (if positive) set by CMP, SUB, ADD, TEST, AND, OR, XOR instructions. If SF = 0, then jump | None |
| JNZ | Jump if not zero | JNZ D | Short jump if not zero (not equal) set by CMP, SUB, ADD, TEST, AND, OR, XOR instructions. If ZF = 0, then jump | None |
| JO | Jump if overflow | JO D | Short jump if overflow. If OF = 1, then jump | None |
| JP | Jump if parity | JP D | Short jump if parity (even). Only 8 low bits of results are checked set by CMP, SUB, ADD, TEST, AND, OR, XOR instructions. If PF = 1, then jump | None |
| JPE | Jump if parity even | JPE D | Short jump if parity even. Only 8 low bits of results are checked set by CMP, SUB, ADD, TEST, AND, OR, XOR instructions. If PF = 1, then jump | None |
| JPO | Jump if parity odd | JPO D | Short jump if parity odd. Only 8 low bits of resultsare checked set by CMP, SUB, ADD, TEST, AND, OR, XOR instructions. If PF = 0, then jump | None |
| JS | Jump if sign | JS D | Short jump if signed (if negative) set by CMP, SUB, ADD, TEST, AND, OR, XOR instructions. If SF = 1, then jump | None |
| JZ | Jump if zero | JZ D | Short jump if zero (equal) set by CMP, SUB,ADD, TEST, AND, OR, XOR instructions. If ZF = 1, then jump | None |

Destination D label for the above instructions must be in the range of −128 to +127 bytes from the address of the instruction after above conditional jump instructions.

2. **CALL:** Saves procedure linking information on the stack and branches to the procedure (called procedure) specified with the destination (target) operand. The target operand specifies the address of the first instruction in the called procedure. This operand can be an immediate value, a general-purpose register or a memory location.

This instruction can be used to execute four different types of calls:

- Near call—A call to a procedure within the current code segment (the segment currently pointed to by the CS register), sometimes referred to as an intrasegment call.

- Far call—A call to a procedure located in a different segment than the current code segment, sometimes referred to as an intersegment call.

- Inter-privilege-level far call—A far call to a procedure in a segment at a different privilege level than that of the currently executing program or procedure.

The LCALL instruction calls intersegment (far) procedures using a full pointer. LCALL causes the procedure named in the operand to be executed.

LCALL ptr1 6:{ 16|32} uses a four-byte or six-byte operand as a long pointer to the called procedure. LCALL m16 :{ 16|32} fetches the long pointer from the specified memory location. In Real Address Mode or Virtual 8086 Mode, the long pointer provides 16 bits for the CS register and 16 or 32 bits for the EIP register. Both forms of the LCALL instruction push the CS and IP registers as a return address.

| Mnemonic | Syntax | Description | Operation | Flag affected |
|---|---|---|---|---|
| CALL | CALL rel16 | Call near, relative, displacement relative to next instruction | Transfers control to procedure, return address is (IP) is pushed to stack. | None |
| | CALL rel32 | Call near, relative, displacement relative to next instruction | *4-byte address* may be entered in this form: | |
| | CALL r/m16 | Call near, absolute indirect, address given in r/m16 | 1234H:5678H, first value is a segment, second value is an offset (this is a far call, so CS is also pushed | |
| | CALL r/m32 | Call near, absolute indirect, address given in r/m32 | to stack). | |
| | CALL ptr16:16 | Call far, absolute, address given in operand | | |
| | CALL ptr16:32 | Call far, absolute, address given in operand | | |
| | CALL m16:16 | Call far, absolute indirect, address given in m16:16 | | |
| | CALL m16:32 | Call far, absolute indirect, address given in m16:32 | | |

**Example:**

| | |
|---|---|
| CALL BX | BX = Offset of the first instruction of procedure |
| | IP ← BX (Indirect within segment call) |
| CALL PROC | Direct within segment call PROC is the name of procedure. |
| | Assembler determines the displacement after CALL. |

3. **RET (Return from procedure call):** The RET instruction transfers control to the return address located on the stack. This address is usually placed on the stack by a call instruction. Issue the RET instruction within the called procedure to resume execution flow at the instruction following the call.

   The optional numeric 16-bit parameter to RET specifies the number of stack bytes or words to be released after the return address is popped from the stack. Typically, these bytes or words are used as input parameters to the called procedure.

| Mnemonic | Description | Syntax | Operation | Flag affected |
|---|---|---|---|---|
| RET | Return from a procedure call | RET | Pop word from stack and place it in IP | None |

**Algorithm:**
- Pop from stack: IP
- If immediate operand is present: SP = SP + operand

For an intersegment (near) return, the address on the stack is a segment offset that is popped onto the instruction pointer. The CS register remains unchanged.

**Example:**

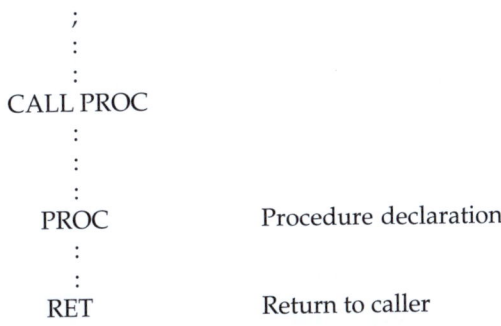

| | |
|---|---|
| CALL PROC | |
| PROC | Procedure declaration |
| RET | Return to caller |

## 5.11 INTERRUPT INSTRUCTIONS

1. **INT:** The INT n instruction generates a call to the interrupt or exception handler specified with the destination operand (see the section titled "Interrupts and Exceptions" in Chapter 4 of the Intel Architecture Software Developer's Manual, Volume 1). The destination operand specifies an interrupt vector number from 0 to 255, encoded as an 8-bit unsigned intermediate value. Each interrupt vector number provides an index to a gate descriptor in the IDT. The first 32 interrupt vector numbers are reserved by Intel for system use. Some of these interrupts are used for internally generated exceptions.

   The INT n instruction is the general mnemonic for executing a software generated call to an interrupt handler. The INTO instruction is a special mnemonic for calling overflow exception (#OF), interrupt vector number 4. The overflow interrupt checks the OF flag in the EFLAGS register and calls the overflow interrupt handler, if the OF flag is set to 1.

| Mnemonic | Description | Syntax | Operation | Flag affected |
|----------|-------------|--------|-----------|---------------|
| INT | Call to interrupt procedure | INT D | Interrupt numbered by immediate byte (0–255) | IF = 0, TF = 0 |

D is immediate 8-bit data.

**Algorithm:**

- Push to stack:

     Flag register

     CS

     IP

     IF = 0

- Transfer control to interrupt procedure

**Example:**    MOV AH, 10H

            INT 20H

Causes sub function number 10H of Interrupt number 20H to be executed. In addition, some sub functions require other values to be passed to the ISR in particular registers.

**Example:** Sub function 09H of Interrupt 21H displays a $-terminated string on the screen. The sub function requires the offset of that string to be passed in the DX register.

2. **INTO:** Interrupt on Overflow or Interrupt 4 if Overflow flag is 1.

| Mnemonic | Description | Syntax | Operation | Flag affected |
|----------|-------------|--------|-----------|---------------|
| INTO | Interrupt on overflow | INTO | Interrupt 4 if overflow flag is 1 | None |

**Algorithm:**

If OF = 1, then INT 4

**Example:**    MOV AL, -5

            SUB AL, 127 ; AL = 7Ch (124)

            INTO        ; Process error

3. **IRET (Interrupt return):** Returns program control from an exception or interrupt handler to a program or procedure that was interrupted by an exception, an external interrupt or a software-generated interrupt.

| Mnemonic | Description | Syntax | Operation | Flag affected |
|----------|-------------|--------|-----------|---------------|
| IRET | Interrupt return | IRET | Pop IP, CS and flag register | All |

**Algorithm:**

Pop from stack:

     IP

     CS

     Flag register

## 5.12 LOOP INSTRUCTIONS

This instruction performs a loop operation using the ECX or CX register as a counter. Each time the LOOP instruction is executed, the count register is decremented, then checked for 0. If the count is 0, the loop is terminated and program execution continues with the instruction following the LOOP instruction. If the count is not zero, a near jump is performed to the destination (target) operand, which is presumably the instruction at the beginning of the loop. If the address-size attribute is 32 bits, the ECX register is used as the count register; otherwise the CX register is used. The target instruction is specified with a relative offset (a signed offset relative to the current value of the instruction pointer in the EIP register). This offset is generally specified as a label in assembly code, but at the machine code level, it is encoded as a signed, 8-bit immediate value, which is added to the instruction pointer. Offsets of −128 to +127 are allowed with this instruction.

Some forms of the loop instruction (LOOPcc) also accept the ZF flag as a condition for terminating the loop before the count reaches zero. With these forms of the instruction, a condition code (cc) is associated with each instruction to indicate the condition being tested for. Here, the LOOPcc instruction itself does not affect the state of the ZF flag; the ZF flag is changed by other instructions in the loop.

| Mnemonic | Description | Syntax | Operation | Flag affected |
|---|---|---|---|---|
| LOOP | Decrement count; jump short if count<>0 | LOOP rel8 | Decrease CX, jump to label if CX not zero, CX = CX − 1 if CX <> 0, then jump else no jump, continue | None |
| LOOPE | Decrement count; jump short if count<>0 and ZF = 1 | LOOPE rel8 | Decrease CX, jump to label if CX not zero and equal (ZF = 1), CX = CX − 1 if (CX <> 0) and (ZF = 1), then jump else no jump, continue | Z |
| LOOPZ | Decrement count; jump short if count<>0 and ZF = 1 | LOOPZ rel8 | Decrease CX, jump to label if CX not zero and ZF = 1 CX = CX − 1 if (CX <> 0) and (ZF = 1), then jump else no jump, continue | Z |
| LOOPNE | Decrement count; jump short if count <> 0 and ZF = 0 | LOOPNE rel8 | Decrease CX, jump to label if CX not zero and not equal (ZF = 0) CX = CX − 1 if (CX <> 0) and (ZF = 0), then jump else no jump, continue | Z |
| LOOPNZ | Decrement count; jump short if count < > 0 and ZF = 0 | LOOPNZ rel8 | Decrease CX, jump to label if CX not zero and ZF = 0, CX = CX − 1 if (CX <> 0) and (ZF = 0), then jump else no jump, continue | Z |

**Example:**

```
          MOV CX, 06
          LABLE
          LOOP LABLE ; CX = CX − 1
                     ; If CX<>0 then jump to LABLE
          MOV BX, 2000H
          MOV CX, 100
LABLE     INC BX
          CMP [BX], FFH ; [BX] ← [BX]-FF
          LOOPE LABLE ; CX ← CX − 1 and Check whether ZF = 1 or not
```

## 5.13 STRING INSTRUCTIONS

The string instructions facilitate operations on sequences of bytes or words. The current byte or word of the source string is at DS: SI and that of the destination string is at ES: DI. Each instruction works on one byte or word and then automatically adjusts SI and DI; if the Direction flag is clear, then the index is incremented, otherwise it is decremented.

To work on an entire string at a time, each string instruction can be accompanied by a repeat prefix, either REP or one of REPE and REPNE (or their synonyms REPZ and REPNZ). These cause the instruction to be repeated the number of times in the count register, CX; for REPE and REPNE, the Zero flag is tested at the end of each operation and the loop is stopped if the condition (Equal or Not Equal to Zero) fails.

There are five string data transfer instructions; these are MOVS, LODS, STOS, INS, OUTS:

1. **MOVS:** One of the more useful string data transfer instruction is MOVS because it transfers data from one memory location to another. This is only instruction that allows memory-to-memory data transfer in 8086. The MOVS instruction transfers a byte or word from the data segment location addressed by the SI to the extrasegment location addressed by DI.

Permissible forms of MOVS instruction are as follows:

| Mnemonic | Description | Syntax | Operation | Flag affected |
|---|---|---|---|---|
| MOVSB | Move byte from one string to another, i.e. Copy byte at DS:[SI] to ES:[DI]. Update SI and DI. | MOVS | ES:[DI] = DS:[SI]; if DF = 0, then SI = SI + 1 and DI = DI +1 otherwise SI = SI – 1 and DI = DI – 1 | None |
| MOVSW | Move a word from one string to another, i.e. Copy word at DS:[SI] to ES:[DI]. Update SI and DI. | MOVSW | ES:[DI] = DS:[SI]; if DF = 0, then SI = SI + 2 and DI = DI + 2 otherwise SI = SI – 2 and DI = DI – 2 | None |

**Example:** MOV SI, FIRST_STRING    ; SI points first string
          MOV DI, SECOND_STRING ; DI points second string
          CLD                       ; DF = 0
          MOV CX, 20             ; CX = 20
          REP MOVSB           ; CX ← CX – 1, copy string bytes from DI to SI until CX = 0

2. **LODS:** The LODS instruction loads AL, AX with data stored at the data segment offset address indicated by the register SI. After loading AL with byte or AX with word, the counters of SI increments if DF = 0 and decrement if DF = 1.

Permissible forms of LODS instruction are as follows:

| Mnemonic | Description | Syntax | Operation | Flag affected |
|---|---|---|---|---|
| LODSB | Load string byte into AL register, i.e. Load byte at DS: [SI] into AL. Update SI. | LODSB | AL = DS:[SI]; if DF = 0, then SI = SI + 1 otherwise SI = SI – 1 | None |
| LODSW | Load string word into AX register, i.e. Load word at DS:[SI] into AX. Update SI. | LODSW | AL = DS:[SI]; if DF = 0, then SI = SI + 2 otherwise SI = SI – 2 | None |

**Example:**   CLD                          ; DF=0, SI is auto incremented
MOV SI, SOURCE_STRING ; SI point's start of string
LODS SOURCE_STRING   ; Copy byte or word from string to AL or AX

3. **STOS:** The STOS instruction stores AL or AX at the extrasegment memory location addressed by the register DI. The STOSB instruction stores the byte in AL at the extra-segment memory location addressed by DI. The STOSW instruction stores AX in the extrasegment memory location addressed by DI.

Permissible forms of STOS instruction are as follows:

| Mnemonic | Description | Syntax | Operation | Flag affected |
|---|---|---|---|---|
| STOSB | Store byte from AL into string, i.e. Store byte in AL into ES: [DI]. Update DI. | STOSB | ES:[DI] = AL; if DF = 0, then DI = DI + 1 otherwise DI = DI – 1 | None |
| STOSW | Store word from AX into string, i.e. Store word in AX into ES: [DI]. Update DI. | STOSW | ES:[DI] = AX; if DF = 0, then DI = DI + 2 otherwise DI = DI – 2 | None |

**Example:**    MOV DI, TARGET_STRING ; DI point's start of destination string

STOSB                          ; Replace byte in string with byte from AL

## 5.13.1 String Comparison Instruction

These are the additional string instructions that allow a section of memory to be tested against a constant or against another section of memory. To perform these tasks, CMPS and SCAS instructions are used.

1. **CMPS:** The CMPS compares two sections of memory data as bytes or words. The contents of data segment memory location addressed by the SI are compared by the contents of the extrasegment memory location addressed by DI.

Permissible forms of CMPS instruction are as follows:

| Mnemonic | Description | Syntax | Operation | Flag affected |
|---|---|---|---|---|
| CMPSB | Compare two string bytes, i.e. Compare bytes: ES: [DI] from DS: [SI]. | CMPSB | DS: [SI] – ES: [DI]; if DF = 0, then SI = SI + 1 and DI = DI + 1 otherwise SI = SI – 1 and DI = DI – 1 | Set flags according to result |
| CMPSW | Compare two string words, i.e. Compare words: ES: [DI] from DS: [SI] SI = SI – 2 and DI = DI – 2 | CMPSW, | DS: [SI] – ES: [DI]; if DF = 0, then SI = SI + 2 and DI = DI + 2 otherwise | Set flags according to result |

**Example:**   MOV SI, FIRST_STRING     ; SI points first string

MOV DI, SECOND_STRING ; DI points second string

CLD                      ; DF=0

MOV CX, 50               ; CX=50

REPE CPMSB               ; Repeat comparison of string  bytes until end of string or compared bytes are not equal or until CX = 0

## 5.14 REPEAT INSTRUCTIONS

| Mnemonic | Description | Syntax | Operation | Flag affected |
|---|---|---|---|---|
| REP | Repeat | REP chain instruction | Repeat following MOVSB, MOVSW, LODSB, LODSW, STOSB, STOSW instructions CX times | None |
| REPE | Repeat while equal, zero | REPE chain instruction | Repeat following CMPSB, CMPSW, SCASB, SCASW instructions while ZF = 1 (result is equal), maximum CX times | Z |
| REPNE | Repeat while not equal (zero) | REPNE chain instruction | Repeat following CMPSB, CMPSW, SCASB, SCASW instructions while ZF = 0 (result is not equal), maximum CX times | Z |
| REPNZ | Repeat while not equal (zero) | REPNZ chain instruction | Repeat following CMPSB, CMPSW, SCASB, SCASW instructions while ZF = 0 (result is not zero), maximum CX times | Z |
| REPZ | Repeat while equal, zero | REPZ chain instruction | Repeat following CMPSB, CMPSW, SCASB, SCASW instructions while ZF = 1 (result is zero), maximum CX times | Z |

**Note:** Chain instruction means any string instruction.

**Example:** REP MOVSB     ; Copy string bytes until the number of bytes loaded into CX has been copied.

REPE/REPZ CMPSB     ; Compare string byte until end of string or until string bytes not equal.

REPNE/REPNZ SCASW ; Scan a string of words until a word in the string matches the word in AX or until all of the strings have been scanned.

## 5.15 ASSEMBLY LANGUAGE PROGRAMMING

**Program 1:** WAP to add two 8 bit numbers and sum is 16-bit.

### PROGRAM

| MOV AL, FF | : | Load FFH in AL register |
| MOV BL, 22 | : | Load 22H in AL register |
| ADD AL, BL | : | Add content of BL to AL |
| HLT | | |

**Program 2:** WAP to add two 16-bit numbers and sum is 16-bit.

### PROGRAM

| MOV AX, 4622 | : | 16-bit data in AX |
| MOV BX, 3244 | : | 16-bit data in BX |
| ADD AX, BX | : | Contents of BX is added to AX |
| HLT | | |

**Program 3:** WAP to add a string of words and sum is 16-bit.

**PROGRAM**

|         |                |   |                                          |
|---------|----------------|---|------------------------------------------|
|         | MOV SI, 0300   | : | Source address in SI                     |
|         | MOV CX, 0005   | : | Count value is loaded in CX              |
| LOOP :  | MOV AX, [SI]   | : | Load AX with data which is located by SI |
|         | ADD AX, BX     | : | Contents of BX in AX                     |
|         | INC SI         | : | Increment SI                             |
|         | INC SI         | : | Increment SI                             |
|         | MOV BX,AX      | : | Contents of BX in AX                     |
|         | DEC CX         | : | Decrement CX                             |
|         | JNZ LOOP       | : | Jump to 0106 if CX ≠ 0                    |
|         | HLT            |   |                                          |

**Program 4:** WAP to convert packed BCD to unpacked BCD.

DATA SEGMENT
NUM DB 45H
DATA ENDS

CODE SEGMENT
ASSUME CS: CODE, DS: DATA

**PROGRAM**

|          |                |
|----------|----------------|
| START :  | MOV AX, DATA   |
|          | MOV DS, AX     |
|          | MOV AX, NUM    |
|          | MOV AH, AL     |
|          | MOV CL, 4      |
|          | SHR AH, CL     |
|          | AND AX, 0F0FH  |
|          | INT 3H         |
|          | CODE ENDS      |
|          | END START      |

**Program 5:** WAP to find factorial of a number.

**PROGRAM**

|         |                |                              |
|---------|----------------|------------------------------|
|         | MOV AX, 05H    |                              |
|         | MOV CX, AX     |                              |
| Back :  | DEC CX         |                              |
|         | MUL CX         |                              |
|         | LOOP back      |                              |
|         | MOV [D000], AX | : To store the result at D000H |
|         | HLT            |                              |

**RESULT**

[D000] = 120H

**Program 6:** WAP to transfer the block of data to new location B001H to B008H from memory location E000H to E007H.

### PROGRAM

(E000H) = 10
(E001H) = 15
(E002H) = 20
(E003H) = 25
(E004H) = 30
(E005H) = 35
(E006H) = 40
(E007H) = 45

```
            MOV BL, 08H
            MOV CX, E000H
            MOV EX, B001H
Loop :      MOV DL, [CX]
            MOV [EX], DL
            DEC BL
            JNZ loop
            HLT
```

### RESULT

(B000H) = 10
(B001H) = 15
(B002H) = 20
(B003H) = 25
(B004H) = 30
(B005H) = 35
(B006H) = 40
(B007H) = 45

**Program 7:** WAP to separate odd and even numbers.

### PROGRAM

(1600H) = 2
(1601H) = 5
(1602H) = 7
(1603H) = 6
(1604H) = 12
(1605H) = 15

```
            MOV CL, 06       : Set counter in CL register
            MOV SI, 1600     : Set source index as 1600
            MOV DI, 1500     : Set destination index as memory address 1500
Loop :      LODSB            : Load data from source memory
            ROR AL, 01       : Rotate AL once to right
```

| | | |
|---|---|---|
| | JB Loop | : If bit is one jump to loop |
| | ROL AL, 01 | : Rotate AL once to left |
| | MOV [DI], AL | : Move result to destination |
| | INC DI | : Increment destination index |
| | DEC CL | : Decrement the count |
| | JNZ Loop | : Jump if not zero to loop |
| | HLT | |

**RESULT**

(1500H) = 5

(1501H) = 7

(1503H) = 15

**Program 8:** WAP to convert the binary number stored at 6000H into its equivalent BCD number. Store the result from memory location 6100H.

**PROGRAM**

(6000) H = 8AH

| | | |
|---|---|---|
| | LXI H, 6200H | : Initialize lookup table pointer |
| | LXI D, 6000H | : Initialize source memory pointer |
| | LXI B, 7000H | : Initialize destination memory pointer |
| BACK : | LDAX D | : Get the number |
| | MOV L, A | : A point to the 7-segment code |
| | MOV A, M | : Get the 7-segment code |
| | STAX B | : Store the result at destination memory location |
| | INX D | : Increment source memory pointer |
| | INX B | : Increment destination memory pointer |
| | MOV A, C | |
| | CPI O5H | : Check for last number |
| | JNZ BACK | : If not repeat |
| | HLT | : End of program |
| | **RESULT** | |

**Program 9:** WAP to find the number of negative elements (most significant bit 1) in a block of data. The length of the block is in memory location 2200H and the block itself begins in memory location 2201H. Store the number of negative elements in memory location 2300H.

**PROGRAM**

(2200H) = 04H

(2201H) = 56H

(2202H) = A9H

(2203H) = 73H

(2204H) = 82H

|  | LDA 2200H |  |
|---|---|---|
|  | MOV C, A | : Initialize count |
|  | MVI B, 00 | : Negative number = 0 |
|  | LXI H, 2201H | : Initialize pointer |
| BACK : | MOV A, M | : Get the number |
|  | ANI 80H | : Check for MSB |
|  | JZ SKIP | : If MSB = 1 |
|  | INR B | : Increment negative number count |
| SKIP : | INX H | : Increment pointer |
|  | DCR C | : Decrement count |
|  | JNZ BACK | : If count 0, repeat |
|  | MOV A, B |  |
|  | STA 2300H | : Store the result |
|  | HLT | : Terminate program execution |

**RESULT**

(2300H)=02 since 2202H and 2204H contain numbers with an MSB of 1.

**Program 10:** WAP to count number of l's in the contents of D register and store the count in the B register.

**PROGRAM**

(2200H) = 04H
(2201H) = 34H
(2202H) = A9H
(2203H) = 78H
(2204H) = 56H

|  | MVI B, 00H |  |
|---|---|---|
|  | MVI C, 08H |  |
|  | MOV A, D |  |
| BACK : | RAR |  |
|  | JNC SKIP |  |
|  | INR B |  |
| SKIP : | DCR C |  |
|  | JNZ BACK |  |
|  | HLT |  |

**RESULT**

(2202H) = A9H

**Program 11:** WAP to add two multi-byte numbers and store the result as the third number.

DATA SEGMENT
BYTES EQU 08H
NUM1 DB 05H, 5AH, 6CH, 55H, 66H, 77H, 34H, 12H
NUM2 DB 04H, 56H, 04H, 57H, 32H, 12H, 19H, 13H

NUM3 DB 0AH DUP (00)

DATA ENDS

CODE SEGMENT

ASSUME CS: CODE, DS: DATA

**PROGRAM**

| | |
|---|---|
| START : | MOV AX, DATA |
| | MOV DS, AX |
| | MOV CX, BYTES |
| | LEA SI, NUM1 |
| | LEA DI, NUM2 |
| | LEA BX, NUM3 |
| | MOV AX, 00 |
| NEXT : | MOV AL, [SI] |
| | ADC AL, [DI] |
| | MOV [BX], AL |
| | INC SI |
| | INC DI |
| | INC BX |
| | DEC CX |
| | JNZ NEXT |
| | INT 3H |
| | CODE ENDS |
| | END START |

**Program 12:** WAP to subtract two multi-byte numbers and store the result as the third number.

DATA SEGMENT

BYTES EQU 08H

NUM2 DB 05H, 5AH, 6CH, 55H, 66H, 77H, 34H, 12H

NUM1 DB 04H, 56H, 04H, 57H, 32H, 12H, 19H, 13H

NUM3 DB 0AH DUP (00)

DATA ENDS

CODE SEGMENT

ASSUME CS: CODE, DS: DATA

**PROGRAM**

| | |
|---|---|
| START : | MOV AX, DATA |
| | MOV DS, AX |
| | MOV CX, BYTES |
| | LEA SI, NUM1 |
| | LEA DI, NUM2 |
| | LEA BX, NUM3 |
| | MOV AX, 00 |

```
NEXT :        MOV AL, [SI]
              SBB AL, [DI]
              MOV [BX], AL
              INC SI
              INC DI
              INC BX
              DEC CX
              JNZ NEXT
              INT 3H
              CODE ENDS
              END START
```

**Program 13:** WAP to multiply two multi-byte numbers and store the result as the third number.

**DATA SEGMENT**

```
BYTES EQU 08H
NUM1 DB 05H, 5AH, 6CH, 55H, 66H, 77H, 34H, 12H
NUM2 DB 04H, 56H, 04H, 57H, 32H, 12H, 19H, 13H
NUM3 DB 0AH DUP (00)
DATA ENDS
```

**CODE SEGMENT**

ASSUME CS: CODE, DS: DATA

**PROGRAM**

```
START :       MOV AX, DATA
              MOV DS, AX
              MOV CX, BYTES
              LEA SI, NUM1
              LEA DI, NUM2
              LEA BX, NUM3
              MOV AX, 00
NEXT :        MOV AL, [SI]
              MOV DL, [DI]
              MUL DL
              MOV [BX], AL
              MOV [BX+1], AH
              INC SI
              INC DI
              INC BX
              INC BX
              DEC CX
              JNZ NEXT
              INT 3H
              CODE ENDS
              END START
```

**Program 14:** WAP to add two ASCII numbers.

CODE SEGMENT

ASSUME CS: CODE

**PROGRAM**

```
START :        MOV AL, '5'
               MOV BL, '9'
               ADD AL, BL
               AAA
               OR AX, 3030H
               INT 3H
               CODE ENDS
               END START
```

**Program 15:** WAP to subtract two ASCII numbers.

CODE SEGMENT

ASSUME CS: CODE

**PROGRAM**

```
START :        MOV AL, '9'
               MOV BL, '5'
               SUB AL, BL
               AAS
               OR AX, 3030H
               INT 3H
               CODE ENDS
               END START
```

**Program 16:** WAP to multiply two ASCII numbers.

CODE SEGMENT

ASSUME CS: CODE

**PROGRAM**

```
START :        MOV AL, 5
               MOV BL, 9
               MUL BL
               AAM
               OR AX, 3030H
               INT 3H
               CODE ENDS
               END START
```

**Program 17:** WAP to convert BCD to ASCII.

DATA SEGMENT

NUM DB 45H

DATA ENDS

CODE SEGMENT

ASSUME CS: CODE, DS: DATA

**PROGRAM**

```
START :      MOV AX, DATA
             MOV DS, AX
             MOV AX, NUM
             MOV AH, AL
             MOV CL, 4
             SHR AH, CL
             AND AX, 0F0FH
             OR AX, 3030H
             INT 3H
             CODE ENDS
             END START
```

**Program 18:** WAP to move a Block using strings.

DATA SEGMENT

SRC DB 'MICROPROCESSOR'

DB 10 DUP (?)

DST DB 20 DUP (0)

DATA ENDS

CODE SEGMENT

ASSUME CS: CODE, DS: DATA, ES: DATA

**PROGRAM**

```
START :      MOV AX, DATA
             MOV DS, AX
             MOV ES, AX
             LEA SI, SRC
             LEA DI, DST
             MOV CX, 20
             CLD
             REP MOVSB
             INT 3H
             CODE ENDS
             END START
```

**Program 19:** WAP to perform string reversal.

DATA SEGMENT

ORG 2000H

SRC DB 'MICROPROCESSOR$'

COUNT EQU ($-SRC)

DEST DB ?
DATA ENDS

CODE SEGMENT
ASSUME CS:CODE, DS:DATA

**PROGRAM**

```
START :      MOV AX, DATA
             MOV DS, AX
             MOV CX, COUNT
             LEA SI, SRC
             LEA DI, DEST
             ADD SI, CX
             DEC CX
BACK :       MOV AL, [SI]
             MOV [DI], AL
             DEC SI
             INC DI
             DEC CX
             JNZ BACK
             INT 3H
             CODE ENDS
             END START
```

**Program 20:** WAP to compare two strings.

```
PRINTSTRING MACRO MSG
MOV AH, 09H
LEA DX, MSG
INT 21H
ENDM

DATA SEGMENT
ORG 2000H
STR1 DB 'MICROPROCESSORS'
LEN EQU ($-STR1)
STR2 DB 'MICROPROCSSOR'
M1 DB "STRINGS R EQUAL$"
M2 DB "STRINGS R NOT EQUAL$"
DATA ENDS

CODE SEGMENT
ASSUME CS: CODE, DS: DATA, ES: DATA
```

**PROGRAM**

```
START :      MOV AX, DATA
             MOV ES, AX
```

```
            LEA SI,STR1
            LEA DI, STR2
            MOV CX, LEN
            CLD
            REPE CMPSB
            JNE FAIL
            PRINTSTRING M1
            INT 3H
FAIL :      PRINTSTRING M2
            INT 3H
            CODE ENDS
            END START
```

**Program 21:** WAP to read a character from keyboard (with Echo).

**Algorithm:**

Step 1 Load DOS function call for reading a key

Step 2 Execute DOS function call

**PROGRAM**

| | | |
|---|---|---|
| MODEL TINY | : | Select tiny model |
| CODE | : | Start CODE Segment |
| STARTUP | : | Start Program |
| MOV AH,01H | : | Select function 01H |
| INT 21H | : | Access DOS to read key |
| EXIT | : | Exit to DOS |
| END | : | End of file |

After execution, ASCII equivalent of typed character will be in AL and typed character can be seen on the screen.

**Program 22:** WAP to initiate 8251 and to check the transmission and reception of character.

**Step 1:** Connect the 8253/8251 study module card to the 8086 kit via a 50-pin FRC. Check the polarity of the cable for proper data transmission.

**Step 2 :** Connect the 8253 kit to the computer input/output port with a RS232 cable.

**Step 3 :** Connect the CLK 0 tag to the 8086 CLK

**Step 4 :** Connect Gate 0 tag to + 5V VCC

**Step 5 :** Connect out O tag to R × C and T × C tags

**Step 6 :** Enter the program given below

**Step 7 :** Enter the program by pressing Reset, Exmem, Next Keys

**Step 8 :** Execute the program using <Reset>, <Go>, <.> Keys

**PROGRAM**

```
            MOV DX, FFD6        : Load Mode Control Word and Send it
            MOV AL, CEH
```

```
                    OUT DX, AL
                    MOV CX, 2
    X :             LOOP X                    : Delay
                    MOV AL, 36                : Load Command Word and Send it
                    OUT DX, AL
    TEST1 :         IN AL, DX                 : Read Status
                    AND AL, 81H               : And Check Status of Data Set Ready and Transmit
                                                Ready
                    CMP AL, 81H               : Is it Ready?
                    JNE TEST 1                : Continue to Poll if not Ready
    Y :             MOV DX, FFD0              : Otherwise Point it Data Address
                    MOV AL, Data to send      : Load Data to Send
                    OUT DX, AL                : Send it
                    MOV AL, 00
                    OUT DX, AL
                    JMP Y
```

**Program 23:** WAP to generate Sinusoidal Wave Using 8255.

```
ASSUME CS:CODE,DS:DATA
SINE DB 0, 11, 22, 33, 43, 54, 63, 72, 81, 90, 97, 104, 109, 115, 119, 122
DB 125, 126, 127, 126, 122, 119, 115, 109, 104, 97, 90, 81, 72, 63, 54, 43, 33, 22, 11
PA EQU 44A0H
CR EQU 44A3H
DATA ENDS
```

**PROGRAM**

```
    START :         MOV AX, DATA
                    MOV DS, AX
                    MOV DX, CR
                    MOV AL, 80H
                    OUT DX, AL
    REPEAT :        MOV DX, PA
                    LEA SI, SINE
                    MOV CX, 36
    NEXT :          MOV AL, [SI]
                    ADD AL, 128
                    OUT DX, AL
                    INC SI
                    LOOP NEXT
                    MOV CX, 36
                    LEA SI, SINE
    NEXT1 :         MOV AL, 128
                    MOV AH, [SI]
```

```
        SUB AL, AH
        OUT DX, AL
        INC SI
        LOOP NEXT1
        JMP REPEAT
        MOV AH, 4CH
        INT 21H
        CODE ENDS
        END START
```

## PROBLEMS

1. Write opcode for instructions MOV AL, BL.

2. Write opcode for the instruction MUL CL.

3. What segment register may not be popped from stack?

4. Which register moves onto the stack with the PUSH instruction?

5. Which is more efficient, a MOV with an offset or an LEA instruction?

6. Which instruction sets and clears direction flag?

7. What is wrong with the INC [BX] instruction?

8. What is the difference between IMUL and MUL instructions?

9. What is the difference between IDIV and DIV instructions?

10. Illustrate the difference between SUB and CMP instructions.

11. Which instructions are used with BCD and ASCII arithmetic operations?

12. Describe the difference between AND and TEST instructions.

13. Differentiate between NOT and NEG instructions.

14. What is the use of DF flag bit?

15. Explain JMP instruction in detail. Also discuss various jumps.

16. Explain how the near RET instruction work?

17. What is the use of shift and rotate instructions?

18. What does the CMPS and SCASB instructions do?

19. When two 16-bit numbers are divided, where are the remainder and the quotient found?

20. What is the function of REP and REPE prefix in string instructions?

21. Write an assembly language program to add two 16-bit numbers.

22. Write an assembly language program to divide 16-bit number by another 16-bit number.

23. Write an assembly language program to arrange a given data array in ascending order.

# 6 Programmable Peripheral Interface 8255A

## 6.1 PROGRAMMABLE PERIPHERAL INTERFACE 8255A

The Intel 82C55A is a high-performance, CHMOS version of the industry standard 8255A general purpose programmable I/O device which is designed for use with all Intel and most other microprocessors. It provides 24 I/O pins which may be individually programmed in two groups of 12 and used in three major modes of operation. The 82C55A is pin compatible with the NMOS 8255A and 8255A-5. In MODE 0, each group of 12 I/O pins may be programmed in sets of 4 and 8 to be inputs or outputs. In MODE 1, each group may be programmed to have 8 lines of input or output. Three of the remaining 4 pins are used for handshaking and interrupt control signals. MODE 2 is a strobed bi-directional bus configuration. The 82C55A is fabricated on Intel's advanced CHMOS III technology which provides low power consumption with performance equal to or greater than the equivalent NMOS product. The 82C55A is available in 40-pin DIP and 44-pin plastic leaded chip carrier (PLCC) packages.

## 6.2 FEATURES OF 8255

1. Three 8-bit Peripheral Ports—Ports A, B and C
2. Three programming modes for Peripheral Ports: Mode 0 (Basic Input/Output), Mode 1 (Strobed Input/Output) and Mode 2 (Bi-directional)
3. Total of 24 programmable I/O lines
4. 8-bit bi-directional system data bus with standard microprocessor interface controls.

## 6.3 ARCHITECTURE OF 8255A

The 82C55A is a programmable peripheral interface device designed for use in Intel microcomputer systems. Its function is that of a general purpose I/O component to interface peripheral equipment to the microcomputer system bus. The functional configuration of the 82C55A is programmed by the system software so that normally no external logic is necessary to interface peripheral devices or structures (Fig. 6.1).

### 6.3.1 Data Bus Buffer

This 3-state bi-directional 8-bit buffer is used to interface the 8255A to the system data bus. Data is transmitted or received by the buffer upon execution of input or output instructions by the CPU. Control words and status information are also transferred through the data bus buffer.

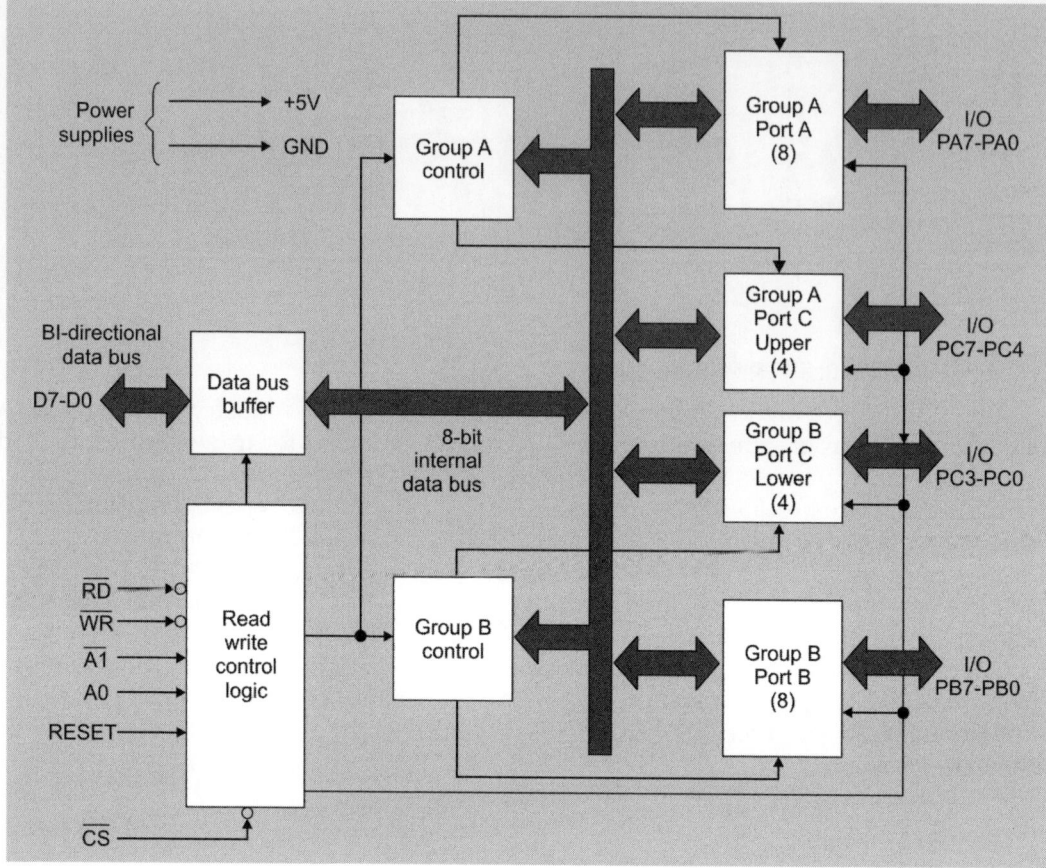

**Fig. 6.1:** Architecture of 8259

### 6.3.2  Read Write Control Logic

The function of this block is to manage all of the internal and external transfers of both Data and Control or Status words. It accepts inputs from the CPU Address and Control buses and in turn, issues commands to both of the Control Groups.

### 6.3.3  Group A and Group B Control

The functional configuration of each port is programmed by the systems software. In essence, the CPU "outputs" a control word to the 8255A. The control word contains information, such as "mode", "bit set", "bit reset", etc. that initializes the functional configuration of the 8255A. Each of the Control blocks (Group A and Group B) accepts "commands" from the Read/Write Control Logic, receives "control words" from the internal data bus and issues the proper commands to its associated ports.

   Control Group A - Port A and Port C upper (C7-C4)

   Control Group B - Port B and Port C lower (C3-C0)

   The control word register can be both written and read as shown in the address decode table in the pin description. Figure 6.2 shows the control word format for both Read and Write operations. When the control word is read, bit D7 will always be a logic "1", as this implies control word mode information.

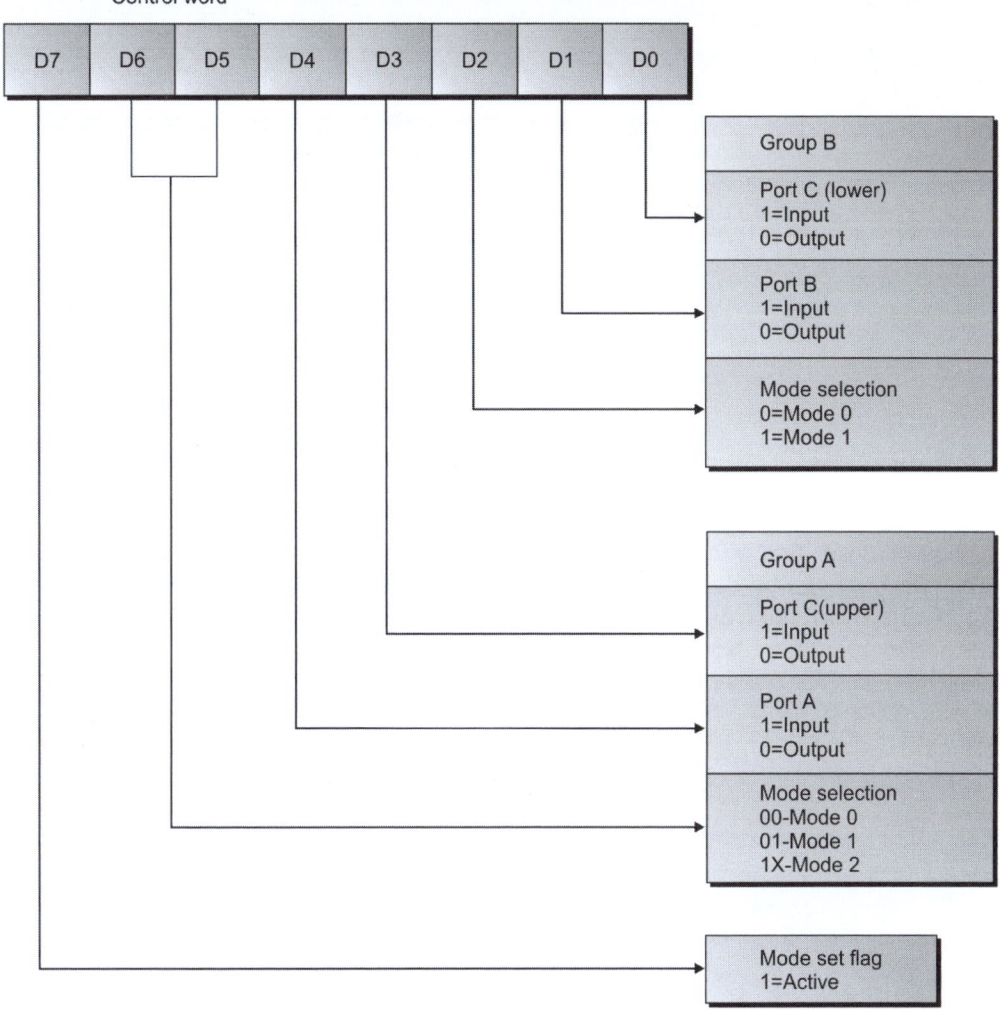

**Fig. 6.2:** Control word format

### 6.3.4 Port A, Port B and Port C

The 8255A contains three 8-bit ports (A, B and C). All can be configured in a wide variety of functional characteristics by the system software but each has its own special features or "personality" to further enhance the power and flexibility of the 8255A.

### *Port A*

This has an 8-bit latched/buffered O/P and 8-bit input latch. It can be programmed in three mode: mode 0, mode 1, mode 2. Both "pull-up" and "pull-down" bus-hold devices are present on Port A (Fig. 6.3).

### *Port B*

This has an 8-bit latched/buffered O/P and 8-bit input latch. It can be programmed in mode 0, mode1. Only "pull-up" bus-hold devices are present on Port B.

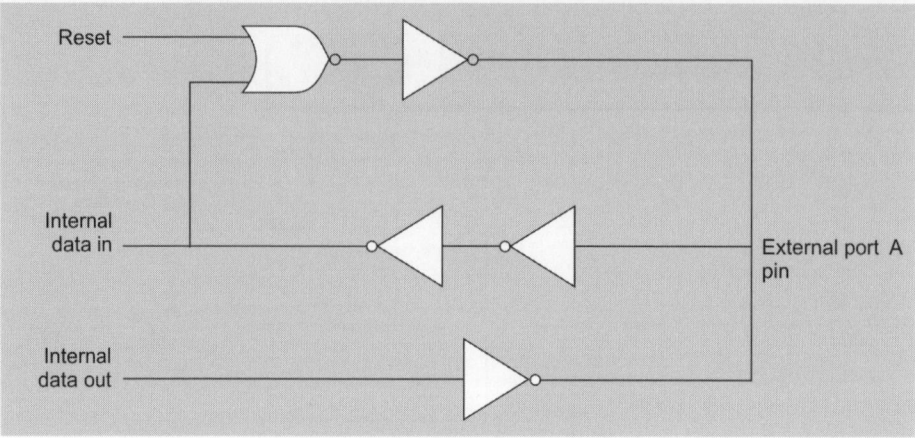

**Fig. 6.3:** Bus-hold circuit for Port A

## Port C

This has one 8-bit data output latch/buffer and one 8-bit data input buffer (no latch for input). This port can be divided into two 4-bit ports under the mode control. Each 4-bit port contains a 4-bit latch and it can be used for the control signal outputs and status signal inputs in conjunction with ports A and B. It can be programmed in mode 0. Only "pull-up" bus-hold devices are present on Port C.

See Fig. 6.4(a, b) for the bus-hold circuit configuration for Ports A, B and C.

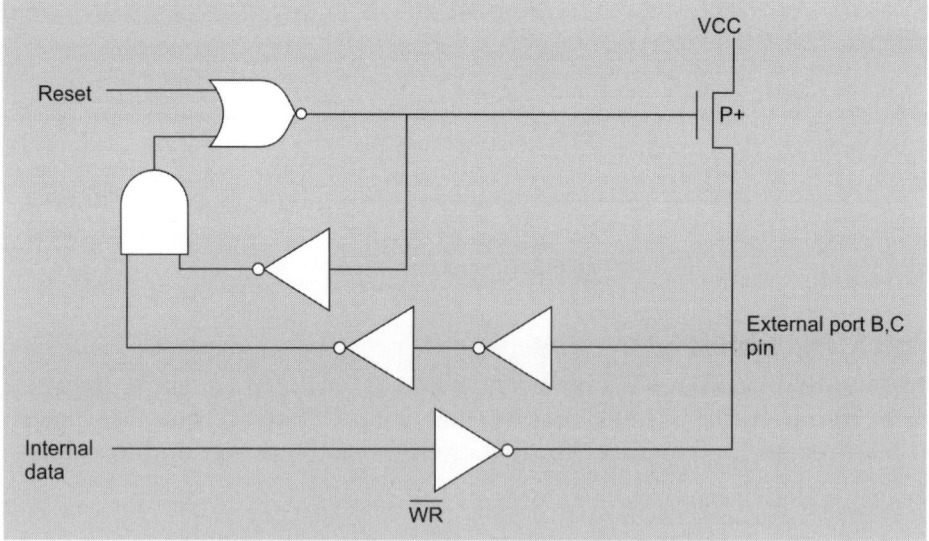

**Fig. 6.4:** Bus-hold circuit for Ports B and C

## 6.4 PIN CONFIGURATION OF 8255A

Pin configuration of 8255A PPI is shown in Fig. 6.5. It is available in 40-pin DIP package. The signal description of 8255A is briefly presented as follows:

**PA7-PA0:** These are eight port A lines that act as either latched output or buffered input lines depending upon the control word loaded into the control word register.

**PC7-PC4:** These are upper nibbles of port C lines. They may act as either output latches or input buffer lines. This port can also be used for generation of handshake lines in mode 1 or mode 2.

**PC3-PC0:** These are the lower port C lines; other details are the same as PC7-PC4 lines.

**PB0-PB7:** These are the eight port B lines which are used as latched output lines or buffered input lines in the same way as port A.

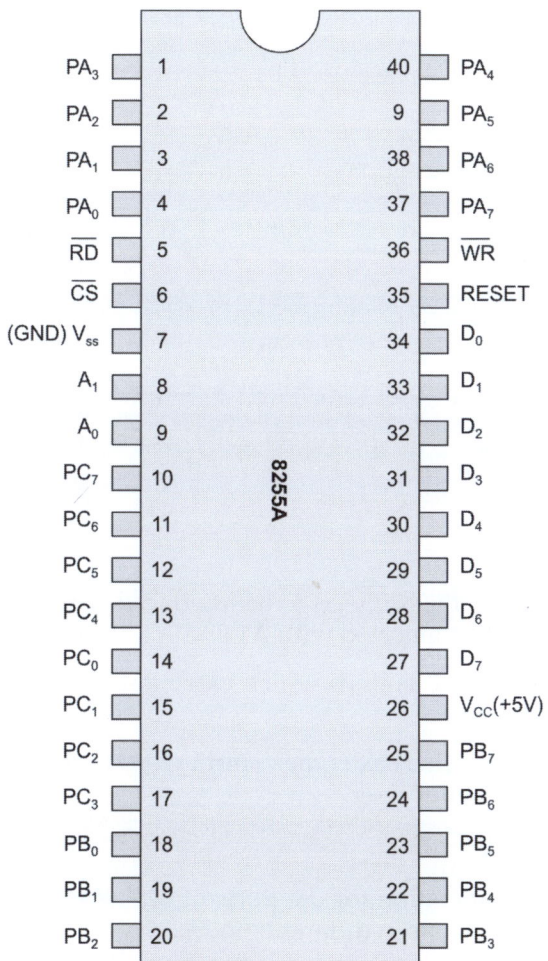

**Fig. 6.5:** Pin diagram of 8255A

$\overline{RD}$: This is the input line driven by the microprocessor and should be low to indicate read operation to 8255A.

$\overline{WR}$: This is an input line driven by the microprocessor. A low on this line indicates write operation.

$\overline{CS}$: This is a chip select line. If this line goes low, it enables the 8255A to respond to $\overline{RD}$ and $\overline{WR}$ signals, otherwise $\overline{RD}$ and $\overline{WR}$ signals are neglected.

**A1-A0:** These are the address input lines and are driven by the microprocessor. These lines A1-A0 perform the following operations for 8255A. These address lines are used for addressing any one of the four registers, i.e. three ports and a control word register as given in Tables 6.1 to 6.3.

**Table 6.1**

| $\overline{RD}$ | $\overline{WR}$ | $\overline{CS}$ | A1 | A0 | Input read cycle |
|---|---|---|---|---|---|
| 0 | 1 | 0 | 0 | 0 | Port A to data bus |
| 0 | 1 | 0 | 0 | 1 | Port B to data bus |
| 0 | 1 | 0 | 1 | 0 | Port C to data bus |
| 0 | 1 | 0 | 1 | 1 | CWR to data bus |

**Table 6.2**

| $\overline{RD}$ | $\overline{WR}$ | $\overline{CS}$ | A1 | A0 | Output write cycle |
|---|---|---|---|---|---|
| 1 | 0 | 0 | 0 | 0 | Data bus to port A |
| 1 | 0 | 0 | 0 | 1 | Data bus to port B |
| 1 | 0 | 0 | 1 | 0 | Data bus to port C |
| 1 | 0 | 0 | 1 | 1 | Data bus to CWR |

**Table 6.3**

| $\overline{RD}$ | $\overline{WR}$ | $\overline{CS}$ | A1 | A0 | Function |
|---|---|---|---|---|---|
| X | X | 1 | X | X | Data bus tri-stated |
| 1 | 1 | 0 | X | X | Data bus tri-stated |

In case of 8086 systems, if the 8255A is to be interfaced with lower order data bus, the A0 and A1 pins of 8255A are connected with A1 and A2, respectively.

**D0-D7:** These are the data bus lines which carry data or control word to/from the microprocessor.

**RESET:** Logic high on this line clears the control word register of 8255A. All ports are set as input ports by default after reset.

## 6.5 OPERATIONS OF 8255A

There are three modes of operation for the ports of 8255: Mode 0, Mode 1 and Mode 2. Figure 6.6 shows different signals in different modes.

### 6.5.1 Mode 0 Operation

It is basic or simple I/O. It does not use any handshake signals. It is used for interfacing an I/P device or an O/P device. It is used when timing characteristics of I/O devices is well known.

### 6.5.2 Mode 1 Operation

It uses handshake I/O. Three lines are used for handshaking. It is used for interfacing an I/P device or an O/P device. Mode 1 operation is used when timing characteristics of I/O devices are not well known or when I/O devices supply or receive data at

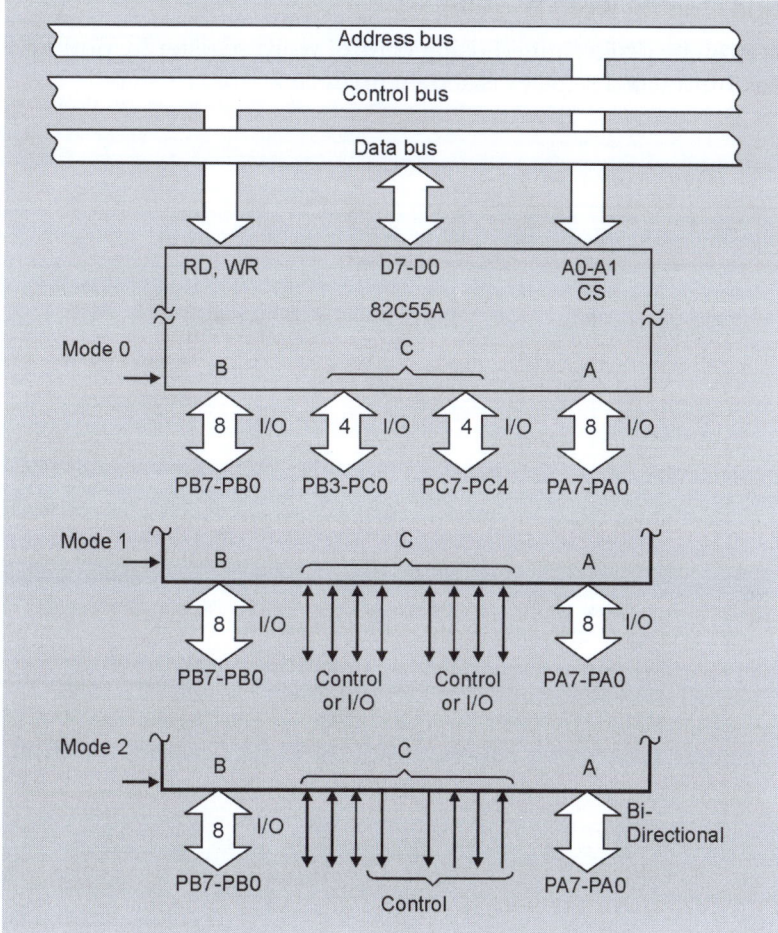

**Fig. 6.6:** Basic mode definitions and bus interface

irregular intervals. Handshake signals of the port inform the processor that the data is available, data transfer complete, etc. More details about mode 1 operation are provided later.

### 6.5.3 Mode 2 Operation

It is bi-directional handshake I/O. Mode 2 operation uses five lines for handshaking. It is used with an I/O device that receives data sometimes and sends data sometimes, e.g. hard disk drive. Mode 2 operation is useful when timing characteristics of I/O devices are not well known or when I/O devices supply or receive data at irregular intervals.

Port A can work in Mode 0, Mode 1 or Mode 2

Port B can work in Mode 0 or Mode 1

Port C can work in Mode 0 only, if at all Port A, Port B and Port C can work in Mode 0

Port A and Port B can work in Mode 1

Only Port A can work in Mode 2

### 6.5.4 I/O Mode Control Word Register

The I/O modes can be programmed using control word register by putting D7 at logic 1. The format of control word register is shown in Fig. 6.7.

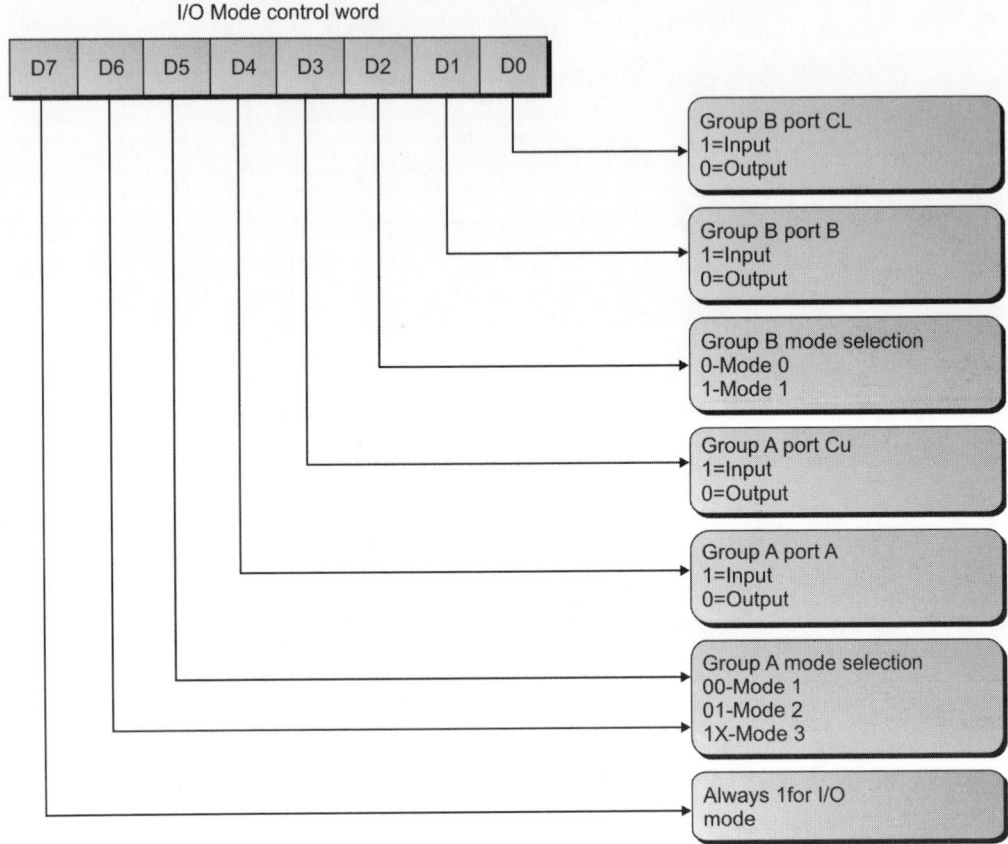

**Fig. 6.7:** I/O mode control word register

### 6.5.5 Bit Set/Reset Mode (BSR Mode)

In this mode, any of the 8-bits of port C can be set or reset depending on D0 of the control word. The bit to be set or reset is selected by bit select flags D3, D2 and D1of the CWR. The format of CWR for BSR mode is shown in Fig. 6.8. This feature reduces the software requirements in control based applications.

### 6.5.6 Interrupt Control Functions

When the 8255A is programmed to operate in mode 1 or mode 2, control signals are provided that can be used as interrupt request inputs to the CPU. The interrupt request signals, generated from port C, can be inhibited or enabled by setting or resetting the associated INTE flip-flop, using the bit set/reset function of port C. This function allows the Programmer to disallow or allow a specific I/O device to interrupt the CPU without affecting any other device in the interrupt structure.

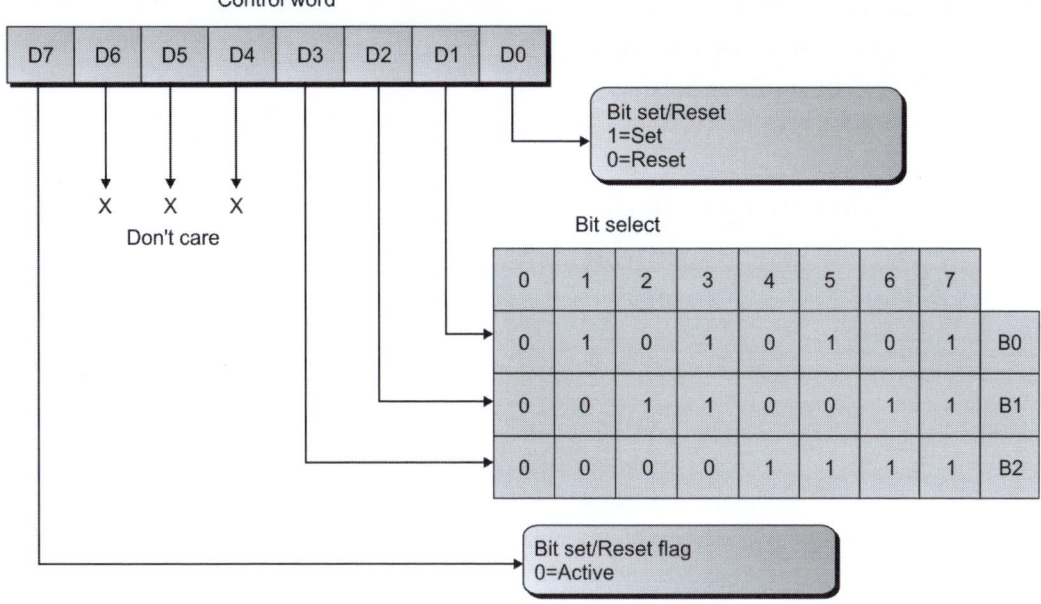

**Fig. 6.8:** BSR mode control word

**INTE flip-flop definition:**

(BIT-SET)-INTE is SET-Interrupt enable

(BIT-RESET)-INTE is RESET-Interrupt disable

## 6.6 I/O OPERATING MODES

### 6.6.1 Mode 0 (Basic Input/Output)

This mode is also called basic input/output mode shown in Fig. 6.9. This mode provides simple input and output capabilities using each of the three ports. Data can be simply read from and written to the input and output ports, respectively after appropriate initialization. No "handshaking" is required, data is simply written to or read from a specified port.

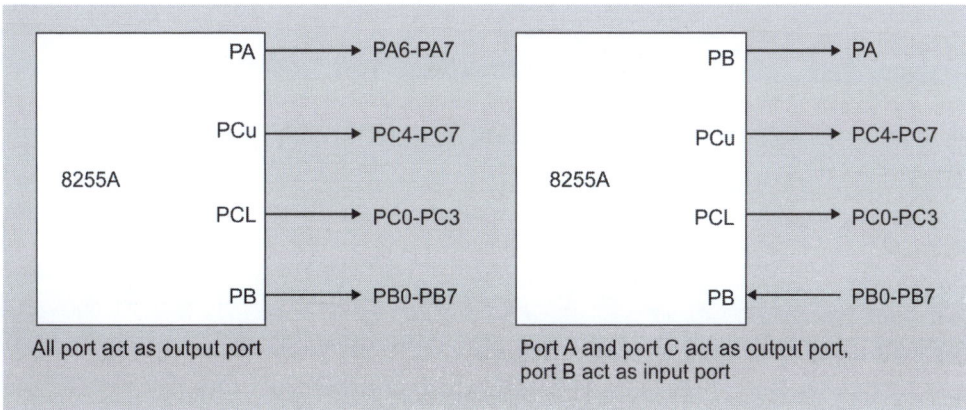

**Fig. 6.9:** Mode 0 operation

### 6.6.1.1 Features of Mode 0

1. Two 8-bit ports and two 4-bit ports
2. Any port can be input or output
3. Outputs are latched
4. Inputs are not latched
5. 16 different input/output configurations are possible in this mode.

**Table 6.4:** Mode 0 port definition

| A | | B | | Group A | | | Group B | |
| D4 | D3 | D1 | D0 | Port A | Port C (Upper) | # | Port B | Port C (Lower) |
|---|---|---|---|---|---|---|---|---|
| 0 | 0 | 0 | 0 | Output | Output | 0 | Output | Output |
| 0 | 0 | 0 | 1 | Output | Output | 1 | Output | Input |
| 0 | 0 | 1 | 0 | Output | Output | 2 | Input | Output |
| 0 | 0 | 1 | 1 | Output | Output | 3 | Input | Input |
| 0 | 1 | 0 | 0 | Output | Input | 4 | Output | Output |
| 0 | 1 | 0 | 1 | Output | Input | 5 | Output | Input |
| 0 | 1 | 1 | 0 | Output | Input | 6 | Input | Output |
| 0 | 1 | 1 | 1 | Output | Input | 7 | Input | Input |
| 1 | 0 | 0 | 0 | Input | Output | 8 | Output | Output |
| 1 | 0 | 0 | 1 | Input | Output | 9 | Output | Input |
| 1 | 0 | 1 | 0 | Input | Output | 10 | Input | Output |
| 1 | 0 | 1 | 1 | Input | Output | 11 | Input | Input |
| 1 | 1 | 0 | 0 | Input | Input | 12 | Output | Output |
| 1 | 1 | 0 | 1 | Input | Input | 13 | Output | Input |
| 1 | 1 | 1 | 0 | Input | Input | 14 | Input | Output |
| 1 | 1 | 1 | 1 | Input | Input | 15 | Input | Input |

### 6.6.1.2 Timing Diagram in Mode 0 (Figs 6.10 and 6.11)

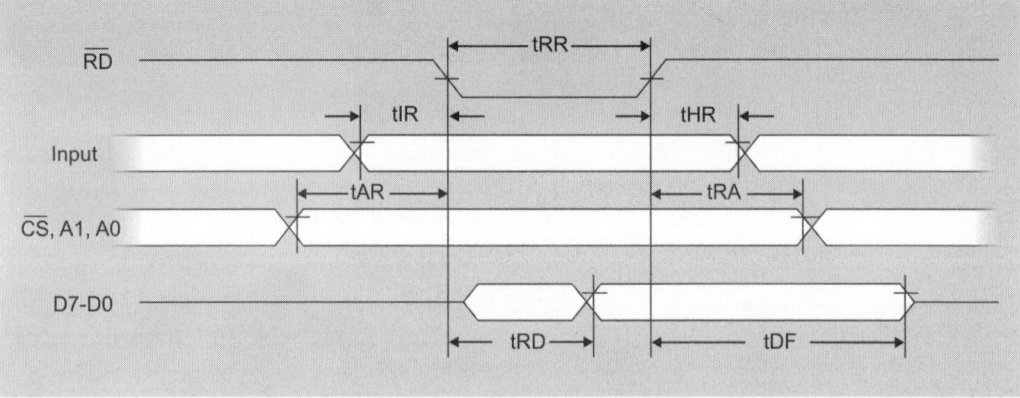

**Fig. 6.10:** Timing diagram for basic input in mode 0

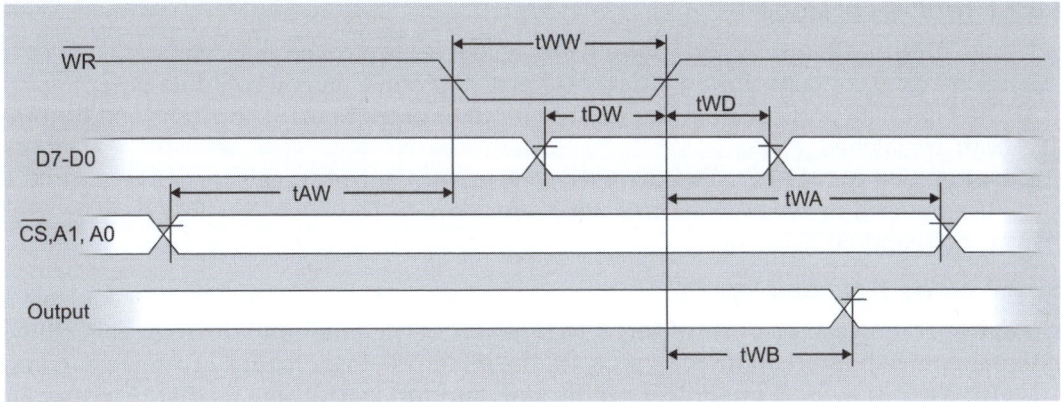

**Fig. 6.11:** Timing diagram for basic output in mode 0

## 6.6.2 Mode 1: (Strobed Input/Output)

In this mode, the handshaking controls the input and output action of the specified port. Port C lines PC0-PC2 provides strobe or handshake lines for port B. This group which includes port B and PC0-PC2 is called group B for strobed data input/output. Port C lines PC3-PC5 provides strobe lines for port A shown in Fig. 6.12. This group includes port A and PC3-PC5 from group A. Thus, port C is utilized for generating handshake signals.

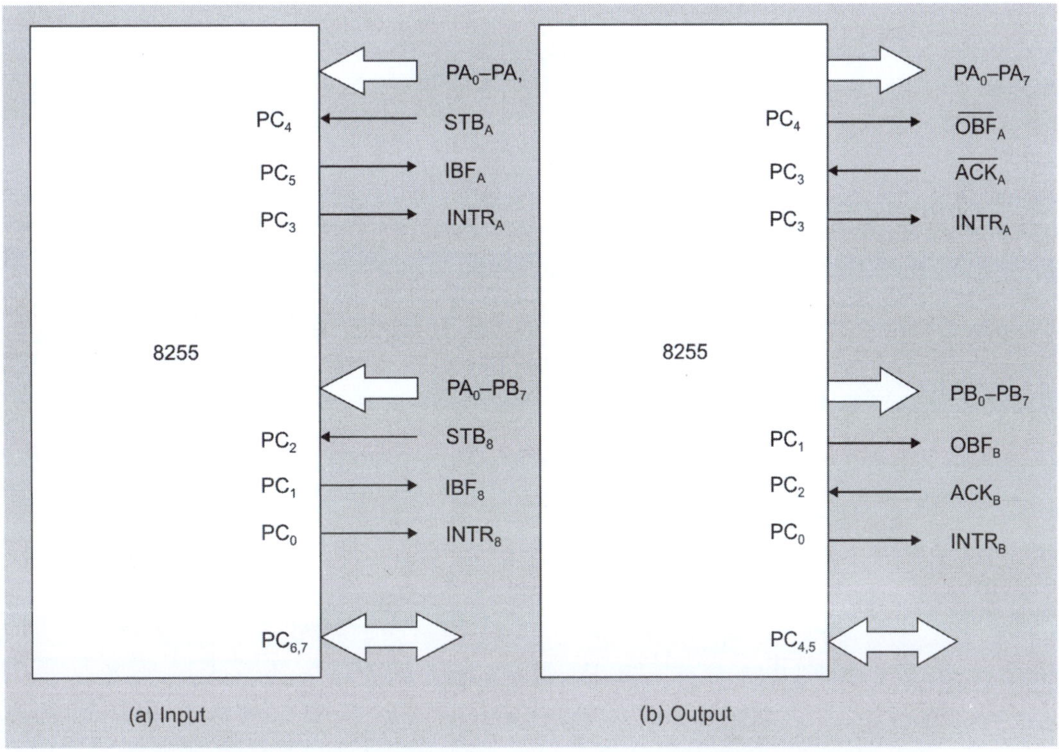

**Fig. 6.12:** Different signals in mode1 for port A as (a) I/P port and (b) O/P port

### 6.6.2.1 Features of Mode 1

1. Two groups—group A and group B are available for strobed data transfer.
2. Each group contains one 8-bit data I/O port and one 4-bit control/data port.
3. The 8-bit data port can be either used as input or output port. The inputs and outputs both are latched.
4. Out of 8-bit port C, PC0-PC2 are used to generate control signals for port B and PC3-PC5 are used to generate control signals for port A. The lines PC6, PC7 may be used as independent data lines.

### 6.6.2.2 Mode 1: Strobed Input

Mode 1 operation causes port A/port B to function as latching input devices. This allows external data to be stored into the port until the microprocessor is ready to receive it. Port C is also used in mode 1 operation, not for data but for control signals that help operate either one or both port A and port B as strobed input ports. Figure 6.13 shows the port A and port B internal structure, their control word register and timing waveforms, respectively for Mode 1 strobed input operation.

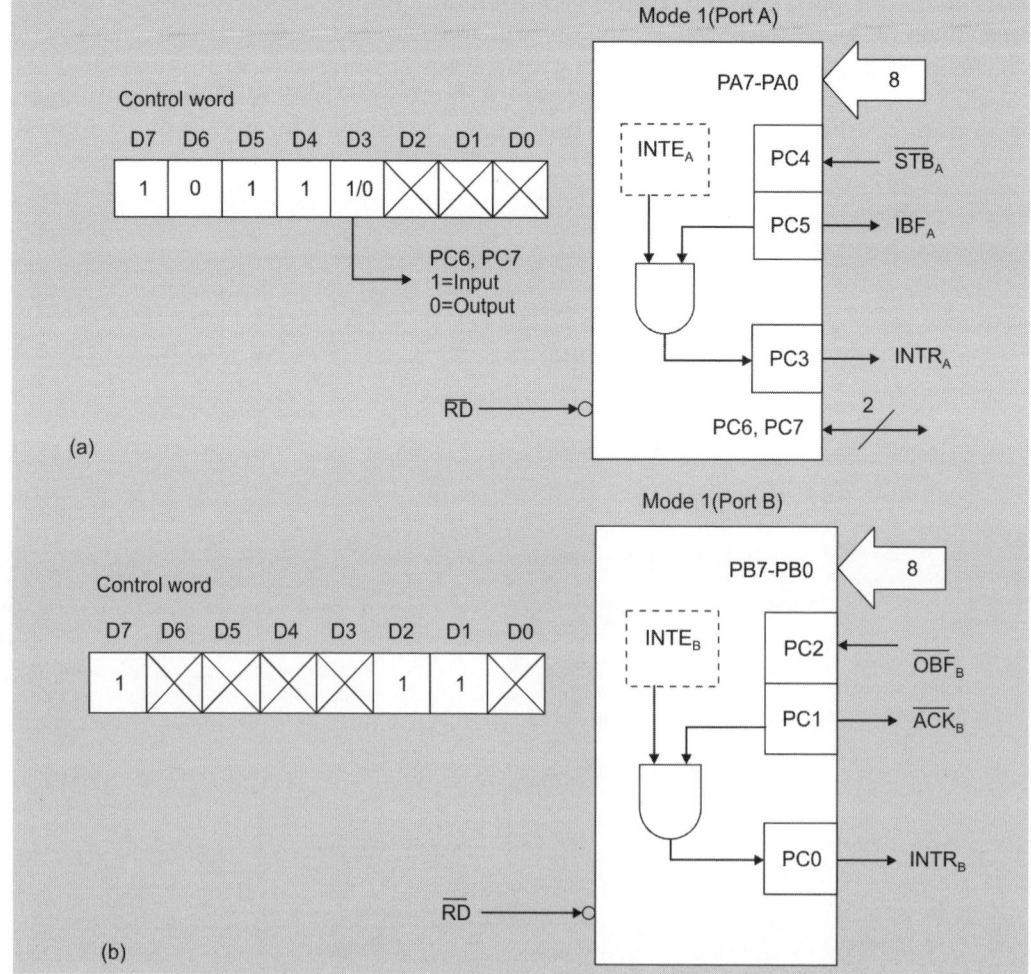

**Fig. 6.13:** Mode 1 control word register for (a) port A and (b) port B

*Signal definitions for mode1 strobed input*

**STB (Strobed input):** A "low" on this input loads data into the input latch.

**IBF (Input buffer full F/F):** A "high" on this output indicates that the data has been loaded into the input latch; in essence, an acknowledgement. IBF is set by STB input being low and is reset by the rising edge of the RD input.

*Control word register*

| D7 | D6 | D5 | D4 | D3 | D2 | D1 | D0 |
|----|----|----|----|----|----|----|----|
| I/O | I/O | $IBF_A$ | $INTE_A$ | $INTR_A$ | $INTE_B$ | $IBF_B$ | $INTR_B$ |
| | | Group A | | | | Group B | |

**INTR (Interrupt request):** A "high" on this output can be used to interrupt the CPU when an input device is requesting service. INTR is set by the  is a "one", IBF is a "one" and INTE is a "one". It is reset by the falling edge of. This procedure allows an input device to request service from the CPU by simply strobing its data into the port.

**$INTE_A$:** Controlled by bit set/reset of PC4

**$INTE_B$:** Controlled by bit set/reset of PC2

### 6.6.2.3 *Timing Diagram* (Fig. 6.14)

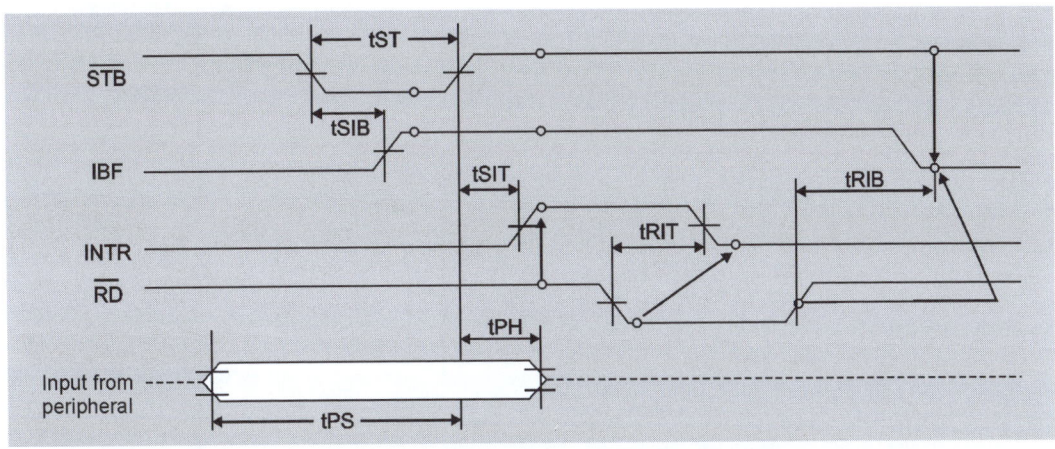

**Fig. 6.14:** Timing diagram for mode 1 (strobed input)

### 6.6.2.4 *Mode 1: Strobed Output*

Figure 6.15 illustrates the internal structure, control word register and timing waveforms, respectively of 8255A when it is operated as a strobed output device under Mode 1.

*Signal definitions for mode 1 strobed output*

**OBF (Output buffer full F/F):** The $\overline{OBF}$ output will go "low" to indicate that the CPU has written data out to the specified port. The $\overline{OBF}$ F/F will be set by the rising edge of the $\overline{WR}$ input and reset by $\overline{ACK}$ input being low.

**ACK (Acknowledge input):** A "low" on this input informs the 82A55A that the data from Port A or Port B has been accepted; in essence, a response from the peripheral device indicating that it has received the data output by the CPU.

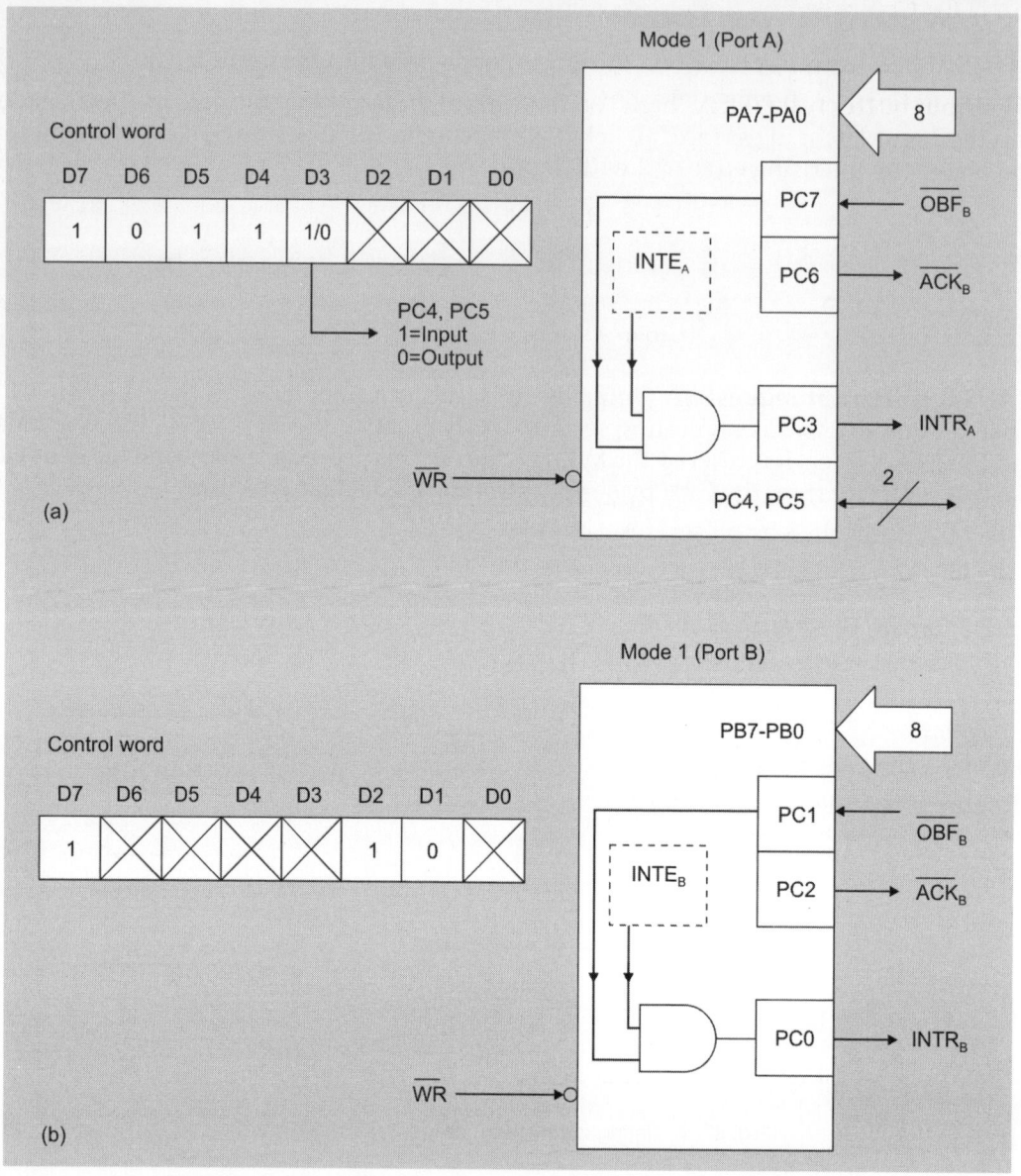

**Fig. 6.15:** Mode 1 control word register for (a) port A and (b) port B

*Control word register*

| D7 | D6 | D5 | D4 | D3 | D2 | D1 | D0 |
|---|---|---|---|---|---|---|---|
| $\overline{OBF}_A$ | $INTE_A$ | I/O | I/O | $INTR_A$ | $INTE_B$ | $\overline{OBF}_B$ | $INTR_B$ |
| | Group A | | | | | Group B | |

**INTR (Interrupt request):** A "high" on this output can be used to interrupt the CPU when an output device has accepted data transmitted by the CPU. INTR is set when $\overline{ACK}$ is a "one", $\overline{OBF}$ is a "one" and INTE is a "one". It is reset by the falling edge of $\overline{WR}$.

**INTE_A:** Controlled by bit set/reset of PC6

**INTE_B:** Controlled by bit set/reset of PC2

### 6.6.2.5 *Timing Diagram* (Fig. 6.16)

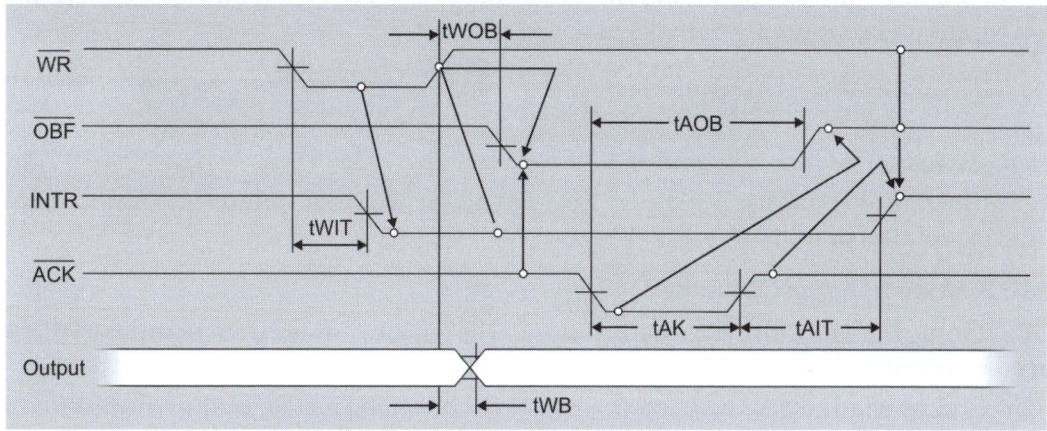

**Fig. 6.16:** Timing diagram for mode 1 (strobed output)

## 6.6.3 Mode 2: Strobed Bi-directional Bus I/O

This mode of operation of 8255A is also called as strobed bi-directional I/O and is shown in Fig. 6.17. This mode of operation provides 8255A with an additional feature for communicating with a peripheral device on an 8-bit data bus. Handshaking signals are provided to maintain proper data flow and synchronization between the data transmitter and receiver. The interrupt generation and other functions are similar to mode 1.

In this mode, 8255A is a bi-directional 8-bit port with handshake signals. The RD and WR signals decide whether the 8255A is going to operate as an input port or output port.

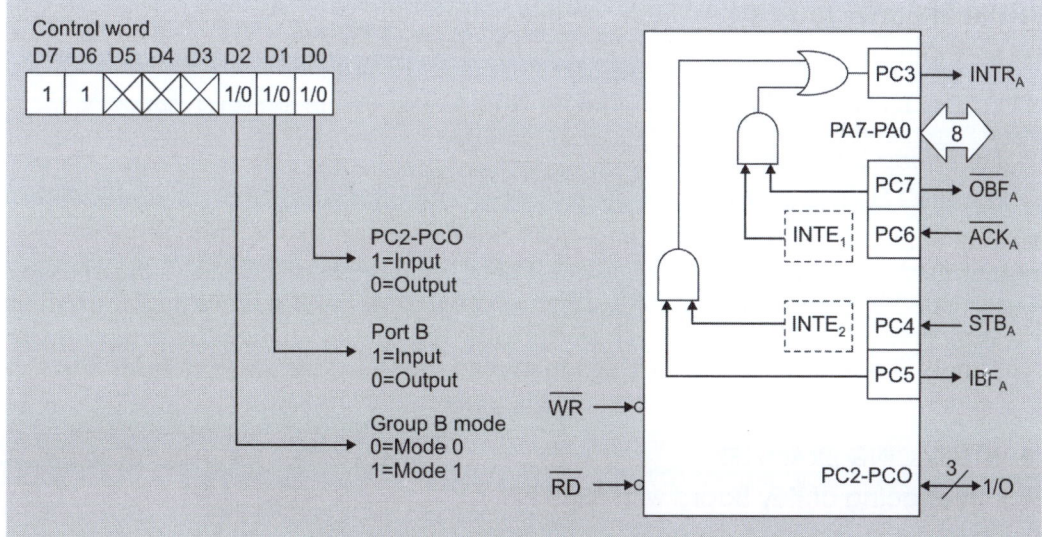

**Fig. 6.17:** Mode 2 control word register

### 6.6.3.1 Features of Mode 2

1. The single 8-bit port in group A is available.
2. The 8-bit port is bi-directional and additionally a 5-bit control port is available.
3. Three I/O lines are available at port C (PC2 – PC0).
4. Inputs and outputs are both latched.
5. The 5-bit control port C (PC3-PC7) is used for generating/accepting handshake signals for the 8-bit data transfer on port A.

*Signal definition for mode 2 control*

**INTR (interrupt request):** A high on this output can be used to interrupt the CPU for input or output operations.

### 6.6.3.2 Mode 2 Output Operations

**$\overline{OBF}$ (Output buffer full):** The $\overline{OBF}$ output will go"low" to indicate that the CPU has written data out to port A.

**$\overline{ACK}$ (Acknowledge):** A "low" on this input enables the tri-state output buffer of Port A to send out the data. Otherwise, the output buffer will be in the high impedance state.

*Control word register*

| D7 | D6 | D5 | D4 | D3 | D2 | D1 | D0 |
|---|---|---|---|---|---|---|---|
| $\overline{OBF}_A$ | $INTE_1$ | $IBF_A$ | $INTE_2$ | $INTR_A$ | X | X | X |
| | Group A | | | | Group B | | |

**$INTE_1$ (The INTE flip-flop associated with $\overline{OBF}$ ):** Controlled by bit set/reset of PC6.

### 6.6.3.3 Mode 2 Input Operations

**$\overline{STB}$ (Strobe input):** A "low" on this input loads data into the input latch.

**IBF (Input buffer full F/F):** A "high" on this output indicates that data has been loaded into the input latch.

**$INTE_2$ (The INTE flip-flop associated with IBF):** Controlled by bit set/reset of PC4.

### 6.6.3.4 Timing Diagram (Fig. 6.18)

### 6.7 READING PORT C STATUS

In Mode 0, Port C transfers data to or from the peripheral device. When the 8255A is programmed to function in Modes 1 or 2, Port C generates or accepts "hand-shaking" signals with the peripheral device. Reading the contents of Port C allows the programmer to test or verify the "status" of each peripheral device and change the program flow accordingly. There is no special instruction to read the status information from Port C. A normal read operation of Port C is executed to perform this function.

### 6.8 INTERFACING EXAMPLES

### 6.8.1 Interfacing of Key Board with 8255 PPI

The keyboard encoder de-bounces the key switches and provides a strobe signal whenever a key is pressed and data output contain the ASCII coded key code. Figure 6.19 shows a

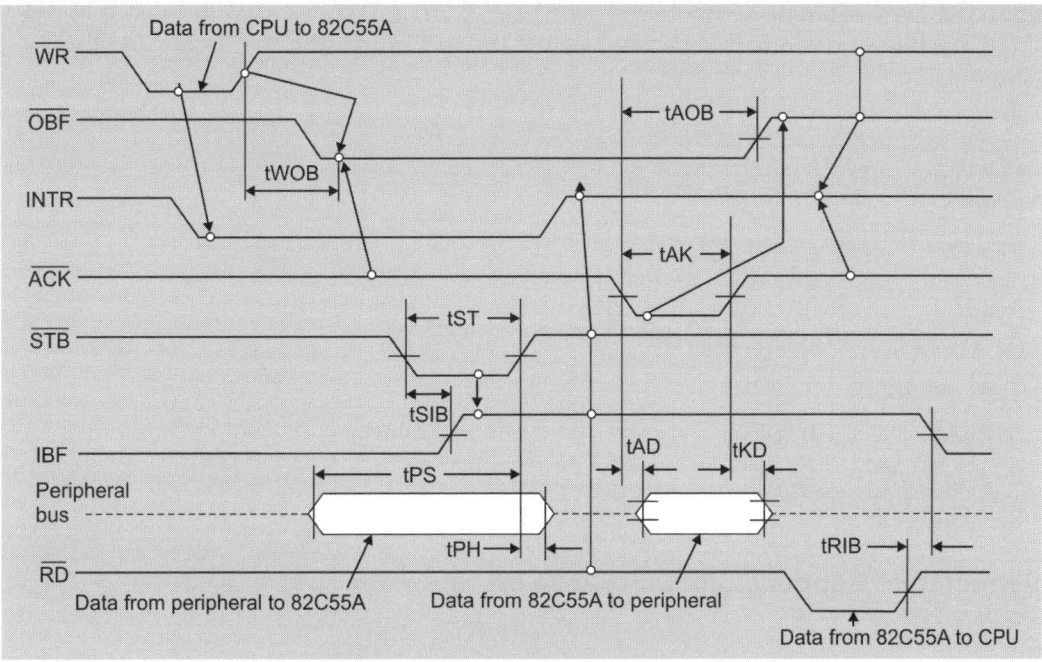

**Fig. 6.18:** Timing diagram for mode 2

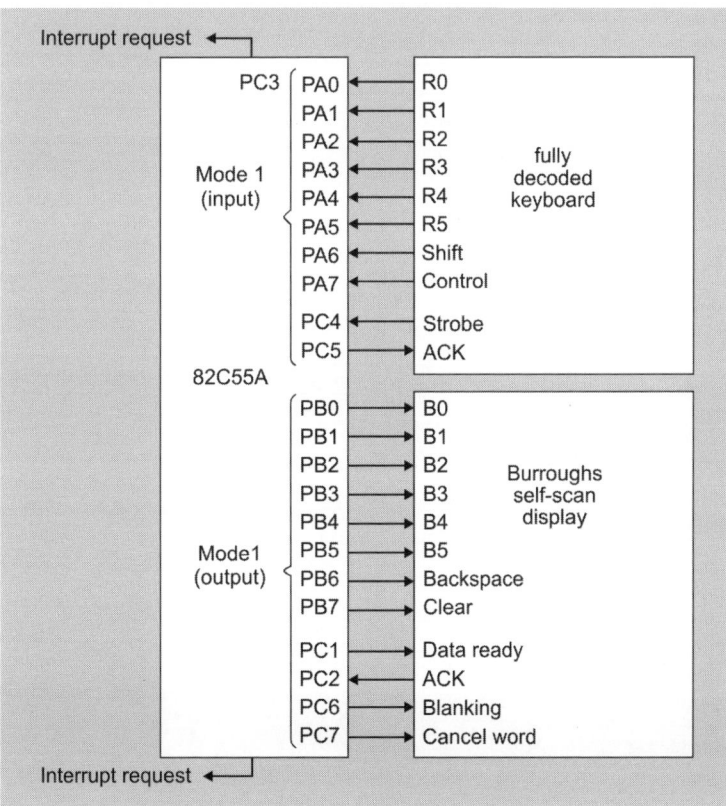

**Fig. 6.19:** Keyboard and display interface

keyboard connected to a strobed input port A. The data available is connected to PC4. The code that reads the keyboard encoder and return the ASCII key code in AL is written as:

```
BIT5 EQU    20H
PORTC       EQU   22H
PORTA       EQU   20H
READ        PROC  NEAR
```

**Read:**

```
IN AL, PORTC ; Read port C
TEST AL, BIT5 ; Test IBF
JZ Read        ; If IBF=0
IN AL, PORTA ; Read Data
READ ENDP
```

## Interfacing of Stepper Motor with 8255A PPI (Fig. 6.20)

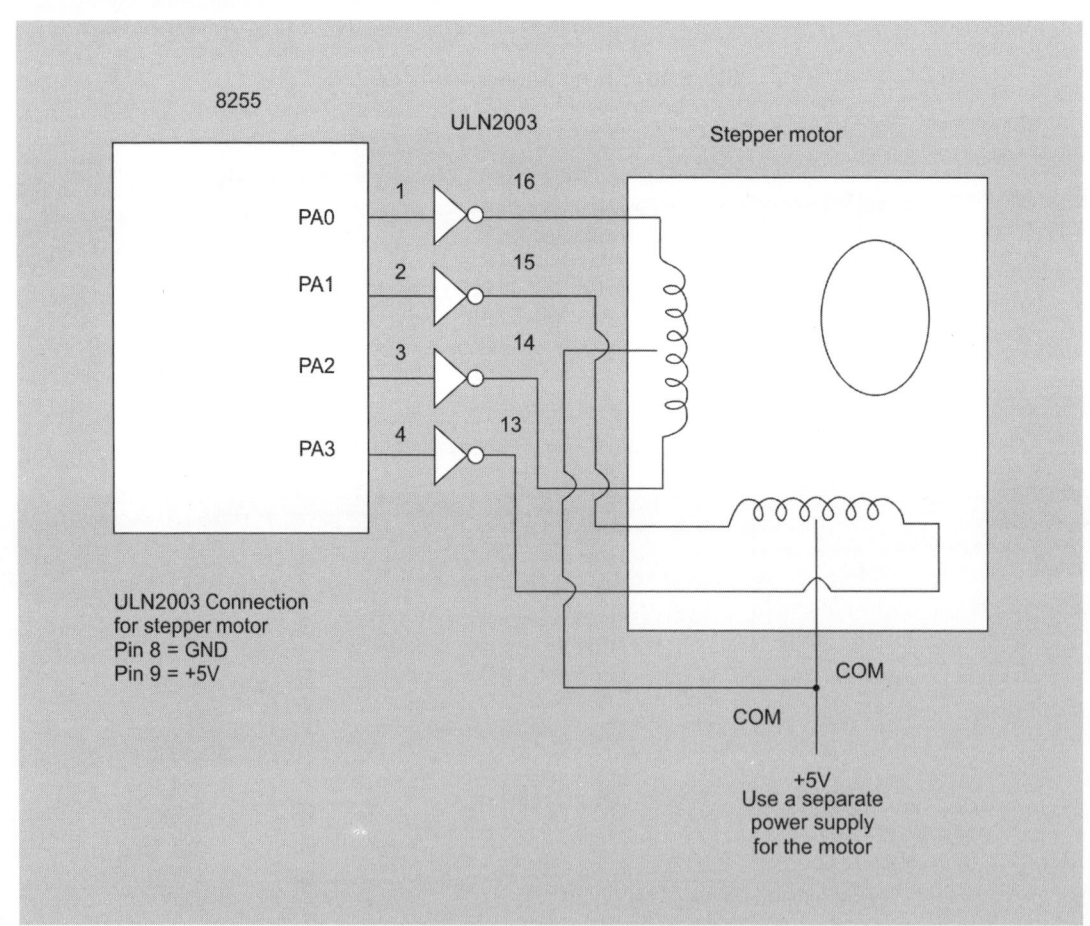

**Fig. 6.20:** Interfacing of stepper motor

# Interfacing of LCD with 8255A PPI (Figs 6.21–6.23)

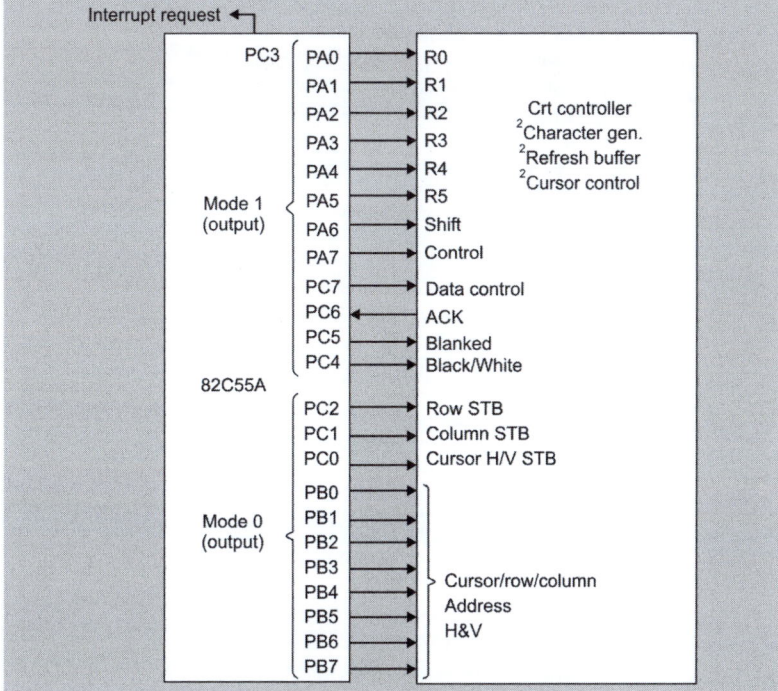

**Fig. 6.21:** Basic CRT controller interface

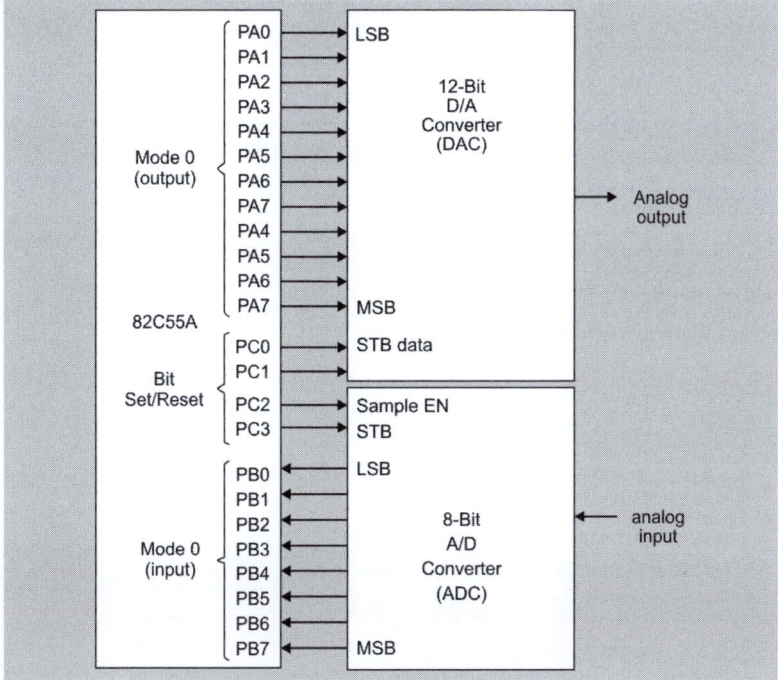

**Fig. 6.22:** Digital to analog, analog to digital

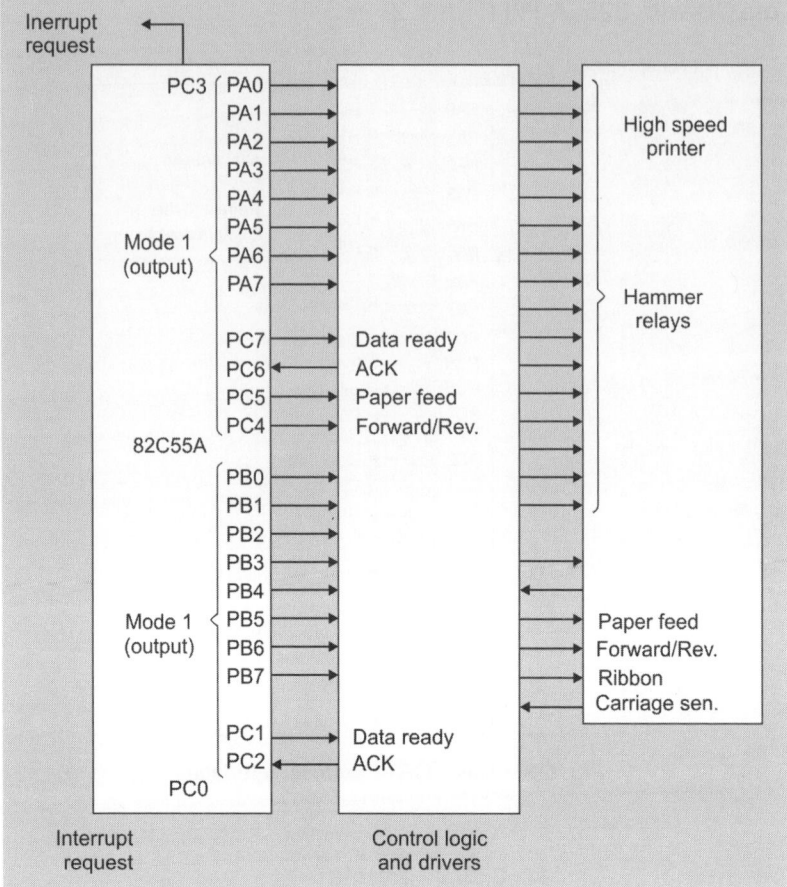

**Fig. 6.23:** Printer interface

## PROBLEMS

1. Explain the internal structure of 8255A in detail.

2. Explain the pin diagram of 8255A in detail.

3. Describe the operating modes of 8255A.

4. Illustrate control word register in I/O modes.

5. Which pins are general purpose I/O pins during mode 2 operation of 8255A PPI?

6. Describe briefly I/O operating modes of 8255A.

7. Which group of pins is used during bi-directional operation of 8255A?

8. In strobed input operation, what is the function of $\overline{ACK}$ signal?

9. What clears the $\overline{OBF}$ signal in strobed output operation of 8255A?

10. Write the software required to decide whether PC4 is at logic 1 when the 8255A is operated in the strobed output mode.

# 7

# Programmable Interval Timer/Counter 8253/54

## 7.1 INTRODUCTION

The Intel 82C54 is a high-performance, CHMOS version of the industry standard 8254 counter/timer which is designed to solve the timing control problems common in microcomputer system design. It provides three independent 16-bit counters, each capable of handling clock inputs up to 10 MHz. All modes are software programmable. The 82C54 is pin compatible with the HMOS 8254, and is a superset of the 8253. Six programmable timer modes allow the 82C54 to be used as an event counter, elapsed time indicator, programmable one-shot and in many other applications. The 82C54 is fabricated on Intel's advanced CHMOS III technology which provides low power consumption with performance equal to or greater than the equivalent HMOS product. The 82C54 is available in 24-pin DIP and 28-pin plastic leaded chip carrier (PLCC) packages.

The 8254 is a superset of the 8253. The 8254 uses HMOS technology and comes in a 24-pin plastic or CERDIP package. The 8254 is useful wherever the microprocessor must control real time events.

## 7.2 8253 VERSUS 8254

| 8253 | 8254 |
|---|---|
| Operating frequency 0–2.6 MHz | Operating frequency 0–10 MHz |
| Uses NMOS technology | Uses HMOS technology |
| Read back command not available | Read back command available |
| Reads and writes of the same counter cannot be interleaved | Reads and writes of the same counter can be interleaved |

## 7.3 FEATURES OF 8254

1. Three independent 16-bit down counters.
2. 8254 can handle inputs from DC to 10 MHz (5MHz 8254-5 8MHz 8254 10MHz 8254-2), whereas 8253 can operate up to 2.6 MHz.
3. Three counters are identical pre-settable and can be programmed for either binary or BCD count.
4. Counter can be programmed in six different modes.
5. Compatible with all Intel and most other microprocessors.
6. 8254 has powerful command called READ BACK command which allows the user to check the count value, programmed mode, current mode and current status of the counter.

## 7.4  PIN CONFIGURATION OF 8254

Figure 7.1 shows the pin configuration of the 8254 and a general definition of the pins is illustrated as follows:

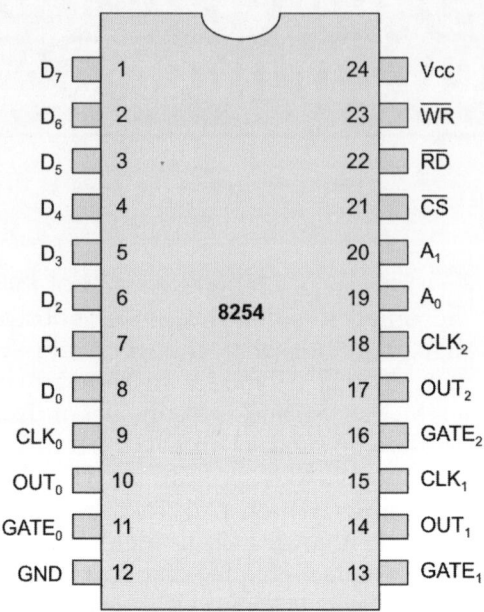

**Fig. 7.1:** Pin diagram of 8254

This 24-pin chip (above) has three distinct blocks as shown in Fig. 7.1.

The first block consists of the **counter output** (OUT0, OUT1, OUT2), **enabling gate** (GATE1, GATE2, GATE3) and **input** (CLK0, CLK1, CLK2) pins. Peripherals (like motors and switches) can be wire up to these pins.

The second block is the **eight data lines D0-D7.** Processor uses to transfer (reading and writing) data to and from the 8254.

The third block is the chip select ($\overline{CS}$), chip reading and writing ($\overline{RD}$ and $\overline{WR}$) and programming mode select lines (A0 and A1). Processor uses these lines to program the 8254 when reading or writing counter values.

**D0-D7 (pin-1-8, type-i/o):** The D0-D7 data bus of the 8254 is a bi-directional bus connected to D0-D7 of the system data bus.

**CLK 0 (pin-9, type-I):**  Clock input of Counter 0.

**OUT 0 (pin-10, type-O):** Output of Counter 0.

**GATE 0 (pin-11, type-I):** Gate input of Counter 0.

**GND (pin-12):** Power supply connection.

**VCC (pin-24):** 5V power supply connection.

$\overline{WR}$ **(pin-23, type-I):** This input is low during CPU write operations.

$\overline{RD}$ **(pin-22, type-I):** This input is low during CPU read operations.

$\overline{CS}$ **(pin-21, type-I):** A low on this input enables the 8254 to respond to $\overline{RD}$ and $\overline{WR}$ signals. Otherwise, $\overline{RD}$ and $\overline{WR}$ signals are ignored.

**A1, A0 (pin-20–19, type-I):** Used to select one of the three Counters or the Control Word Register for read or write operations. Normally connected to the system address bus.

**CLK 2 (pin-18, type-I):** Clock input of Counter 2.

**OUT 2 (pin-17, type-O):** Output of Counter 2.

**GATE 2 (pin-16, type-I):** Gate input of Counter 2.

**CLK 1 (pin-15, type-I):** Clock input of Counter 1.

**GATE 1 (pin-14, type-I):** Gate input of Counter 1.

**OUT 1 (pin-13, type-O):** Output of Counter 1.

## 7.5  ARCHITECTURE OF 8254

Figure 7.2 shows the block diagram of 8253/54. It includes three counters: a data bus buffer, Read/Write control logic and a control word register. Each counter has two input signals CLOCK and GATE and one output signal OUT.

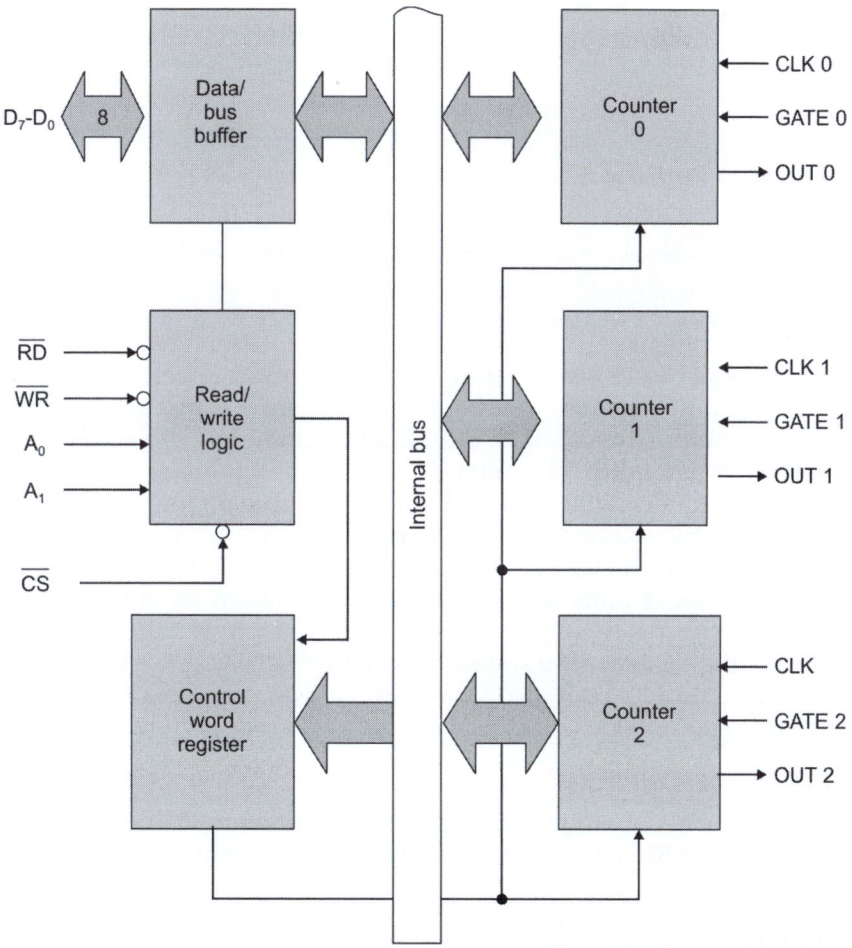

**Fig. 7.2:** Architecture of 8254

### 7.5.1  Data Bus Buffer

This tri-state, bi-directional, 8-bit buffer is used to interface the 8253/54 to the system data bus. The data bus buffer has three basic functions.

1. Programming the modes of 8253/54.
2. Loading the count registers.
3. Reading the count values.

### 7.5.2 Read/Write Control Logic

The read/write control logic has five signals: RD, WR, CS and the address lines A0 and A1. In the peripheral I/O mode, the RD and WR signals are connected to IOR and IOW, respectively. In memory-mapped I/O, these are connected to MEMR and MEMW. Address lines A0 and A1 of the CPU are usually connected to lines A0 and A1 of the 8253/54 and CS is tied to a decoded address. The control word register and counters are selected according to the signals on lines A0 and A1.

| A1 | A0 | Selection |
| --- | --- | --- |
| 0 | 0 | Counter 0 |
| 0 | 1 | Counter 1 |
| 1 | 0 | Counter 2 |
| 1 | 1 | Control word register |

### 7.5.3 Control Word Register

This register is accessed when lines A0 and A1 are at logic 1. It is used to write a command word which specifies the counter to be used (binary or BCD), its mode and either a read or write operation.

### 7.5.4 Counter

These three functional blocks are identical in operation. Each counter consists of a single, 16-bit, pre-settable, down counter. The counter can operate in either binary or BCD and its input, gate and output are configured by the selection of modes stored in the control word register. The counters are fully independent. The programmer can read the contents of any of the three counters without disturbing the actual count in process.

The Control Logic allows one register at a time to be loaded from the internal bus. Both bytes are transferred to the CE simultaneously. $CR_M$ and $CR_L$ are cleared when the Counter is programmed. In this way, if the Counter has been programmed for one byte counts (either most significant byte only or least significant byte only), the other byte will be zero. Note that the CE cannot be written into; whenever a count is written, it is written into the CR. The Control Logic is also shown in the diagram. CLKn, GATEn and OUTn are all connected to the outside world through the Control Logic (Fig. 7.3).

### 7.6 OPERATIONAL DESCRIPTION

The complete functional definition of the 8253/54 is programmed by the system software. Once programmed, the 8253/54 is ready to perform whatever timing tasks it is assigned to accomplish.

### 7.6.1 Programming the 8253/54

Each counter of the 8253/54 is individually programmed by writing a control word into the control word register (A0 – A1 = 11). Figure 7.4 shows the control word format. Bits SC1 and SC0 select the counter, bits RW1 and RW0 select the read, write or latch command, bits M2, M1 and M0 select the mode of operation and bit BCD decides whether it is a BCD counter or binary counter.

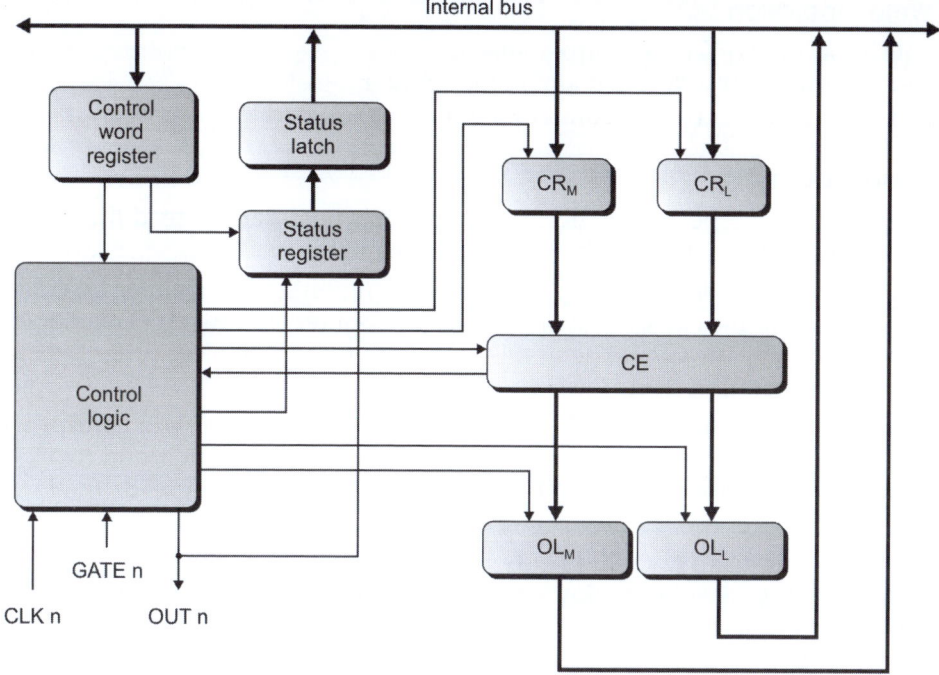

**Fig. 7.3:** Internal block diagram of counter

| $D_7$ | $D_6$ | $D_5$ | $D_4$ | $D_3$ | $D_2$ | $D_1$ | $D_0$ |
|-------|-------|-------|-------|-------|-------|-------|-------|
| $SC_1$ | $SC_0$ | $RW_1$ | $RW_0$ | $M_2$ | $M_1$ | $M_0$ | BDC |

SC - Select counter

| $SC_1$ | $SC_0$ | |
|--------|--------|---|
| 0 | 0 | Select counter 0 |
| 0 | 1 | Select counter 1 |
| 1 | 0 | Select counter 2 |
| 1 | 1 | Illegal for 8253<br>Read-back command for 8254<br>(See Read operations) |

M - Mode

| $M_2$ | $M_1$ | $M_0$ | |
|-------|-------|-------|---|
| 0 | 0 | 0 | Mode 0 |
| 0 | 0 | 1 | Mode 1 |
| x | 1 | 0 | Mode 2 |
| x | 1 | 1 | Mode 3 |
| 1 | 0 | 0 | Mode 4 |
| 1 | 0 | 1 | Mode 5 |

RW-Read/Write

| $RW_1$ | $RW_0$ | |
|--------|--------|---|
| 0 | 0 | Counter latch command<br>(See Read operations) |
| 0 | 1 | Read/write least significant byte only |
| 1 | 0 | Read/write least significant byte only |
| 1 | 1 | Read/write least significant byte first,<br>then most significant byte |

BCD:

| 0 | Binary counter 16-bits |
|---|---|
| 1 | Binary coded decimal (BCD)<br>counter (4 decades) |

**Fig. 7.4:** Control word format

## 7.6.2 Write Operation

1. Write a control word into control register.
2. Load the low-order byte of a count in the counter register.
3. Load the high-order byte of count in the counter register.

## 7.6.3 Read Operation

In some applications, especially in event counters, it is necessary to read the value of the count in process. This can be done by three possible methods:

1. **Simple read:** It involves reading a count after inhibiting the counter by controlling the gate input or the clock input of the selected counter, and two I/O read operations are performed by the CPU. The first I/O operation reads the low-order byte and the second I/O operation reads the high-order byte.

2. **Counter latch command:** In the second method, an appropriate control word is written into the control register to latch a count in the output latch, and two I/O read operations are performed by the CPU. The first I/O operation reads the low-order byte, and the second I/O operation reads the high-order byte.

3. **Read-Back command (available only for 8254):** The third method uses the Read-Back command. This command allows the user to check the count value, programmed mode and current status of the OUT pin and Null count flag of the selected counter(s). Figure 7.5 shows the format of the control word register for read-back command.

$A_0 A_1 = 11 \; \overline{CS} = 0 \; \overline{RD} = 1 \; \overline{WR} = 0$

| $D_7$ | $D_6$ | $D_5$ | $D_4$ | $D_3$ | $D_2$ | $D_1$ | $D_0$ |
|---|---|---|---|---|---|---|---|
| 1 | 1 | Count | Status | $CNT_2$ | $CNT_1$ | $CNT_0$ | 0 |

$D_5$ : 0=Latch count of selected counter(s)
$D_4$ : 0=Latch status of selected counter(s)
$D_3$ : 0=Select counter 2
$D_2$ : 0=Select counter 1
$D_1$ : 0=Select counter 0
$D_1$ : Reserved for future expansion: must be 0.

**Fig. 7.5:** Control word register for read-back command

The read-back command may be used to latch multiple counter output latches by setting the COUNT bit D5 = 0 and selecting the desired counter(s). Each counter's latch count is held until it is read (or the counter is reprogrammed). That counter is automatically unlatched when read.

## 7.7 OTHER FEATURES OF READ-BACK COMMAND (AVAILABLE ONLY FOR 8254)

The read-back command may also be used to latch status information of selected counter(s) by setting STATUS bit D4 = 0. The contents of the counter must be latched before reading. The status of a counter is then accessed by a read from that counter. Figure 7.6 shows the counter status format.

Bit D5 - D0 contains the counter's programmed mode exactly as written in the last mode control word. Bit D7 contains the current status of the output pin. In 8254, it is not possible to read count from the counter, if the count is not loaded into the counting element (CE). The Bit D6 indicates whether the counting element has count or not. If D6 = 0, counting element has count, otherwise null count.

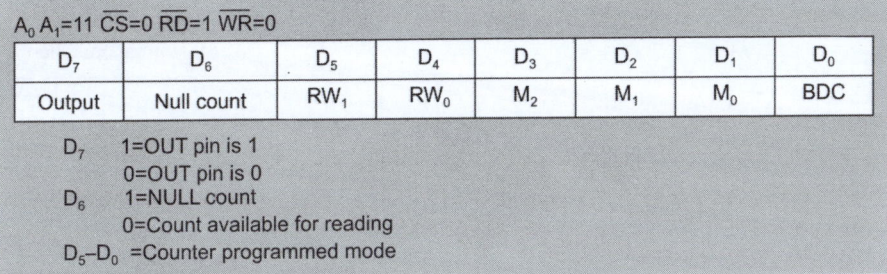

$A_0 A_1 = 11$ $\overline{CS} = 0$ $\overline{RD} = 1$ $\overline{WR} = 0$

| $D_7$ | $D_6$ | $D_5$ | $D_4$ | $D_3$ | $D_2$ | $D_1$ | $D_0$ |
|--------|------------|--------|--------|--------|--------|--------|--------|
| Output | Null count | $RW_1$ | $RW_0$ | $M_2$ | $M_1$ | $M_0$ | BDC |

$D_7$   1=OUT pin is 1
0=OUT pin is 0
$D_6$   1=NULL count
0=Count available for reading
$D_5$–$D_0$ =Counter programmed mode

**Fig. 7.6:** Counter status format

## 7.8 INTERLEAVED READ AND WRITE

Another feature of the 8254 is that reads and writes of the same counter may be interleaved. For example, if the counter is programmed for the two byte counts, the following sequence is valid:

1. Read least significant byte.
2. Write new least significant byte.
3. Read most significant byte.
4. Write new most significant byte.

## 7.9 MODE DEFINITIONS

### 7.9.1 Mode 0: Interrupt on Terminal Count (Fig. 7.7)

a. **Normal operation**
1. The output will be initially low after the mode set operation.
2. After the count is loaded into the selected count register, the output will remain low and the counter will count.
3. When the terminal count is reached, the output will go high and remain high until the selected count is reloaded.

b. **Gate disable**
1. Gate = 1 enables counting.
2. Gate = 0 disables counting.

**Note:** Gate has no effect on OUT.

c. **New count:** If a new count is written to the counter, it will be loaded on the next CLK pulse and counting will continue from the new count.

**In case of two byte count:**
1. Writing the first byte disables counting.
2. Writing the second byte loads the new count on the next CLK pulse and counting will continue from the new count.

### 7.9.2 Mode 1: Hardware Retriggerable One Shot (Fig. 7.8)

a. **Normal operation**
1. The output will be initially high.
2. The output will go low on the CLK pulse following the rising edge at the gate input.

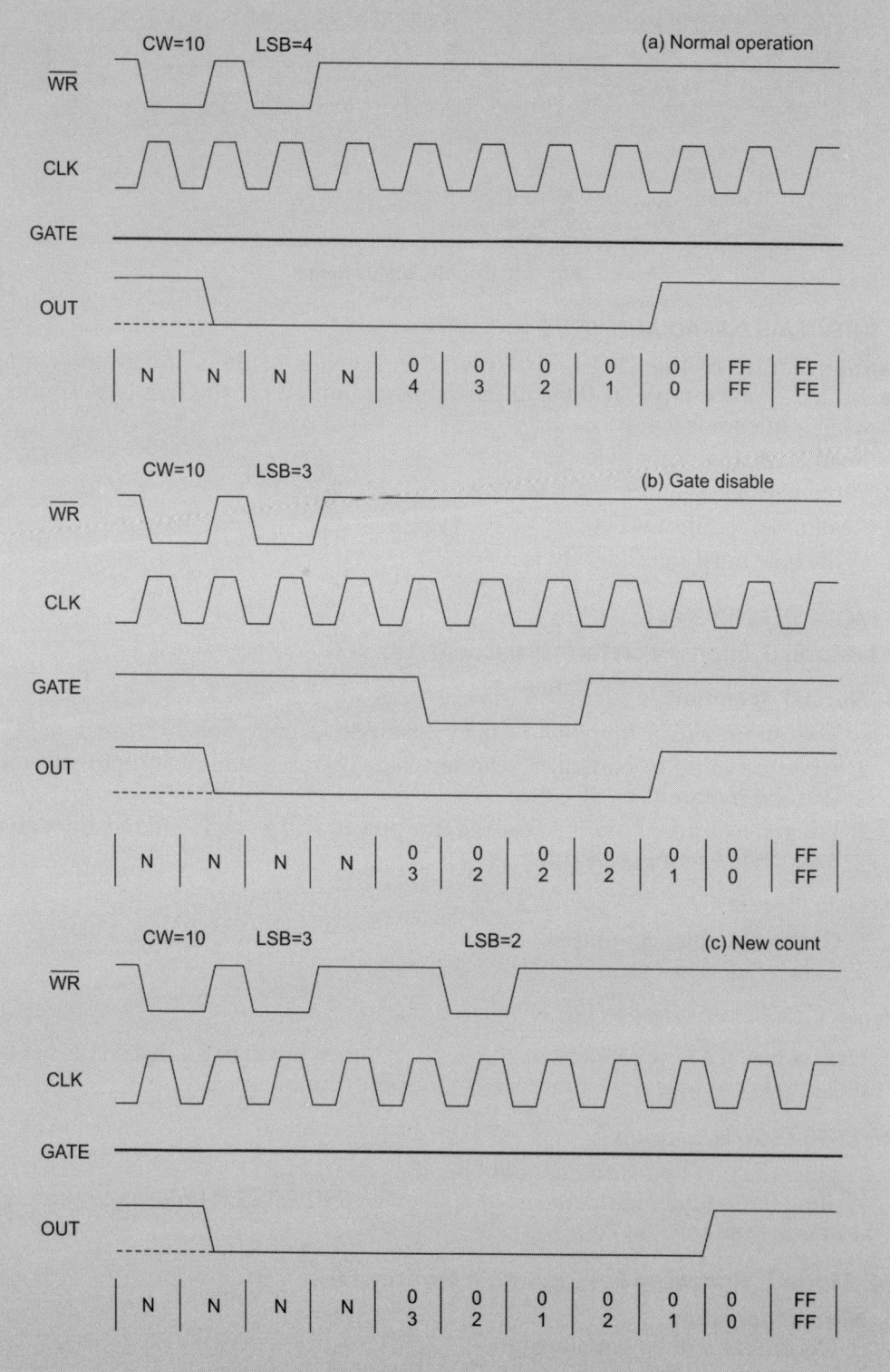

**Fig. 7.7:** Mode 0 interrupt on terminal count

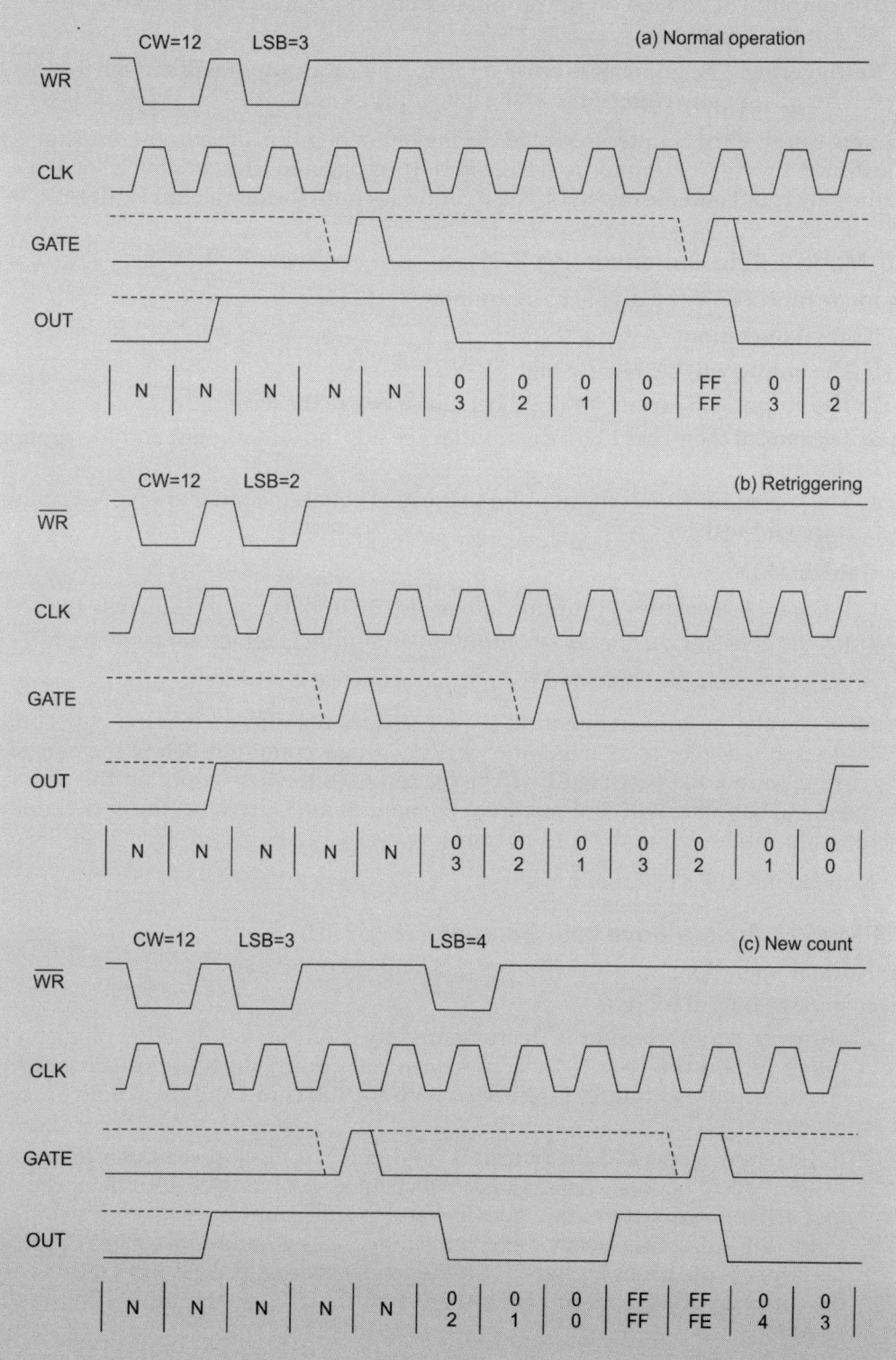

**Fig. 7.8:** Mode 1 hardware retriggerable one shot

3. The output will go high on the terminal count and remain high until the next rising edge at the gate input.

b. **Retriggering:** The one shot is retriggerable, hence the output will remain low for the full count after any rising edge of the gate input.

c. **New count:** If the counter is loaded during one shot pulse, the current one shot is not affected unless the counter is retriggered. If retriggered, the counter is loaded with the new count and the one-shot pulse continues until the new count expires.

### 7.9.3 Mode 2: Rate Generator (Fig. 7.9)

This mode functions like a divide by-N counter.

a. **Normal operation**
1. The output will be initially high.
2. The output will go low for one clock pulse before the terminal count.
3. The output then goes high, the counter reloads the initial count and the process is repeated.
4. The period from one output pulse to the next equals the number of input counts in the count register.

b. **Gate disable**
1. If Gate = 1, it enables a counting, otherwise it disables counting (Gate = 0).
2. If Gate goes low during a low output pulse, output is set immediately high.

A trigger reloads the count and the normal sequence is repeated.

c. **New count:** The current counting sequence does not affect when the new count is written. If a trigger is received after writing a new count but before the end of the current period, the new count will be loaded with the new count on the next CLK pulse and counting will continue from the new count. Otherwise, the new count will be loaded at the end of the current counting cycle.

**Note:** In mode 2, a count of 1 is illegal.

### 7.9.4 Mode 3: Square Wave Rate Generator (Fig. 7.10)

a. **Normal operation:**
1. Initially output is high.
2. For even count, counter is decremented by 2 on the falling edge of each clock pulse. When the counter reaches terminal count, the state of the output is changed and the counter is reloaded with the full count and the whole process is repeated.
3. If the count is odd and the output is high, the first clock pulse (after the count is loaded) decrements the count by 1. Subsequent clock pulses decrement the clock by 2. After timeout, the output goes low and the full count is reloaded. The first clock pulse (following the reload) decrements the count by 3 and subsequent clock pulse decrements the count by two. Then the whole process is repeated. In this way, if the count is odd, the output will be high for $(n + 1)/2$ counts and low for $(n - 1)/2$ counts.

b. **Gate disable:** If Gate is 1, counting is enabled, otherwise it is disabled. If Gate goes low while output is low, output is set high immediately. After this, when Gate goes

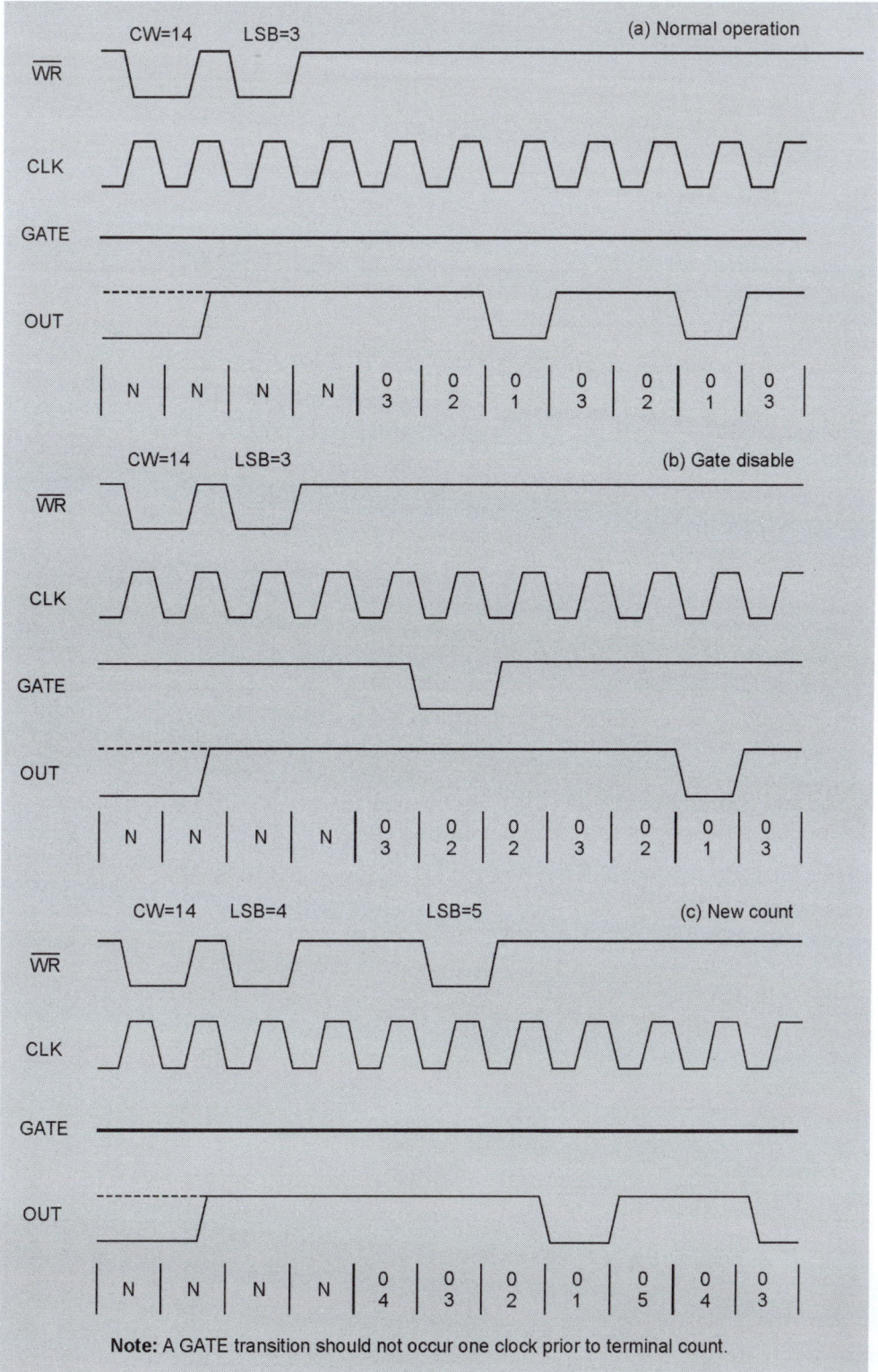

Note: A GATE transition should not occur one clock prior to terminal count.

Fig. 7.9: Mode 2 rate generator

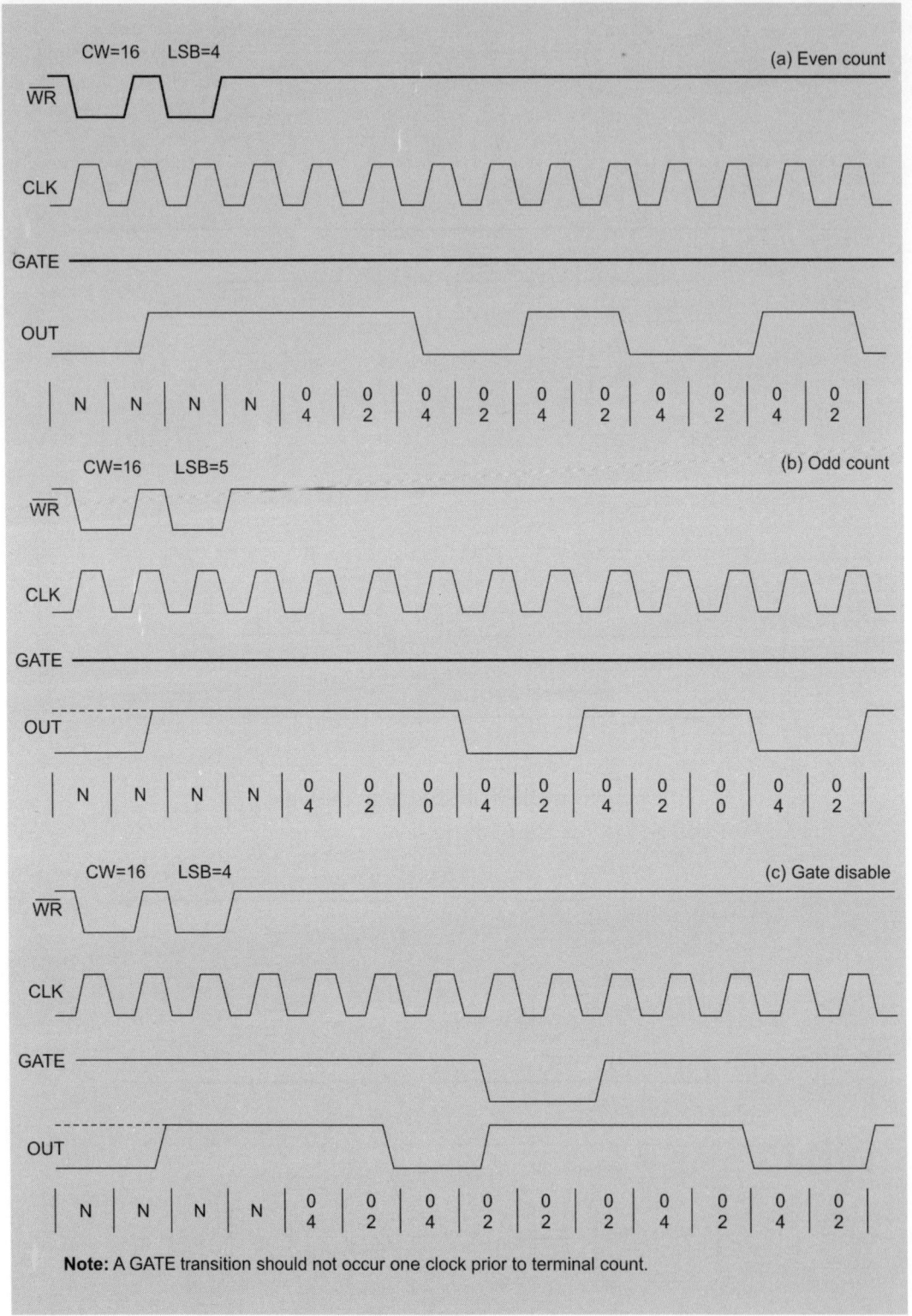

**Note:** A GATE transition should not occur one clock prior to terminal count.

**Fig. 7.10:** Mode 3 square wave mode

high, the counter is loaded with the initial count on the next clock pulse and the sequence is repeated.

c. **New count:** The current counting sequence does not affect when the new count is written. If a trigger is received after writing a new count but before the end of the current half-cycle of the square wave, the counter will be loaded with the new count on the next CLK pulse and counting will continue from the new count, otherwise the new count will be loaded at the end of the current half-cycle.

### 7.9.5 Mode 4: Software Triggered Strobe (Fig. 7.11)

**a. Normal operation**

1. The output will be initially high

2. The output will go low for one CLK pulse after the terminal count (TC).

b. **Gate disable:** If Gate is one, the counting is enabled, otherwise it is disabled. The Gate has no effect on the output.

c. **New count:** If a new count is written during counting, it will be loaded on the next CLK pulse and counting will continue from the new count. If the count is two byte, then

1. Writing the first byte has no effect on counting.

2. Writing the second byte allows the new count to be loaded on the next CLK pulse.

### 7.9.6 Mode 5: Hardware Triggered Strobe (Retriggerable) (Fig. 7.12)

**a. Normal operation**

1. The output will be initially high.

2. The counting is triggered by the rising edge of the Gate.

3. The output will go low for one CLK pulse after the terminal count (TC).

b. **Retriggering:** If the triggering occurs on the Gate input during the counting, the initial count is loaded on the next CLK pulse and the counting will be continued until the terminal count is reached.

c. **New count:** If a new count is written during counting, the current counting sequence will not be affected. If the trigger occurs after the new count is written but before the terminal count, the counter will be loaded with the new count on the next CLK pulse and counting will continue from there.

### 7.10 PROGRAMMING EXAMPLE

**Example:** Write a program to initialize counter 2 in mode 0 with a count of C030H. Assume address for control register = 0BH, counter 0 = 08H, counter 1 = 09H and counter 2 = 0AH.

**Solution:** Control word

| $D_7$ | $D_6$ | $D_5$ | $D_4$ | $D_3$ | $D_2$ | $D_1$ | $D_0$ | |
|-------|-------|-------|-------|-------|-------|-------|-------|---|
| $SC_1$ | $SC_2$ | $RW_1$ | $RW_0$ | $M_2$ | $M_1$ | $M_0$ | BCD | |
| 1 | 0 | 1 | 1 | 0 | 0 | 0 | 0 | = B0H |

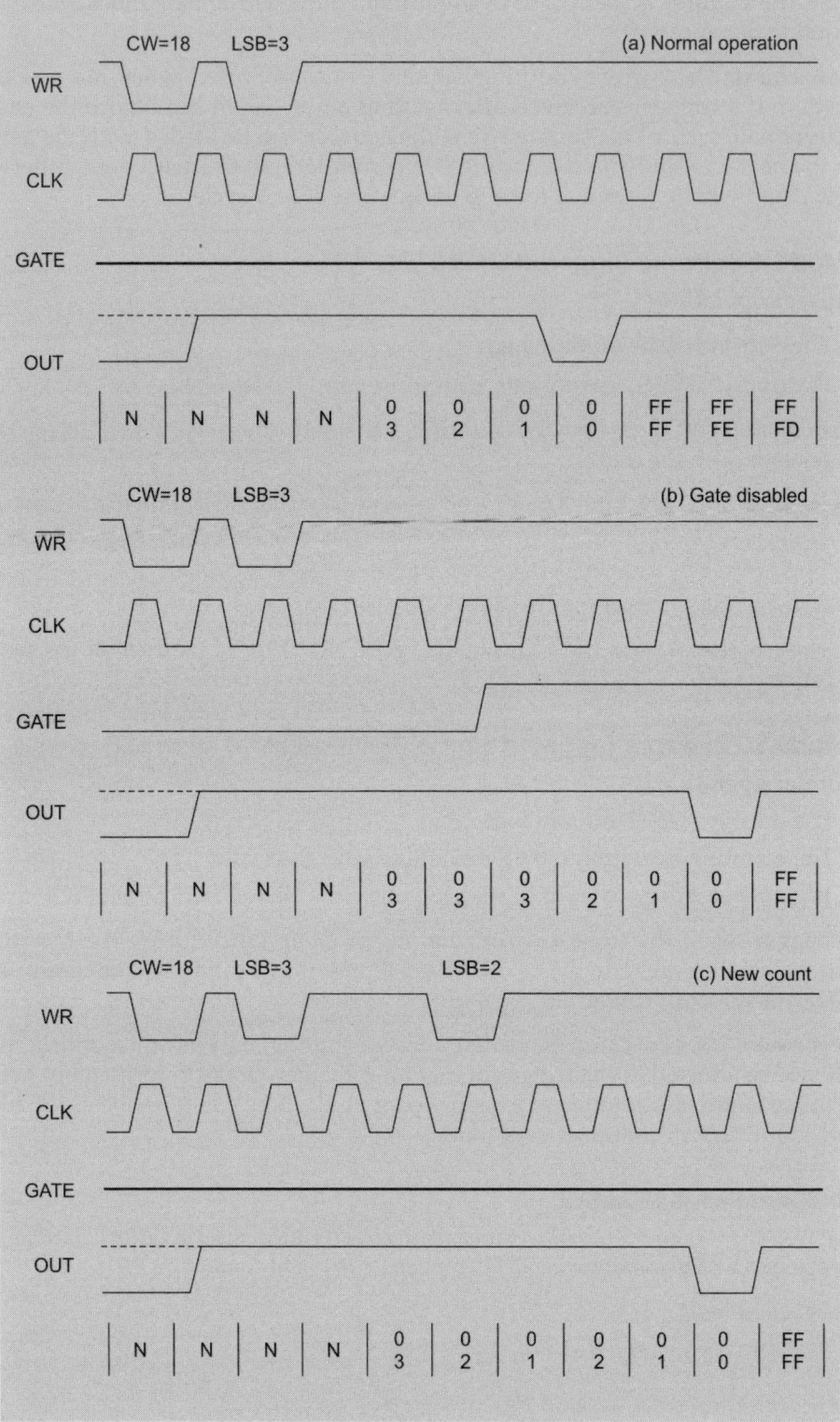

**Fig. 7.11:** Mode 4 software triggered strobe

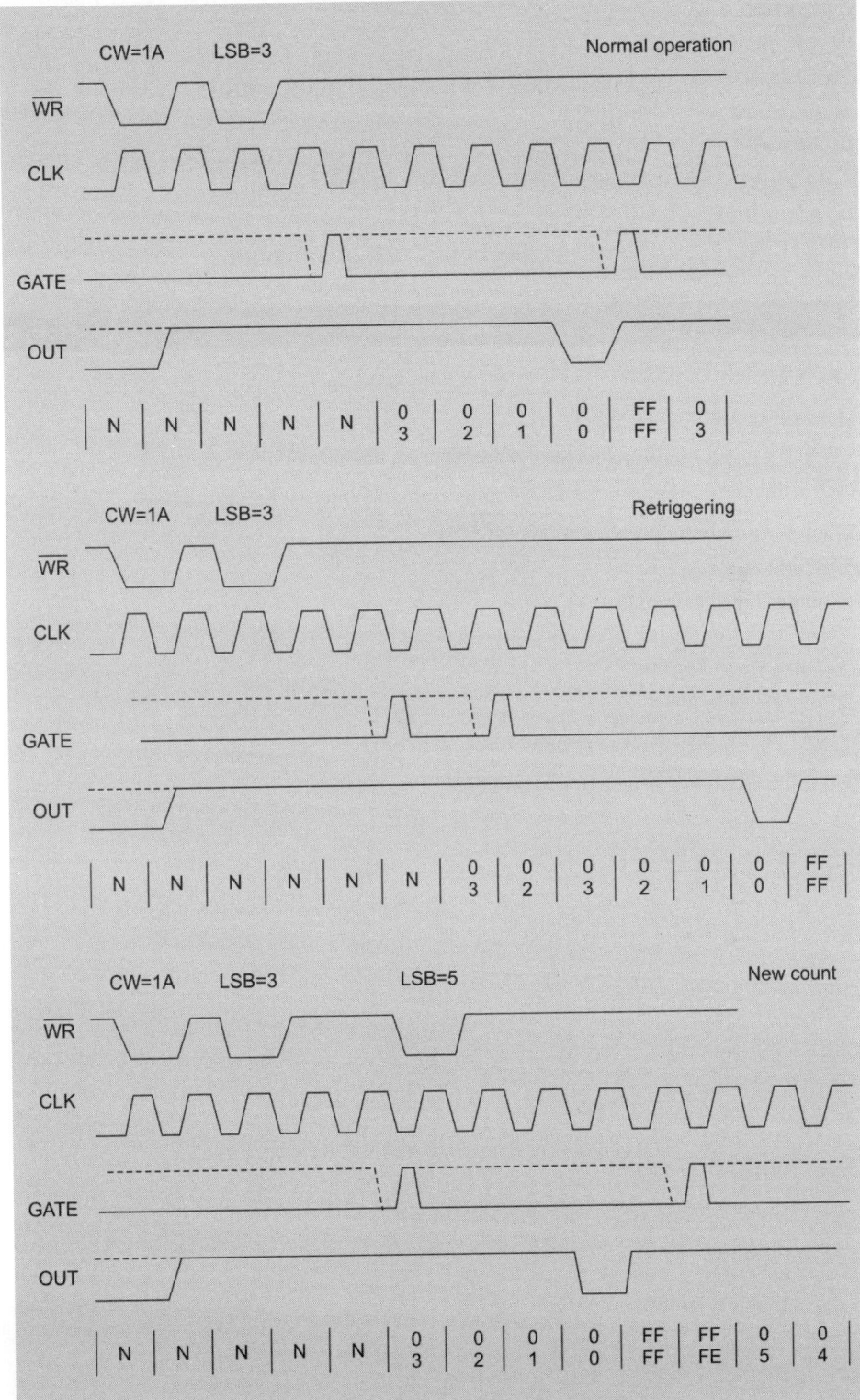

**Fig. 7.12:** Mode 5 hardware triggered strobe

**Source program**

```
MOV AL, B0H
OUT 0BH, AL      : Loads control word (B0H) in the control
                 : register.
MOV AL, 30H
OUT 0AH, AL      : Loads lower byte of (30H) the count
MOV AL, 0C0H
OUT 0AH, AL      : Loads higher byte (C0H) of the count.
```

## PROBLEMS

1. What is the need of programmable interval timer/counter?

2. Explain the features of 8254 PIT.

3. Explain the major building blocks of 8254 PIT with the help of block diagram.

4. Describe the pin configuration of 8254 programmable interval timer/counter.

5. Explain the different operating modes of 8254.

6. Write short notes on
   a. Counter Latch Command
   b. Read-Back Command
   c. Control Word Register
   d. Status Word Register

7. Explain how the read-back command function in 8254.

8. What is the difference between 8253 and 8254?

# 8
# Programmable Interrupt Controller 8259A

## 8.1 INTRODUCTION

The Intel 8259A Programmable Interrupt Controller handles up to eight vectored priority interrupts for the CPU. It is cascadable for up to 64 vectored priority interrupts without additional circuitry. It is packaged in a 28-pin DIP, uses NMOS technology and requires a single A 5V supply. Circuitry is static, requiring no clock input. The 8259A is designed to minimize the software and real time overhead in handling multi-level priority interrupts. It has several modes, permitting optimization for a variety of system requirements. The 8259A is fully upward compatible with the Intel 8259. Software originally written for the 8259 will operate the 8259A in all 8259 equivalent modes (MCS-80/85, non-buffered and edge triggered).

## 8.2 FEATURES OF 8259A

1. 8 levels of interrupts.
2. Can be cascaded in master-slave configuration to handle 64 levels of interrupts.
3. Internal priority resolver.
4. Fixed priority mode and rotating priority mode.
5. Individually maskable interrupts.
6. Modes and masks can be changed dynamically.
7. Accepts IRQ, determines priority, checks whether incoming priority > current level being serviced, issues interrupt signal.
8. In 8085 mode, provides 3 byte CALL instruction. In 8086 mode, provides 8-bit vector number.
9. Polled and vectored mode.
10. Starting address of ISR or vector number is programmable.
11. No clock required.

## 8.3 ARCHITECTURE OF 8259

The 8259A is a device specifically designed for use in real time, interrupt driven microcomputer systems. It manages eight levels or requests and has built-in features for expandability to other 8259A's (up to 64 levels). It is programmed by the system's software as an I/O peripheral. A selection of priority modes is available to the programmer so that the manner in which the requests are processed by the 8259A can be configured to match his system requirements.

The priority modes can be changed or reconfigured dynamically at any time during the main program. This means that the complete interrupt structure can be defined as

227

required, based on the total system environment. The functional block diagram of 8259A is shown in Fig. 8.1 and is explained below.

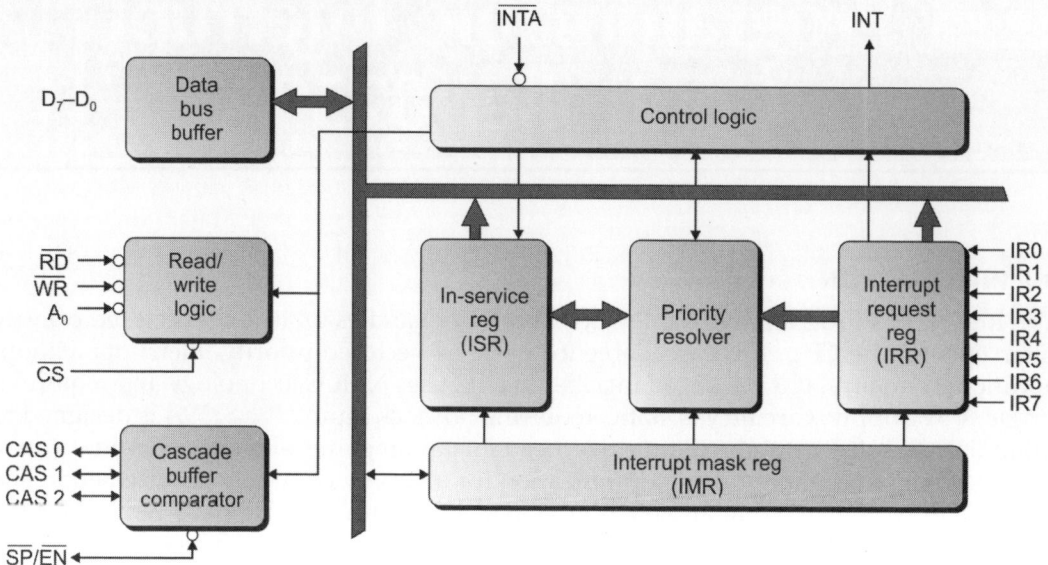

**Fig. 8.1:** Functional block diagram of 8259A

### 8.3.1 Interrupt Request Register (IRR) and In-Service Register (ISR)

The interrupts at the IR input lines are handled by two registers in cascade, the Interrupt Request Register (IRR) and the In-Service Register (ISR). The IRR is used to store all the interrupt levels which are requesting service; and the ISR is used to store all the interrupt levels which are being serviced.

### 8.3.2 Priority Resolver

This logic block determines the priorities of the bits set in the IRR. The highest priority is selected and strobed into the corresponding bit of the ISR during $\overline{\text{INTA}}$ pulse.

### 8.3.3 Interrupt Mask Register (IMR)

The IMR stores the bits which mask the interrupt lines to be masked. The IMR operates on the IRR. Masking of a higher priority input will not affect the interrupt request lines of lower quality.

### 8.3.4 INT (Interrupt)

This output goes directly to the CPU interrupt input. The $V_{OH}$ level on this line is designed to be fully compatible with the 8080A, 8085A and 8086 input levels.

### 8.3.5 $\overline{\text{INTA}}$ (Interrupt Acknowledge)

INTA pulses will cause the 8259A to release vectoring information onto the data bus. The format of this data depends on the system mode ($\mu$PM) of the 8259A.

### 8.3.6 Data Bus Buffer

This 3-state, bi-directional 8-bit buffer is used to interface the 8259A to the system Data Bus. Control words and status information are transferred through the Data Bus Buffer.

### 8.3.7 Read/Write Control Logic

The function of this block is to accept Output commands from the CPU. It contains the Initialization Command Word (ICW) registers and Operation Command Word (OCW) registers which store the various control formats for device operation. This function block also allows the status of the 8259A to be transferred onto the Data Bus.

### 8.3.8 $\overline{CS}$ (Chip Select)

A LOW on this input enables the 8259A. No reading or writing of the chip will occur unless the device is selected.

### 8.3.9 $\overline{WR}$ (Write)

A LOW on this input enables the CPU to write control words (ICWs and OCWs) to the 8259A.

### 8.3.10 $\overline{RD}$ (Read)

A LOW on this input enables the 8259A to send the status of the Interrupt Request Register (IRR), In-Service Register (ISR), the Interrupt Mask Register (IMR) or the Interrupt Level onto the Data Bus.

### 8.3.11 $A_0$

This input signal is used in conjunction with $\overline{WR}$ and $\overline{RD}$ signals to write commands into the various command registers, as well as reading the various status registers of the chip. This line can be tied directly to one of the address lines.

### 8.3.12 Cascaded Buffer/Comparator

This function block stores and compares the IDs of all 8259As used in the system. The associated three I/O pins (CAS0-2) are outputs when the 8259A is used as a master and are inputs when the 8259A is used as a slave. As a master, the 8259A sends the ID of the interrupting slave device onto the CAS0-2 lines. The slave thus selected will send its pre-programmed subroutine address onto the Data Bus during the next one or two consecutive $\overline{INTA}$ pulses.

### 8.4 PIN CONFIGURATION OF 8259

The 8259A is available in 28-pin DIP package as shown in Fig. 8.2.

**VCC (pin-28, type-input):** SUPPLY: A 5V Supply.

**GND (pin-14, type-input):** GROUND

$\overline{CS}$ **(pin-1, type-input):** CHIP SELECT: A LOW on this pin enables $\overline{RD}$ and $\overline{WR}$ communication between the CPU and the 8259A. INTA functions are independent of $\overline{CS}$.

$\overline{WR}$ **(pin-2, type-input):** WRITE: A LOW on this pin when CS is low enables the 8259A to accept command words from the CPU.

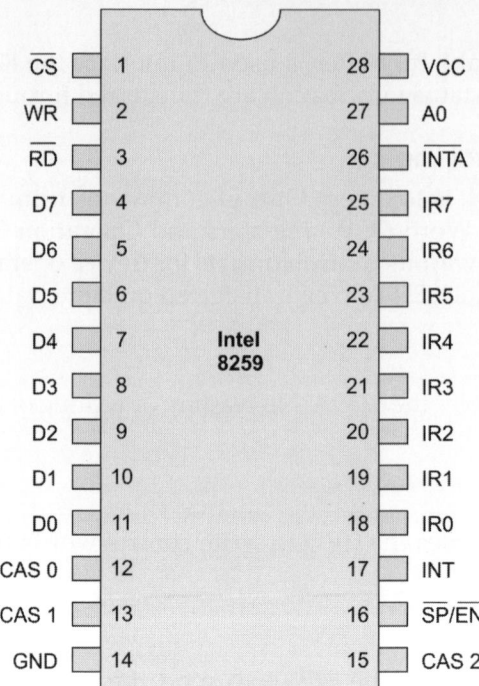

**Fig. 8.2:** Pin diagram of 8259A

$\overline{\text{RD}}$ **(pin-3, type-input):** READ: A LOW on this pin when CS is low enables the 8259A to release status onto the data bus for the CPU.

**D7-D0 (pin-4-11, type-I/O):** BIDIRECTIONAL DATA BUS: Control, status and interrupt-vector information is transferred via this bus.

**CAS0-CAS2 (pin-12, 13, 15, type-I/O):** CASCADE LINES: The CAS lines form a private 8259A bus to control a multiple 8259A structure. These pins are outputs for a master 8259A and inputs for a slave 8259A.

$\overline{\text{SP}}/\overline{\text{EN}}$ **(pin-16, type-I/O):** SLAVE PROGRAM/ENABLE BUFFER: This is a dual function pin. When in the Buffered Mode, it can be used as an output to control buffer transceivers (EN). When not in the buffered mode, it is used as an input to designate a master (SP = 1) or slave (SP = 0).

**INT (pin-17, type-output):** INTERRUPT: This pin goes high whenever a valid interrupt request is asserted. It is used to interrupt the CPU, thus it is connected to the CPU's interrupt pin.

**IR0-IR7 (pin-18-25, type-input):** INTERRUPT REQUESTS: Asynchronous inputs. An interrupt request is executed by raising an IR input (low to high), and holding it high until it is acknowledged (Edge Triggered Mode) or just by a high level on an IR input (Level Triggered Mode).

$\overline{\text{INTA}}$ **(pin-26, type-input):** INTERRUPT ACKNOWLEDGE: This pin is used to enable 8259A interrupt-vector data onto the data bus by a sequence of interrupt acknowledge pulses issued by the CPU.

**A₀ (pin-27, type-input):** ADDRESS LINE: This pin acts in conjunction with the $\overline{CS}$, $\overline{WR}$ and $\overline{RD}$ pins. It is used by the 8259A to decipher various Command Words the CPU writes and status the CPU wishes to read. It is typically connected to the CPU A0 address line (A1 for 8086, 8088).

## 8.5 INTERRUPT SEQUENCE (Fig. 8.3)

1. One or more of the INTERRUPT REQUEST lines (IR0 - IR7) are raised high, setting the corresponding IRR bit(s).

2. The 82C59A evaluates those requests in the priority resolver and sends an interrupt (INT) to the CPU, if appropriate.

3. The CPU acknowledges the lNT and responds with an INTA pulse.

4. The 82C59A does not drive the data bus during the first INTA pulse.

5. The 80C86/88/286 CPU will initiate a second INTA pulse. During this INTA pulse, the appropriate ISR bit is set and the corresponding bit in the IRR is reset. The 82C59A outputs the 8-bit pointer onto the data bus to be read by the CPU.

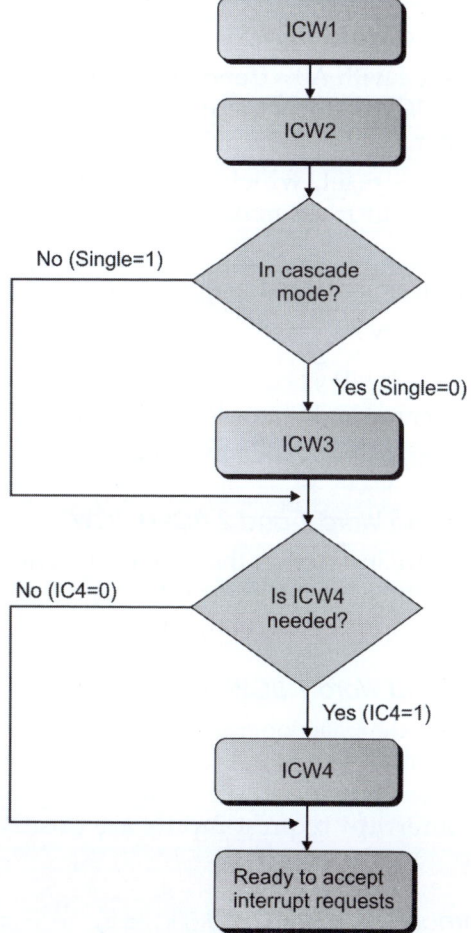

**Fig. 8.3:** Flow chart of interrupt sequence

6. This completes the interrupt cycle. In the AEOI mode, the ISR bit is reset at the end of the second INTA pulse. Otherwise, the ISR bit remains set until an appropriate EOI command is issued at the end of the interrupt subroutine.

## 8.6 PROGRAMMING OF 8259

The 8259A accepts two types of command words generated by the CPU:

1. **Initialization Command Words (ICWs):** Before normal operation can begin, each 8259A in the system must be brought to a starting point by a sequence of 2 to 4 bytes timed by WR pulses.

2. **Operation Command Words (OCWs):** These are the command words which command the 8259A to operate in various interrupt modes. These modes are:

   a. Fully nested mode
   b. Rotating priority mode
   c. Special mask mode
   d. Polled mode

The OCWs can be written into the 8259A anytime after initialization.

### 8.6.1 Initialization Command Word (ICWS)

Whenever a command is issued with A0 = 0 and D4 = 1, this is interpreted as Initialization Command Word 1 (ICW1). ICW1 starts the initialization sequence during which the following automatically occur.

a. The edge sense circuit is reset, which means that following initialization, an interrupt request (IR) input must make a low-to-high transition to generate an interrupt.

b. The Interrupt Mask Register is cleared.

c. IR7 input is assigned priority 7.

d. The slave mode address is set to 7.

e. Special Mask Mode is cleared and Status Read is set to IRR.

f. If IC4 e 0, then all functions selected in ICW4 are set to zero.

#### 8.6.1.1 Initialization Command Word 1 and 2 (ICW1, ICW2)

In an 8086 system, A15-A11 are inserted in the five most significant bits of the vectoring byte and the 8259A sets the three least significant bits according to the interrupt level. A10-A5 are ignored and ADI (Address Interval) has no effect.

#### 8.6.1.2 Initialization Command Word 1 (ICW1)

This ICW is used to program the basic operation of the 8259A to program this ICW for 8086 place logic I in bit IC4. This selects single or cascade operation by programming the SNGL bit. If cascade operation is selected, ICW3 must be programmed. The LTIM bit determines whether the interrupt request inputs are positive edge trigged or level triggered (Fig. 8.4).

#### 8.6.1.3 Initialization Command Word 2 (ICW2) (Fig. 8.5)

This ICW is used to select the vector number used with the interrupt request inputs.

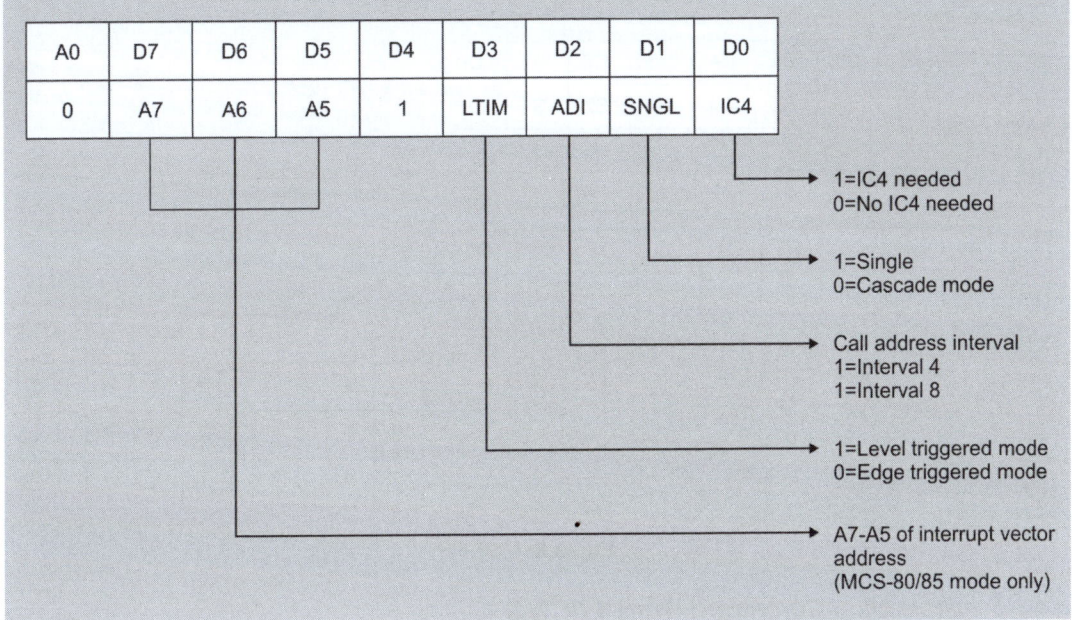

**Fig. 8.4:** ICW1

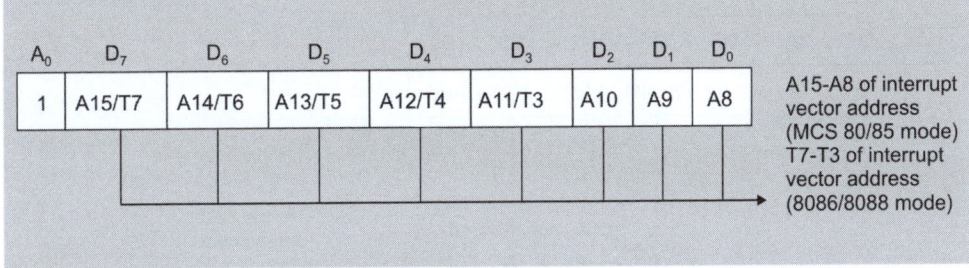

**Fig. 8.5:** ICW2

### 8.6.1.4 Initialization Command Word 3 (ICW3)

This command word is only used when ICW1 indicates that the system is operated in cascaded mode. This ICW indicates where the slave is connected to the master (Fig. 8.6).

The functions of this register are:

a. In the master mode (either when SP = 1 or in buffered mode when M/S = 1 in ICW4), a "1" is set for each slave in the system. The master will then release byte 1 of the call sequence (for MCS-80/85 system) and enable the corresponding slave to release bytes 2 and 3 (byte 2 for 8086 only) through the cascade lines.

b. In the slave mode (either when SP = 0 or if BUF = 1 and M/S = 0 in ICW4), bits 2–0 identify the slave. The slave compares its cascade input with these bits and if they are equal, bytes 2 and 3 of the call sequence (or just byte 2 for 8086) are released by it on the Data Bus.

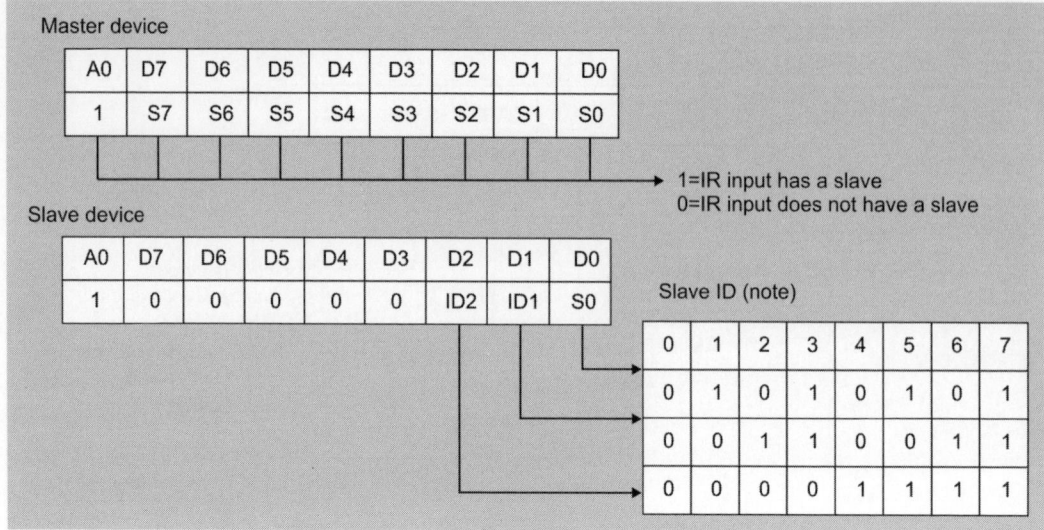

**Fig. 8.6:** ICW3

### 8.6.1.5 Initialization Command Word 4 (ICW4)

The format of ICW4 is given in Fig. 8.7.

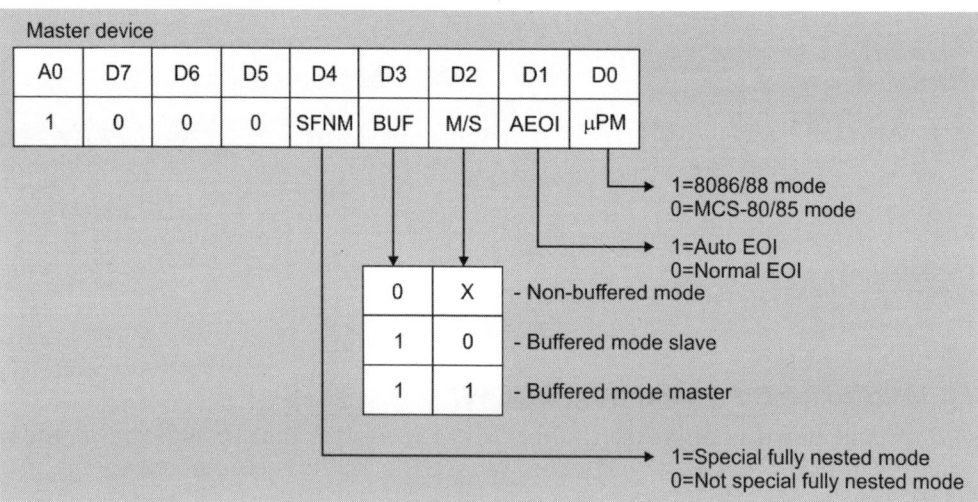

**Fig. 8.7:** ICW4

## 8.6.2 Operational Control Words (OCWS)

After the Initialization Command Words (ICWs) are programmed into the 8259A, the chip is ready to accept interrupt requests at its input lines. However, during the 8259A operation, a selection of algorithms can command the 8259A to operate in various modes through the Operation Command Words (OCWs).

### 8.6.2.1 Operational Control Word 1 (OCW1)

OCW1 sets and clears the mask bits in the Interrupt Mask Register (IMR). M7-M0 represents the eight mask bits. M = 1 indicates the channel is masked (inhibited), M = 0 indicates the channel is enabled (Fig. 8.8a).

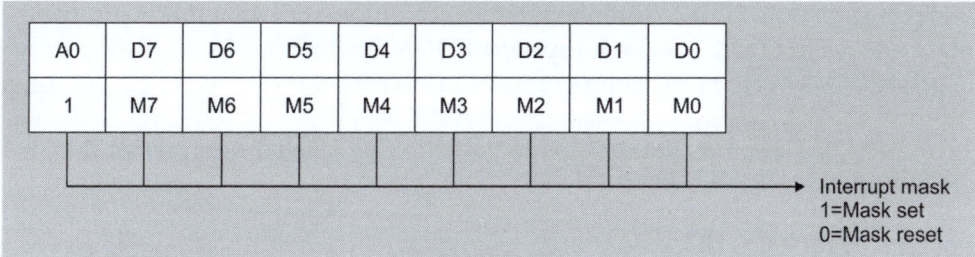

**Fig. 8.8a:** OCW1

### 8.6.2.2 Operational Control Word 2 (OCW2)

OCW2 is programmed only when the AEOI mode is not selected for the 8259A. In this case, this OCW selects the way that the 8259A respond to an interrupt. These modes are:

**Non-specific end of interrupt:** It is a command sent by the interrupt service procedure to signal the end of the interrupt. Interrupt controller automatically determines which interrupt level was active and reset the correct bit of the interrupt status register.

**End of specific interrupt:** It is a command that allows a specific interrupt request to be reset. The exact position is determined with bits L2–10 of the OCW2.

**Rotate on non-specific EOI:** A command that function exactly like the non-specific end of interrupt except that it rotates interrupt priorities after resetting the interrupt status register bit.

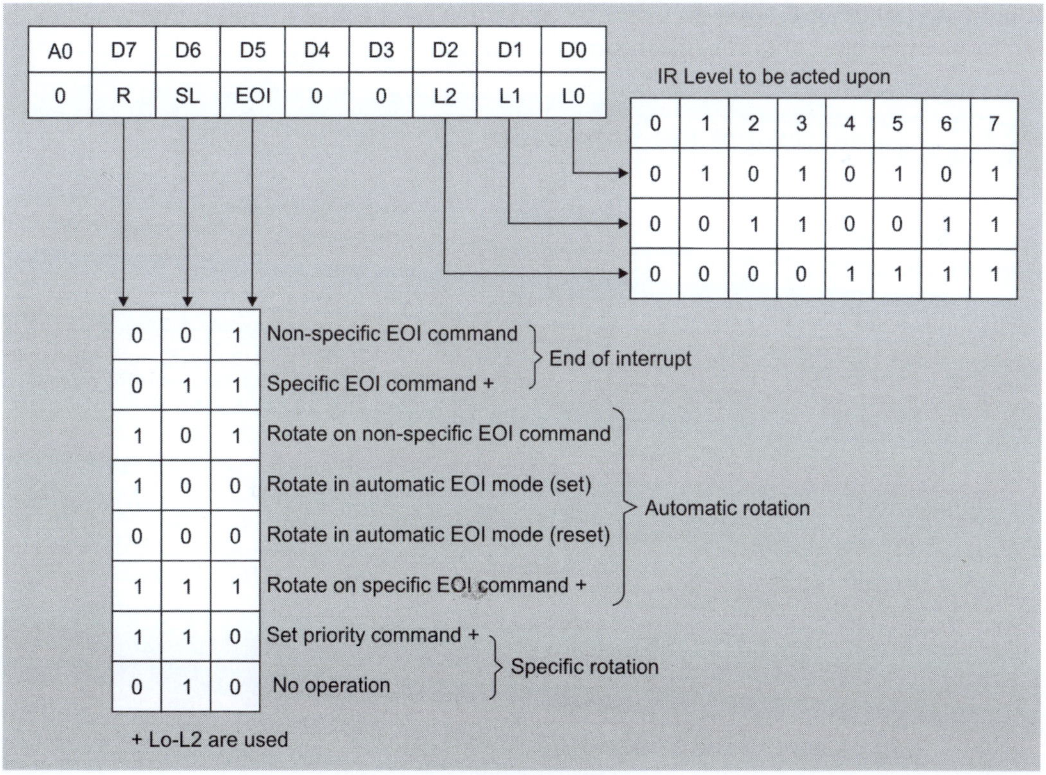

| A0 | D7 | D6 | D5 | D4 | D3 | D2 | D1 | D0 |
|----|----|----|----|----|----|----|----|----|
| 0 | R | SL | EOI | 0 | 0 | L2 | L1 | L0 |

IR Level to be acted upon

| 0 | 1 | 2 | 3 | 4 | 5 | 6 | 7 |
|---|---|---|---|---|---|---|---|
| 0 | 1 | 0 | 1 | 0 | 1 | 0 | 1 |
| 0 | 0 | 1 | 1 | 0 | 0 | 1 | 1 |
| 0 | 0 | 0 | 0 | 1 | 1 | 1 | 1 |

| R | SL | EOI | | |
|---|----|----|---|---|
| 0 | 0 | 1 | Non-specific EOI command | End of interrupt |
| 0 | 1 | 1 | Specific EOI command + | |
| 1 | 0 | 1 | Rotate on non-specific EOI command | |
| 1 | 0 | 0 | Rotate in automatic EOI mode (set) | Automatic rotation |
| 0 | 0 | 0 | Rotate in automatic EOI mode (reset) | |
| 1 | 1 | 1 | Rotate on specific EOI command + | |
| 1 | 1 | 0 | Set priority command + | Specific rotation |
| 0 | 1 | 0 | No operation | |

+ Lo-L2 are used

**Fig. 8.8b:** OCW2

**Rotate on automatic EOI:** A command that selects automatic EOI with rotating priorities.

**Rotate on specific EOI:** Function a specific EOI except that it selects rotating priorities.

**Set priority:** Allows the programmer to set the lowest priority interrupt input using the L2-L1 bits of OCW2.

**L2, L1 and L0:** These bits determine the interrupt level acted upon when the SL bit is active (Table 8.1).

| Table 8.1: L2, L1 and L0 bits | | | |
|---|---|---|---|
| L2 | L1 | L0 | IR level |
| 0 | 0 | 0 | 0 |
| 0 | 0 | 1 | 1 |
| 0 | 1 | 0 | 2 |
| 0 | 1 | 1 | 3 |
| 1 | 0 | 0 | 4 |
| 1 | 0 | 1 | 5 |
| 1 | 1 | 0 | 6 |
| 1 | 1 | 1 | 7 |

### 8.6.2.3 Operational Control Word 3 (OCW3)

This command word is used to select the register to be read, the operation of special mask register and the poll command. If polling is selected, the P bit must be set and then output to the 8259A. The next read operation will read the poll word. The rightmost three bits of the poll word indicate the active interrupt request with the highest priority. The leftmost bit indicates whether there is an interrupt and must be checked to determine whether the rightmost three bits contain valid information (Fig. 8.9).

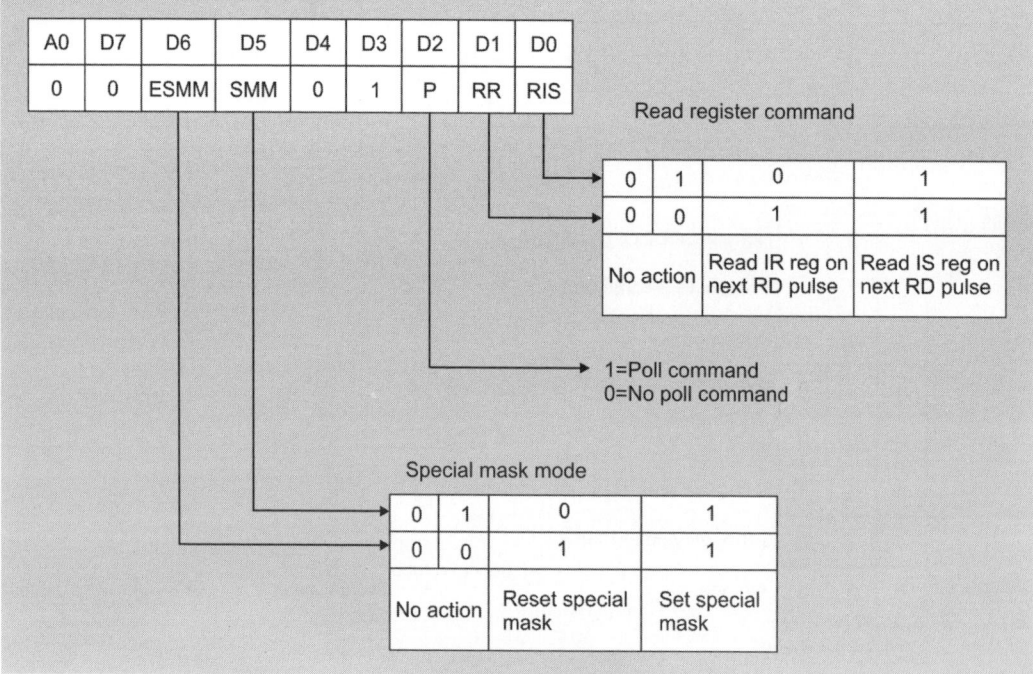

**Fig. 8.9:** OCW3

**ESMM (Enable special mask mode):** When this bit is set to 1, it enables the SMM bit to set or reset the Special Mask Mode. When ESMM = 0, the SMM bit becomes a "don't care".

**SMM (Special mask mode):** If ESMM = 1 and SMM = 1, the 8259A will enter Special Mask Mode. If ESMM = 1 and SMM = 0, the 8259A will revert to normal mask mode. When ESMM = 0, SMM has no effect.

## 8.7 OPERATING MODES

The different modes of operation of 8259A can be programmed by setting or resting the appropriate bits of the ICW or OCW as discussed previously. The different modes of operation of 8259A are explained as follows:

### 8.7.1 Full Nested Mode

This is the default mode of operation of 8259A.This mode is entered after initialization unless another mode is programmed. The interrupt requests are ordered in priority from 0 through 7 (0 highest). When an interrupt is acknowledged, the highest priority request is determined and its vector placed on the bus.

Additionally, a bit of the Interrupt Service register (ISO-7) is set. This bit remains set until the microprocessor issues an End of Interrupt (EOI) command immediately before returning from the service routine or if AEOI (Automatic End of Interrupt) bit is set, until the trailing edge of the last INTA. While the IS bit is set, all further interrupts of the same or lower priority are inhibited, while higher levels will generate an interrupt (which will be acknowledged only if the microprocessor internal Interrupt enable flip-flop has been re-enabled through software). After the initialization sequence, IR0 has the highest priority and IR7 the lowest. Priorities can be changed, as will be explained, in the rotating priority mode.

### 8.7.2 End of Interrupt (EOI) Mode

The ISR bit can be reset either with AEOI bit of ICW1 or by EOI command, issued before returning from the interrupt service routine. There are two types of EOI commands: specific and non-specific. When 8259A is operated in the modes that preserve fully nested structure, it can determine which ISR bit is to be reset on EOI.

When non-specific EOI command is issued to 8259A, it will be automatically reset the highest ISR bit out of those already set. When a mode that may disturb the fully nested structure is used, the 8259A is no longer able to determine the last level acknowledged. In this case, a specific EOI command is issued to reset a particular ISR bit. An ISR bit that is masked by the corresponding IMR bit will not be cleared by non-specific EOI of 8259A, if it is in special mask mode.

### 8.7.3 Automatic Rotation Mode

This is used in the applications where all the interrupting devices are of equal priority. In this mode, an interrupt request IR level receives priority after it is served while the next device to be served gets the highest priority in sequence. Once all the device is served like this, the first device again receives the highest priority.

### 8.7.4 Automatic EOI Mode

Till AEOI = 1 in ICW4, the 8259A operates in AEOI mode. In this mode, the 8259A performs a non-specific EOI operation at the trailing edge of the last INTA pulse automatically. This mode should be used only when a nested multi-level interrupt structure is not required with a single 8259A.

### 8.7.5 Specific Rotation Mode

In this mode, a bottom priority level can be selected, using L2, L1 and L0 in OCW2 and R = 1, SL = 1, EOI = 0. The selected bottom priority fixes other priorities. If IR5 is selected as a bottom priority, then IR5 will have least priority and IR4 will have a next higher priority. Thus, IR6 will have the highest priority. These priorities can be changed during an EOI command by programming the rotate on specific EOI command in OCW2.

### 8.7.6 Specific Mask Mode

In specific mask mode, when a mask bit is set in OCW1, it inhibits further interrupts at that level and enables interrupt from other levels, which are not masked.

### 8.7.7 Edge and Level Triggered Mode

This mode decides whether the interrupt should be edge triggered or level triggered. If bit LTIM of ICW1 = 0, they are edge triggered, otherwise the interrupts are level triggered.

### 8.7.8 Reading 8259 Status

The status of the internal registers of 8259A can be read using this mode. The OCW3 is used to read IRR and ISR while OCW1 is used to read IMR. Reading is possible only in no polled mode.

### 8.7.9 Poll Command

In polled mode of operation, the INT output of 8259A is neglected, though it functions normally, by not connecting INT output or by masking INT input of the microprocessor. The poll mode is entered by setting P = 1 in OCW3. The 8259A is polled by using software execution by microprocessor instead of the requests on INT input. The 8259A treats the next RD pulse to the 8259A as an interrupt acknowledge. An appropriate ISR bit is set, if there is a request. The priority level is read and a data word is placed on to data bus, after RD is activated. A poll command may give more than 64 priority levels.

### 8.7.10 Special Fully Nested Mode

This mode is used in more complicated system, where cascading is used and the priority has to be programmed in the master using ICW4. This is somewhat similar to the normal nested mode. In this mode, when an interrupt request from a certain slave is in service, this slave can further send request to the master, if the requesting device connected to the slave has higher priority than the one being currently served. In this mode, the master, interrupts the CPU only when the interrupting device has a higher or the same priority than the one current being served. In normal mode, other requests than the one being served are masked out.

When entering the interrupt service routine, the software has to check whether this is the only request from the slave. This is done by sending a non-specific EOI to the master, otherwise no EOI should be sent. This mode is important, since in the absence of this mode, the slave would interrupt the master only once and hence the priorities of the slave inputs would have been disturbed.

### 8.7.11 Cascade Mode

The 8259A can be easily interconnected in a system of one master with up to eight slaves to handle up to 64 priority levels. The master controls the slaves through the 3 line cascade bus. The cascade bus acts like chip selects to the slaves during the INTA sequence. In a cascade configuration, the slave interrupt outputs are connected to the master interrupt request inputs. When a slave request line is activated and afterwards acknowledged, the master will enable the corresponding slave to release the device routine address during bytes 2 and 3 of INTA.

The cascade bus lines are normally low and will contain the slave address code from the trailing edge of the first INTA pulse to the trailing edge of the third pulse. Each 8259A in the system must follow a separate initialization sequence and can be programmed to work in a different mode. An EOI command must be issued twice: once for the master and once for the corresponding slave. An address decoder is required to activate the Chip Select (CS) input of each 8259A. The cascade lines of the Master 8259A are activated only for slave inputs, non-slave inputs leave the cascade line inactive (low) (Fig. 8.10).

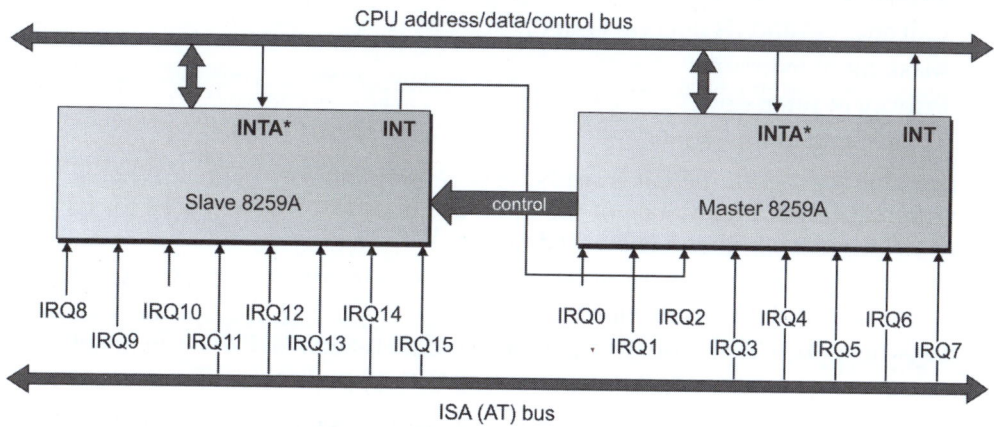

**Fig. 8.10:** Cascade mode of 8259A

### 8.7.12 Buffered Mode

When the 8259A is used in a large system where bus driving buffers are required on the data bus and the cascading mode is used, there exists the problem of enabling buffers.

The buffered mode will structure the 8259A to send an enable signal on SP/EN to enable the buffers. In this mode, whenever the 8259A's data bus outputs are enabled, the $\overline{SP/EN}$ output becomes active. This modification forces the use of software programming to determine whether the 8259A is a master or a slave. Bit 3 in ICW4 programs the buffered mode and bit 2 in ICW4 determines whether it is a master or a slave.

## 8.8 INTERFACING OF 8259 TO MICROPROCESSOR (Fig. 8.11)

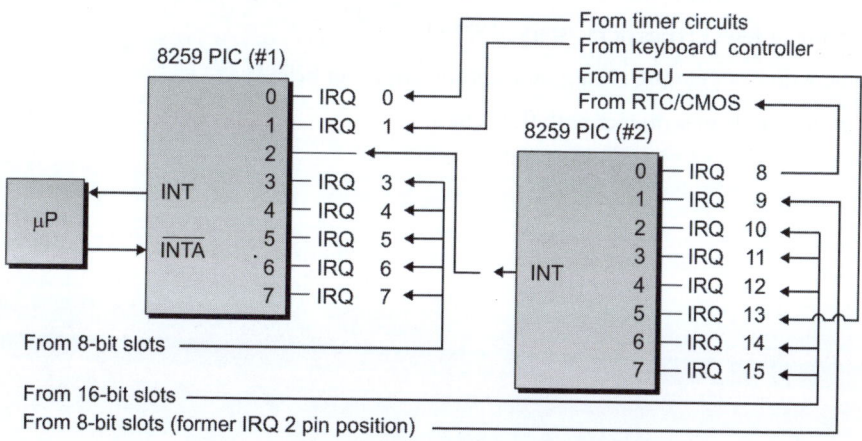

**Fig. 8.11:** Interfacing of 8259 to microprocessor

## 8.9 WORKING OF 8259 WITH MICROPROCESSOR

Step-by-step working of 8259 with microprocessor is given below

1. First the 8259 should be programmed by sending Initialization Command Word (ICW) and Operational Command Word (OCW). These command words will inform 8259 about the following:

   a. Type of interrupt signal (Level triggered/Edge triggered)
   b. Type of processor (8085/8086)
   c. Call address and its interval (4 or 8)
   d. Masking of interrupts
   e. Priority of interrupts
   f. Type of end of interrupts

2. Once 8259 is programmed, it is ready for accepting interrupt signal. When it receives an interrupt through any one of the interrupt lines IR0-IR7, it checks for its priority and also checks whether it is masked or not.

3. If the previous interrupt is completed and if the current request has highest priority and unmasked, then it is serviced.

4. For servicing this interrupt, the 8259 will send INT signal to INTR pin of 8085.

5. In response, it expects an acknowledge INTA (low) from the processor.

6. When the processor accepts the interrupt, it sends three INTA (low) one-by-one.

7. In response to first, second and third INTA (low) signals, the 8259 will supply CALL opcode, low byte of call address and high byte of call address, respectively. Once the processor receives the call opcode and its address, it saves the content of program counter (PC) in stack and loads the CALL address in PC and starts executing the interrupt service routine stored in this call address.

### PROBLEMS

1. What are the common features of 8259A interrupt controller?

2. Draw and explain pin diagram of 8259A programmable interrupt controller.

3. Explain the internal structure of 8259A with neat and clean diagram.

4. What do you mean by command word? Explain the different types of command words of 8259A.

5. Explain the different types of registers of 8259A.

6. Describe various operating modes of interrupt controller 8259A.

7. Explain the interrupt sequence with flow chart.

# 9

# DMA 8237A

## 9.1 INTRODUCTION

Direct Memory Access (DMA) is one of several methods for coordinating the timing of data transfers between an input/output (I/O) device and the core processing unit or memory in a computer. DMA is one of the faster types of synchronization mechanisms, generally providing significant improvement over interrupts, in terms of both latency and throughput. An I/O device often operates at a much slower speed than the core. 1 DMA allows the I/O device to access the memory directly, without using the core. DMA can lead to a significant improvement in performance because data movement is one of the most common operations performed in processing applications. There are several advantages of using DMA which are as follows:

- DMA saves core MIPS because the core can operate in parallel.
- DMA saves power because it requires less circuitry than the core to move data.
- DMA saves pointers because core AGU pointer registers are not needed.
- DMA has no modulo block size restrictions, unlike the core AGU.

## 9.2 BASIC DMA OPERATION

- The direct memory access (DMA) I/O technique provides direct access to the memory while the microprocessor is temporarily disabled.
- A DMA controller temporarily borrows the address bus, data bus and control bus from the microprocessor and transfers the data bytes directly between an I/O port and a series of memory locations.
- The DMA transfer is also used to do high-speed memory-to-memory transfers.
- Two control signals are used to request and acknowledge a DMA transfer in the microprocessor-based system.
- The HOLD signal is a bus request signal which asks the microprocessor to release control of the buses after the current bus cycle.
- The HLDA signal is a bus grant signal which indicates that the microprocessor has indeed released control of its buses by placing the buses at their high-impedance states.
- The HOLD input has a higher priority than the INTR or NMI interrupt inputs.

## 9.3 DMA 8237A

The 8237A Multimode Direct Memory Access (DMA) Controller is a peripheral interface circuit for microprocessor systems. It is designed to improve system performance by allowing external devices to directly transfer information from the system memory.

Memory-to-memory transfer capability is also provided. The 8237A offers a wide variety of programmable control features to enhance data throughput and system optimization and to allow dynamic reconfiguration under program control. The 8237A is designed to be used in conjunction with an external 8-bit address latch. It contains four independent channels and may be expanded to any number of channels by cascading additional controller chips. The three basic transfer modes allow programmability of the types of DMA service by the user. Each channel can be individually programmed to auto initialize to its original condition following an End of Process (EOP). Each channel has a full 64 K address and word count capability.

## 9.4 ARCHITECTURE OF DMA 8237A

The 8237A block diagram shown in Fig. 9.1 includes the major logic blocks and all of the internal registers. The data interconnection paths are also shown. Not shown are the various control signals between the blocks. The 8237A contains 344 bits of internal memory in the form of registers. Table 9.1 lists these registers by name and shows the size of each. A detailed description of the registers and their functions can be found under Register Description.

The 8237A contains three basic blocks of control logic. The Timing Control block generates internal timing and external control signals for the 8237A.

The Program Command Control block decodes the various commands given to the 8237A by the microprocessor prior to servicing a DMA Request. It also decodes the Mode Control word used to select the type of DMA during the servicing. The Priority Encoder block resolves priority contention between DMA channels requesting service simultaneously.

The Timing Control block derives internal timing from the clock input. In 8237A systems, this input will usually be the 2 TTL clock from an 8224 or CLK from an 8085AH or 8284A. 33% duty cycle clock generators, however, may not meet the clock high time requirement of the 8237A of the same frequency. For example, 82C84A-5 CLK output violates the clock high time requirement of 8237A-5. In this case, 82C84A CLK can simply be inverted to meet 8237A-5 clock high and low time requirements. For 8085AH-2 systems above 3.9 MHz, the 8085 CLK (OUT) does not satisfy 8237A-5 clock LOW and HIGH time requirements. In this case, an external clock should be used to drive the 8237A-5.

**Table 9.1:** Registers and their size

| Name of register | Bit size | No. of registers |
|---|---|---|
| Base address register | 16 | 4 |
| Base word count register | 16 | 4 |
| Current address register | 16 | 4 |
| Current word count register | 16 | 4 |
| Temporary address register | 16 | 1 |
| Temporary word count register | 16 | 1 |
| Status register | 8 | 1 |
| Command register | 8 | 1 |
| Temporary register | 8 | 1 |
| Mode register | 6 | 4 |
| Mask register | 4 | 1 |
| Request register | 4 | 1 |

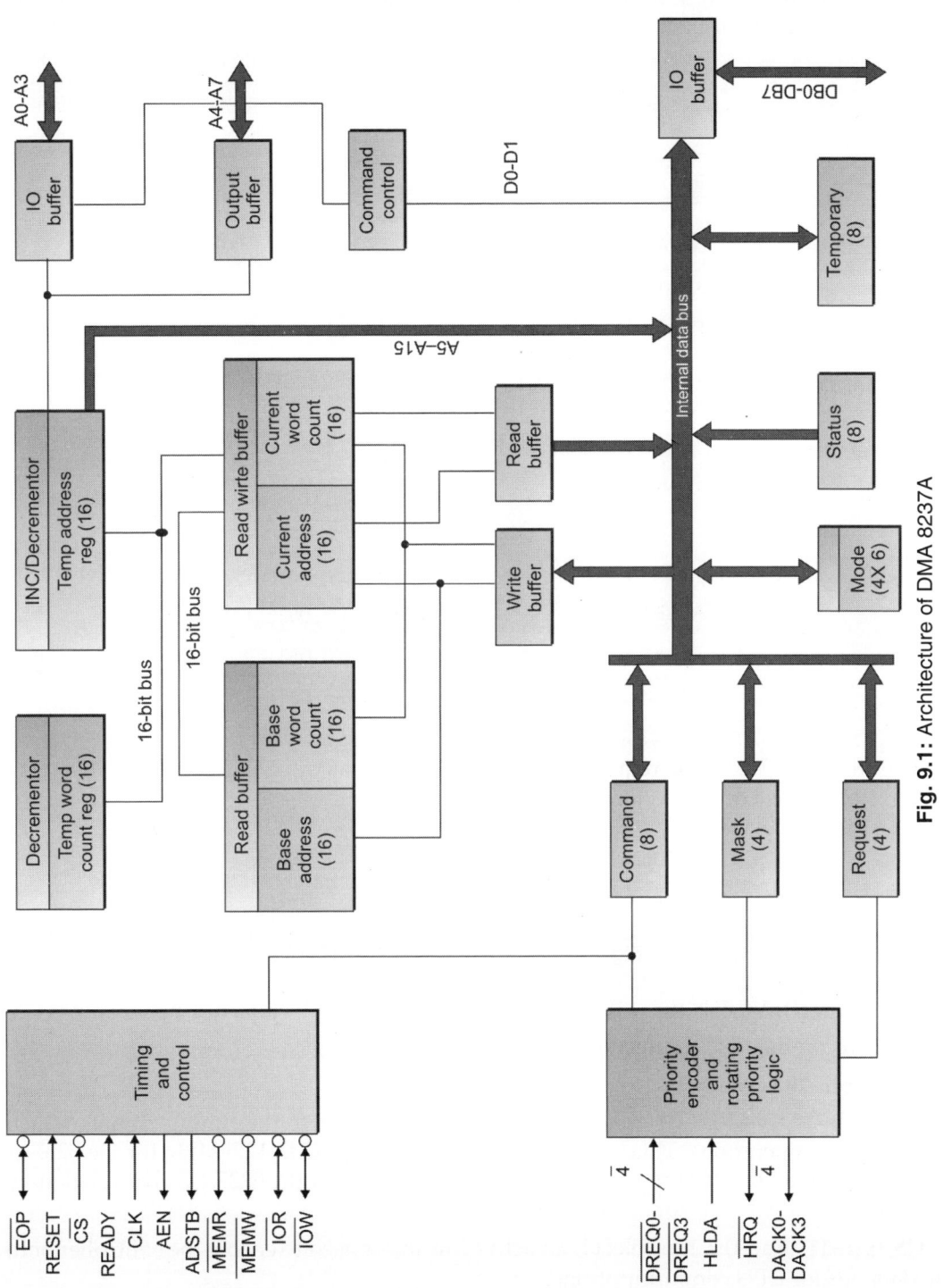

**Fig. 9.1:** Architecture of DMA 8237A

## 9.5 PIN CONFIGURATION OF 8237A

The 8237A is available in 40-pin DIP package as shown in Fig. 9.2. The functions of different pins are described as follows:

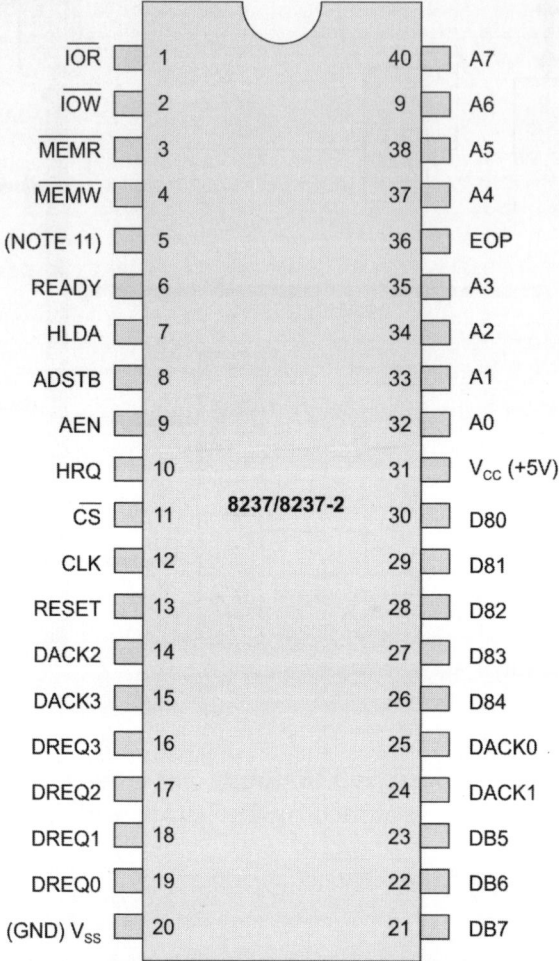

**Fig. 9.2:** Pin diagram of DMA 8237

**VCC (pin-31):** VCC is the +5V power supply pin. A 0.1µF capacitor between pins 20 and 31 is recommended for decoupling.

**GND (pin-20):** Ground.

**CLK (pin-12, type-I):** The Clock Input is used to generate the timing signals which control 8237A operations. This input may be driven from DC to 12.5 MHz for the 8237A-12, from DC to 8 MHz for the 8237A or from DC to 5 MHz for the 8237A-5. The Clock may be stopped in either state for standby operation.

**$\overline{CS}$ (pin-11, type-I):** Chip Select is an active low input used to enable the controller onto the data bus for CPU communications.

**RESET (pin-13, type-I):** This is an active high input which clears the Command, Status, Request and Temporary registers, the First/Last Flip-Flop, and the mode register counter.

The Mask register is set to ignore requests. Following a Reset, the controller is in an idle cycle.

**READY (pin-6, type-I):** This signal can be used to extend the memory read and write pulses from the 8237A to accommodate slow memories or I/O devices. READY must not make transitions during its specified set-up and hold times.

**HLDA (pin-7, type-I):** The active high Hold Acknowledge from the CPU indicates that it has relinquished control of the system buses. HLDA is a synchronous input and must not transit during its specified set-up time. There is an implied hold time (HLDA inactive) of TCH from the rising edge of CLK, during which time HLDA must not transit.

**DREQ0-DREQ3 (pin-16-19, type-I):** The DMA Request (DREQ) lines are individual asynchronous channel request inputs used by peripheral circuits to obtain DMA service. In Fixed Priority, DREQ0 has the highest priority and DREQ3 has the lowest priority. A request is generated by activating the DREQ line of a channel. DACK will acknowledge the recognition of a DREQ signal. Polarity of DREQ is programmable. RESET initializes these lines to active high. DREQ must be maintained until the corresponding DACK goes active. DREQ will not be recognized while the clock is stopped. Unused DREQ inputs should be pulled High or Low (inactive) and the corresponding mask bit set.

**DB0-DB7 (pin-21-23, 26-30, type-I/O):** The Data Bus lines are bi-directional three-state signals connected to the system data bus. The outputs are enabled in the Program condition during the I/O Read to output the contents of a register to the CPU. The outputs are disabled and the inputs are read during an I/O Write cycle when the CPU is programming the 8237A control registers. During DMA cycles, the most significant 8-bits of the address are output onto the data bus to be strobed into an external latch by ADSTB. In memory-to-memory operations, data from the memory enters the 8237A on the data bus during the read-from-memory transfer, and then during the write-to-memory transfer, the data bus outputs write the data into the new memory location.

**$\overline{\text{IOR}}$ (pin-1, type-I/O):** I/O Read is a bi-directional active low three-state line. In the Idle cycle, it is an input control signal used by the CPU to read the control registers. In the Active cycle, it is an output control signal used by the 8237A to access data from the peripheral during a DMA Write transfer.

**$\overline{\text{IOW}}$ (pin-2, type-I/O):** I/O Write is a bi-directional active low three-state line. In the Idle cycle, it is an input control signal used by the CPU to load information into the 8237A. In the Active cycle, it is an output control signal used by the 8237A to load data to the peripheral during a DMA Read transfer.

**EOP (pin-36, type-I/O):** End of Process (EOP) is an active low bi-directional signal. Information concerning the completion of DMA services is available at the bi-directional EOP pin. The 8237A allows an external signal to terminate an active DMA service by pulling the EOP pin low. A pulse is generated by the 8237A when terminal count (TC) for any channel is reached, except for channel 0 in memory-to-memory mode. During memory-to-memory transfers, EOP will be output when the TC for channel 1 occurs.

The EOP pin is driven by an open drain transistor on-chip and requires an external pull-up resistor to VCC. When an EOP pulse occurs, whether internally or externally generated, the 8237A will terminate the service, and if auto initialize is enabled, the base registers will be written to the current registers of that channel. The mask bit and TC bit in the status word will be set for the currently active channel by EOP unless the channel is programmed for auto initialize. In that case, the mask bit remains clear.

**A0-A3 (pin-32-35, type-I/O):** The four least significant address lines are bi-directional three-state signals. In the Idle cycle, they are inputs and are used by the 8237A to address the control register to be loaded or read. In the Active cycle, they are outputs and provide the lower 4-bits of the output address.

**A4-A7 (pin-37-40, type-O):** The four most significant address lines are three-state outputs and provide 4-bits of address. These lines are enabled only during the DMA service.

**HRQ (pin-10, type-O):** The Hold Request (HRQ) output is used to request control of the system bus. When a DREQ occurs and the corresponding mask bit is clear or a software DMA request is made, the 8237A issues HRQ. The HLDA signal then informs the controller when access to the system buses is permitted. For standalone operation where the 8237A always controls the buses, HRQ may be tied to HLDA. This will result in one S0 state before the transfer.

**DACK0-DACK3 (pin-14, 15, 24, 25, type-O):** DMA acknowledge is used to notify the individual peripherals when one has been granted a DMA cycle. The sense of these lines is programmable. RESET initializes them to active low.

**AEN (pin-9, type-O):** Address Enable enables the 8-bit latch containing the upper 8 address bits onto the system address bus. AEN can also be used to disable other system bus drivers during DMA transfers. AEN is active high.

**ADSTB (pin-8, type-O):** This is an active high signal used to control latching of the upper address byte. It will drive directly the strobe input of external transparent octal latches, such as the 82C82. During block operations, ADSTB will only be issued when the upper address byte must be updated, thus speeding operation through elimination of S1 states. ADSTB timing is referenced to the falling edge of the 8237A clock.

**$\overline{\text{MEMR}}$ (pin-3, type-O):** The Memory Read signal is an active low three-state output used to access data from the selected memory location during a DMA Read or a memory-to-memory transfer.

**$\overline{\text{MEMW}}$ (Pin-4, type-O):** The Memory Write signal is an active low three-state output used to write data to the selected memory location during a DMA write or a memory-to-memory transfer.

**NC (pin-5):** Pin 5 is open and should not be tested for continuity.

## 9.6 INTERNAL REGISTER

### 9.6.1 Current Address Register

Each channel has a 16-bit Current Address register. This register holds the value of the address used during DMA transfers. The address is automatically incremented or decremented after each transfer and the intermediate values of the address are stored in the Current Address register during the transfer. This register is written or read by the microprocessor in successive 8-bit bytes. It may also be reinitialized by an auto initialize back to its original value. Auto initialize takes place only after $\overline{\text{EOP}}$.

### 9.6.2 Current Word Register

Each channel has a 16-bit Current Word Count register. This register determines the number of transfers to be performed. The actual number of transfers will be one more than the number programmed in the Current Word Count register (i.e. programming a count of 100 will result in 101 transfers). The word count is decremented after each transfer. The intermediate value of the word count is stored in the register during the transfer. When the value in the register goes from zero to FFFFH, a TC will be generated. This register is loaded or read in successive 8-bit bytes by the microprocessor in the Program Condition. Following the end of a DMA service, it may also be reinitialized by an auto initialization back to its original value. Auto initialize can occur only when occurs. If it is not auto initialized, this register will have a count of FFFFH after TC.

### 9.6.3 Base Address and Base Word Count Registers

Each channel has a pair of Base Address and Base Word Count registers. These 16-bit registers store the original value of their associated current registers. During auto initialization, these values are used to restore the current registers to their original values. The base registers are written simultaneously with their corresponding current register in 8-bit bytes in the Program Condition by the microprocessor. These registers cannot be read by the microprocessor.

### 9.6.4 Command Register

This 8-bit register controls the operation of the 8237A. It is programmed by the microprocessor in the Program Condition and is cleared by Reset or a Master Clear instruction. See Fig. 9.3 for address coding and function of command bits.

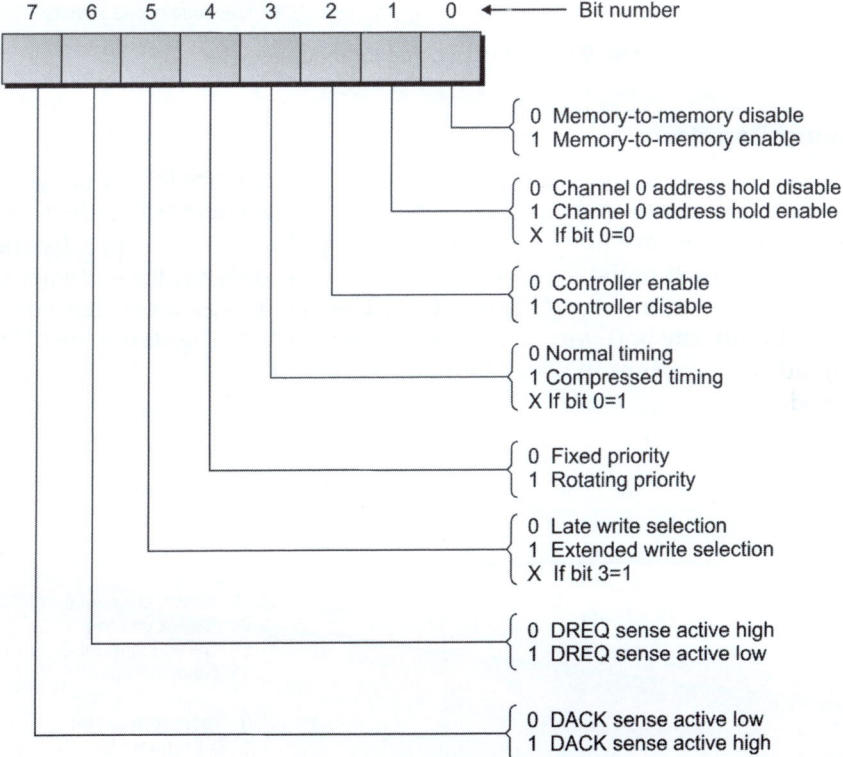

**Fig. 9.3:** Address coding for command register

### 9.6.5 Mode Register

Each channel has a 6-bit Mode register associated with it. When the register is being written to the microprocessor in the Program Condition, bits 0 and 1 determine which channel Mode register is to be written (Fig. 9.4).

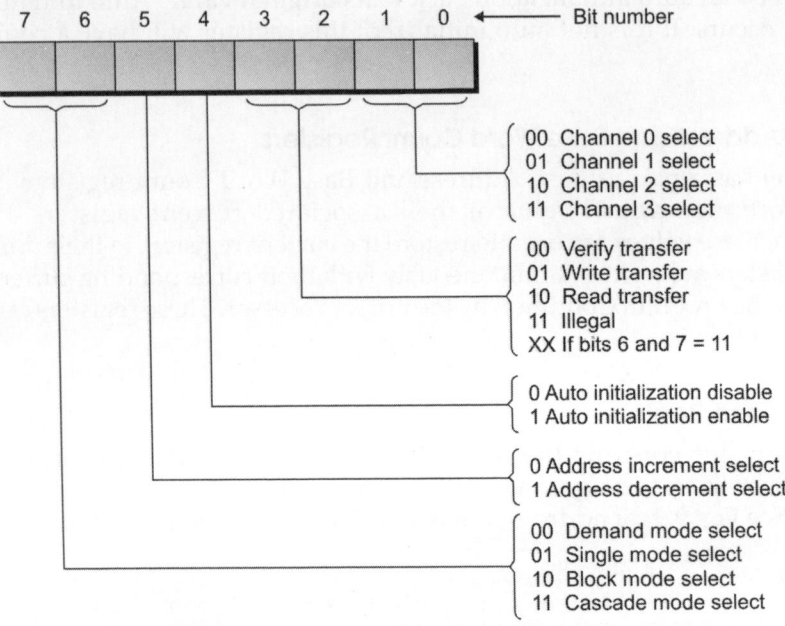

**Fig. 9.4:** Address coding for mode register

### 9.6.6 Request Register

The 8237A can respond to requests for DMA service which are initiated by software as well as by a DREQ. Each channel has a request bit associated with it in the 4-bit Request register. These are non-maskable and subject to prioritization by the Priority Encoder network. Each register bit is set or reset separately under software control or is cleared upon generation of a TC or external. The entire register is cleared by a Reset. To set or reset a bit, the software loads the proper form of the data word. See Fig. 9.5 for register address coding. In order to make a software request, the channel must be in Block Mode.

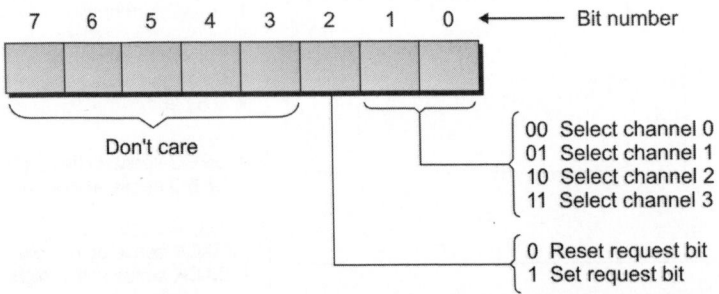

**Fig. 9.5:** Address coding for request register

### 9.6.7 Mask Register

Each channel has associated with it a mask bit which can be set to disable the incoming DREQ. Each mask bit is set when its associated channel produces if the channel is not programmed for auto initialize. Each bit of the 4-bit mask register may also be set or cleared separately under software control. The entire register is also set by a Reset. This disables all DMA requests until a clear Mask register instruction allows them to occur. The instruction to separately set or clear the mask bits is similar in form to that used with the Request register. See Fig. 9.6 for instruction addressing.

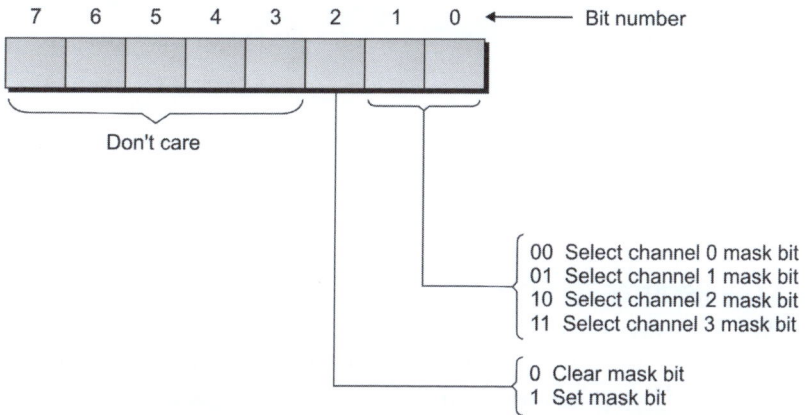

**Fig. 9.6:** Address coding for mask register

All four bits of mask register may also be written with a single command (Fig. 9.7).

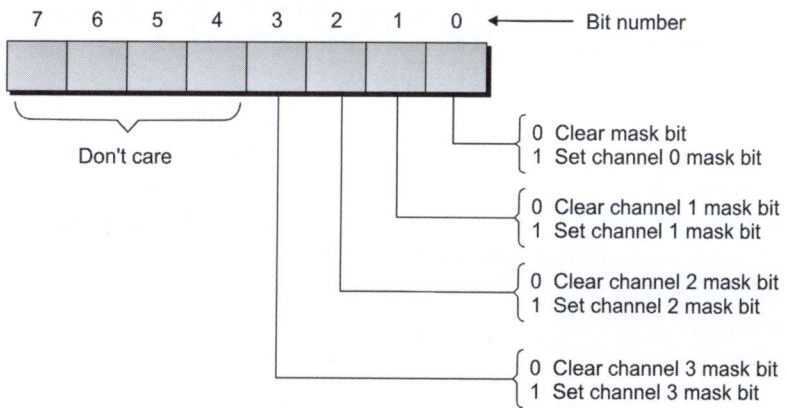

**Fig. 9.7:** Address coding for mask register

### 9.6.8 Status Register

The Status register is available to be read out of the 8237A by the microprocessor. It contains information about the status of the devices at this point. This information includes which channels have reached a terminal count and which channels have pending DMA requests. Bits 0–3 are set every time a TC (terminal count) is reached by that channel or an external is applied. These bits are cleared upon Reset and on each Status Read. Bits 4–7 are set whenever their corresponding channel is requesting service (Fig. 9.8).

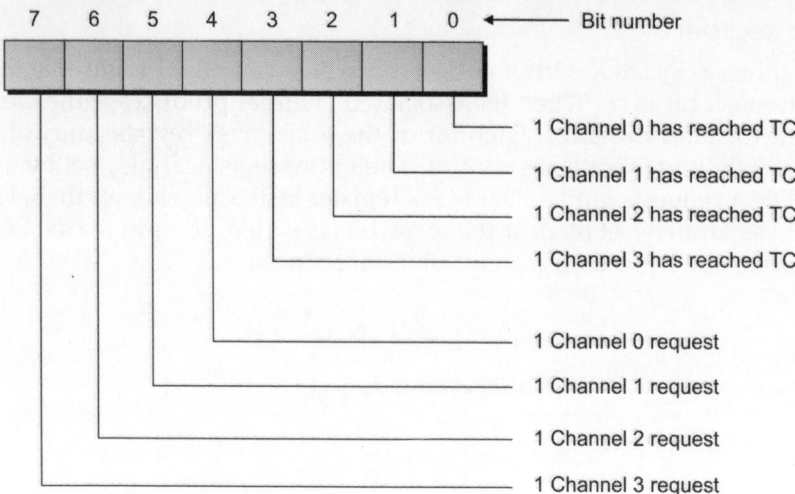

**Fig. 9.8:** Address coding for status register

### 9.6.9 Temporary Register

The Temporary register is used to hold data during memory-to-memory transfers. Following the completion of the transfers, the last word moved can be read by the microprocessor in the Program Condition. The Temporary register always contains the last byte transferred in the previous memory-to-memory operation, unless cleared by a Reset.

### 9.6.10 Software Commands

These are additional special software commands which can be executed in the Program Condition. They do not depend on any specific bit pattern on the data bus. The three software commands are: Clear first, Master clear and Clear mask register (Table 9.2).

**Table 9.2:** Software commands

| A3 | A2 | A1 | A0 | $\overline{IOR}$ | $\overline{IOW}$ | Operation |
|----|----|----|----|-----|-----|-----------|
| 1 | 0 | 0 | 0 | 0 | 1 | Read status register |
| 1 | 0 | 0 | 0 | 1 | 0 | Write command register |
| 1 | 0 | 0 | 1 | 0 | 1 | Illegal |
| 1 | 0 | 0 | 1 | 1 | 0 | Write request register |
| 1 | 0 | 1 | 0 | 0 | 1 | Illegal |
| 1 | 0 | 1 | 0 | 1 | 0 | Write single mask register bit |
| 1 | 0 | 1 | 1 | 0 | 1 | Illegal |
| 1 | 0 | 1 | 1 | 1 | 0 | Write mode register |
| 1 | 1 | 0 | 0 | 0 | 1 | Illegal |
| 1 | 1 | 0 | 0 | 1 | 0 | Clear byte pointer flip/flop |
| 1 | 1 | 0 | 1 | 0 | 1 | Read temporary register |
| 1 | 1 | 0 | 1 | 1 | 0 | Master clear |
| 1 | 1 | 1 | 0 | 0 | 1 | Illegal |
| 1 | 1 | 1 | 0 | 1 | 0 | Clear mask register |
| 1 | 1 | 1 | 1 | 0 | 1 | Illegal |
| 1 | 1 | 1 | 1 | 1 | 0 | Write all mask register bits |

1. **Clear first/last flip-flop:** This command must be executed prior to writing or reading new address or word count information to the 8237A. This initializes the flip-flop to a known state so that subsequent accesses to register contents by the microprocessor will address upper and lower bytes in the correct sequence.

2. **Master clear:** This software instruction has the same effect as the hardware Reset. The Command, Status, Request, Temporary and Internal First/Last Flip-Flop registers are cleared and the Mask register is set. The 8237A will enter the idle cycle.

3. **Clear mask register:** This command clears the mask bits of all four channels, enabling them to accept DMA requests.

## 9.7 TRANSFER TYPES

Each of the three active transfer modes can perform three different types of transfers. These are Read, Write and Verify. Write transfers move data from an I/O device to the memory by activating $\overline{\text{MEMW}}$ and $\overline{\text{IOR}}$. Read transfers move data from memory to an I/O device by activating $\overline{\text{MEMR}}$ and $\overline{\text{IOW}}$. Verify transfers are pseudo transfers. The 8237A operates as in Read or Write transfers generating addresses and responding to $\overline{\text{EOP}}$, etc. However, the memory and I/O control lines all remain inactive. The ready input is ignored in verify mode (Table 9.3).

**Table 9.3:** DMA data transfer

| Types of transfer | Clock cycles per single word transfer |
| --- | --- |
| Internal memory → Internal memory | 2 |
| External memory ↔ Internal memory | 2 + wait states |
| External memory → External memory | 2 + wait states |
| Internal memory ↔ Internal I/O | 2 |
| External memory ↔ Internal I/O | 2 + wait states |
| Internal I/O → Internal I/O | 2 |

*Notes:*
1. Data transfer for one channel takes a minimum of two clock cycles per single word.
2. External memory includes external I/O.

### 9.7.1 Memory-to-Memory

To perform block moves of data from one memory address space to another with a minimum of program effort and time, the 8237A includes a memory-to-memory transfer feature. Programming a bit in the Command register selects channels 0 and 1 to operate as memory-to-memory transfer channels. The transfer is initiated by setting the software DREQ for channel 0.

The 8237A requests a DMA service in the normal manner. After HLDA is true, the device, using four state transfers in Block Transfer mode, reads data from the memory. The channel 0 Current Address register is the source for the address used and is decremented or incremented in the normal manner. The data byte read from the memory is stored in the 8237A internal Temporary register.

Channel 1 then performs a four-state transfer of the data from the Temporary register to memory using the address in its Current Address register and incrementing or decrementing it in the normal manner. The channel 1 current Word Count is decremented. When the Word Count of channel 1 goes to FFFFH, a TC is generated causing an $\overline{\text{EOP}}$ output terminating the service. Channel 0 may be programmed to retain the same address for all transfers. This allows a single word to be written to a block of memory. The 8237A will respond to external $\overline{\text{EOP}}$ signal during memory-to-memory transfers. Data comparators in block search schemes may use this input to terminate the service when a match is found. Memory-to-memory operations can be detected as an active AEN with no DACK outputs.

### 9.7.2 Auto Initialize

By programming a bit in the Mode register, a channel may be set up as an auto initialize channel. During auto initialization, the original values of the Current Address and Current Word Count registers are automatically restored from the Base Address and Base Word Count registers of that channel following $\overline{\text{EOP}}$. The base registers are loaded simultaneously with the current registers by the microprocessor and remain unchanged throughout the DMA service. The mask bit is not altered when the channel is in auto initialization. Following auto initialize, the channel is ready to perform another DMA service, without CPU intervention as soon as a valid DREQ is detected. In order to auto initialize both channels in a memory-to-memory transfer, both Word Counts should be programmed identically. If interrupted externally, pulses should be applied in both bus cycles.

### 9.7.3 Priority

The 8237A has two types of priority encoding available as software selectable options. The first is Fixed Priority which fixes the channels in priority order based upon the descending value of their number. The channel with the lowest priority is 3 followed by 2, 1 and the highest priority channel, 0. After the recognition of any one channel for service, the other channels are prevented from interfering with that service until it is completed. After completion of a service, HRQ will go inactive and the 8237A will wait for HLDA to go low before activating HRQ to service another channel. The second scheme is Rotating Priority. The last channel to get service becomes the lowest priority channel with the others rotating accordingly.

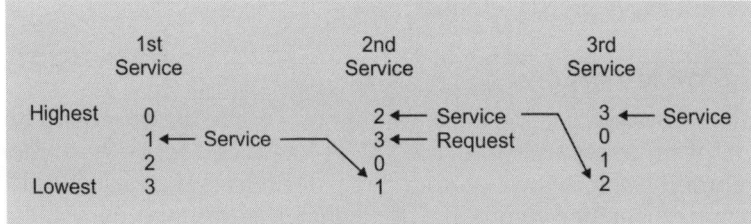

**Fig. 9.9**

With Rotating Priority in a single chip DMA system, any device requesting service is guaranteed to be recognized after no more than three higher priority services have occurred. This prevents any one channel from monopolizing the system.

### 9.7.4 Compressed Timing

In order to achieve even greater throughput where system characteristics permit, the 8237A can compress the transfer time to two clock cycles. From Fig. 9.10, it can be seen that state S3 is used to extend the access time of the read pulse. By removing state S3, the read pulse width is made equal to the write pulse width and a transfer consists only of state S2 to change the address and state S4 to perform the read/write. S1states will still occur when A8-A15 need updating.

### 9.7.5 Address Generation

In order to reduce pin count, the 8237A multiplexes the eight higher order address bits on the data lines. State S1 is used to output the higher order address bits to an external latch from which they may be placed on the address bus. The falling edge of Address Strobe (ADSTB) is used to load these bits from the data lines to the latch. Address Enable (AEN) is used to enable the bits onto the address bus through a three-state enable. The lower order address bits are output by the 8237A directly.

Lines A0-A7 should be connected to the address bus. Figure shows the time relationships between CLK, AEN, ADSTB, and DB0-DB7 and A0-A7. During Block and Demand Transfer mode services, which include multiple transfers, the addresses generated will be sequential. For many transfers, the data held in the external address latch will remain the same. This data need only change when a carry or borrow from A7 to A8 takes place in the normal sequence of addresses. To save time and speed transfers, the 8237A executes S1 states only when updating of A8-A15 in the latch is necessary. This means for long services, S1 states and Address Strobes may occur only once every 256 transfers, a saving of 255 clock cycles for each 256 transfers.

## 9.8 LIST OF REGISTER CODES (Table 9.4)

**Table 9.4:** List of register codes

| Register | Operation | Signals | | | | | | |
|---|---|---|---|---|---|---|---|---|
| | | $\overline{CS}$ | $\overline{IOR}$ | $\overline{IOW}$ | A3 | A2 | A1 | A0 |
| Command | Write | 0 | 1 | 0 | 1 | 0 | 0 | 0 |
| Mode | Write | 0 | 1 | 0 | 1 | 0 | 1 | 1 |
| Request | Write | 0 | 1 | 0 | 1 | 0 | 0 | 1 |
| Mask | Set/reset | 0 | 1 | 1 | 1 | 0 | 1 | 0 |
| Mask | Write | 0 | 1 | 0 | 1 | 0 | 1 | 0 |
| Temporary | Read | 0 | 0 | 1 | 1 | 1 | 0 | 1 |
| Status | Read | 0 | 0 | 1 | 1 | 0 | 0 | 0 |

## 9.9 LIST OF ADDRESS CODES FOR SOFTWARE COMMANDS (Table 9.5)

**Table 9.5:** List of address codes

| Signals | | | | | | Operation |
|---|---|---|---|---|---|---|
| A3 | A2 | A1 | A0 | $\overline{IOR}$ | $\overline{IOW}$ | |
| 1 | 0 | 0 | 0 | 0 | 1 | Read status register |
| 1 | 0 | 0 | 0 | 1 | 0 | Write command register |
| 1 | 0 | 0 | 1 | 0 | 1 | Illegal |
| 1 | 0 | 0 | 1 | 1 | 0 | Write request register |
| 1 | 0 | 1 | 0 | 0 | 1 | Illegal |
| 1 | 0 | 1 | 0 | 1 | 0 | Write single mask register bit |
| 1 | 0 | 1 | 1 | 0 | 1 | Illegal |
| 1 | 0 | 1 | 1 | 1 | 0 | Write mode register |
| 1 | 1 | 0 | 0 | 0 | 1 | Illegal |

*(Contd...)*

**Table 9.5:** List of address codes *(Contd...)*

| Signals | | | | | | Operation |
|---|---|---|---|---|---|---|
| A3 | A2 | A1 | A0 | $\overline{IOR}$ | $\overline{IOW}$ | |
| 1 | 1 | 0 | 0 | 1 | 0 | Clear byte pointer flip-flop |
| 1 | 1 | 0 | 1 | 0 | 1 | Read temporary register |
| 1 | 1 | 0 | 1 | 1 | 0 | Master clear |
| 1 | 1 | 1 | 0 | 0 | 1 | Illegal |
| 1 | 1 | 1 | 0 | 1 | 0 | Clear mask register |
| 1 | 1 | 1 | 1 | 0 | 1 | Illegal |
| 1 | 1 | 1 | 1 | 1 | 0 | Write all mask register bits |

## 9.10  LIST OF DMA CHANNEL I/O PORT ADDRESS (Table 9.6)

**Table 9.6:** List of DMA channel I/O port address

| Channel | Register | Operation | Signals | | | | | | | Internal | Data bus |
|---|---|---|---|---|---|---|---|---|---|---|---|
| | | | $\overline{CS}$ | $\overline{IOR}$ | $\overline{IOW}$ | A3 | A2 | A1 | A0 | FF | DB0-DB7 |
| 0 | Base and | Write | 0 | 1 | 0 | 0 | 0 | 0 | 0 | 0 | A0-A7 |
| | current address | | 0 | 1 | 0 | 0 | 0 | 0 | 0 | 1 | A8-A15 |
| | Current address | Read | 0 | 0 | 1 | 0 | 0 | 0 | 0 | 0 | A0-A7 |
| | | | 0 | 0 | 1 | 0 | 0 | 0 | 0 | 1 | A8-A15 |
| | Base and current | Write | 0 | 1 | 0 | 0 | 0 | 0 | 1 | 0 | W0-W7 |
| | word count | | 0 | 1 | 0 | 0 | 0 | 0 | 1 | 1 | W8-W15 |
| | Current word | Read | 0 | 0 | 1 | 0 | 0 | 0 | 1 | 0 | W0-W7 |
| | count | | 0 | 0 | 1 | 0 | 0 | 0 | 1 | 1 | W8-W15 |
| 1 | Base and current | Write | 0 | 1 | 0 | 0 | 0 | 1 | 0 | 0 | A0-A7 |
| | address | | 0 | 1 | 0 | 0 | 0 | 1 | 0 | 1 | A8-A15 |
| | Current address | Read | 0 | 0 | 1 | 0 | 0 | 1 | 0 | 0 | A0-A7 |
| | | | 0 | 0 | 1 | 0 | 0 | 1 | 0 | 1 | A8-A15 |
| | Base and current | Write | 0 | 1 | 0 | 0 | 0 | 1 | 1 | 0 | W0-W7 |
| | word count | | 0 | 1 | 0 | 0 | 0 | 1 | 1 | 1 | W8-W15 |
| | Current word | Read | 0 | 0 | 1 | 0 | 0 | 1 | 1 | 0 | W0-W7 |
| | count | | 0 | 0 | 1 | 0 | 0 | 1 | 1 | 1 | W8-W15 |
| 2 | Base and current | Write | 0 | 1 | 0 | 0 | 1 | 0 | 0 | 0 | A0-A7 |
| | address | | 0 | 1 | 0 | 0 | 1 | 0 | 0 | 1 | A8-A15 |
| | Current address | Read | 0 | 0 | 1 | 0 | 1 | 0 | 0 | 0 | A0-A7 |
| | | | 0 | 0 | 1 | 0 | 1 | 0 | 0 | 1 | A8-A15 |
| | Base and current | Write | 0 | 1 | 0 | 0 | 1 | 0 | 1 | 0 | W0-W7 |
| | word count | | 0 | 1 | 0 | 0 | 1 | 0 | 1 | 1 | W8-W15 |
| | Current word | Read | 0 | 0 | 1 | 0 | 1 | 0 | 1 | 0 | W0-W7 |
| | count | | 0 | 0 | 1 | 0 | 1 | 0 | 1 | 1 | W8-W15 |
| 3 | Base and current | Write | 0 | 1 | 0 | 0 | 1 | 1 | 0 | 0 | A0-A7 |
| | address | | 0 | 1 | 0 | 0 | 1 | 1 | 0 | 1 | A8-A15 |
| | Current address | Read | 0 | 0 | 1 | 0 | 1 | 1 | 0 | 0 | A0-A7 |
| | | | 0 | 0 | 1 | 0 | 1 | 1 | 0 | 1 | A8-A15 |
| | Base and current | Write | 0 | 1 | 0 | 0 | 1 | 1 | 1 | 0 | W0-W7 |
| | word count | | 0 | 1 | 0 | 0 | 1 | 1 | 1 | 1 | W8-W15 |
| | Current word | Read | 0 | 0 | 1 | 0 | 1 | 1 | 1 | 0 | W0-W7 |
| | count | | 0 | 0 | 1 | 0 | 1 | 1 | 1 | 1 | W8-W15 |

## 9.11 INTERFACING OF DMA WITH MICROPROCESSOR

Figure 9.10 shows a convenient method for configuring a DMA system with the 8237A controller and an 8080A/8085AH microprocessor system. The multimode DMA controller issues a HRQ to the processor whenever there is at least one valid DMA request from a peripheral device. When the processor replies with an HLDA signal, the 8237A takes control of the address bus, the data bus and the control bus. The address for the first transfer operation comes out in two bytes: the least significant 8-bits on the eight address outputs and the most significant 8 bits on the data bus. The contents of the data bus are then latched into an 8-bit latch to complete the full 16-bits of the address bus. The 8282 is a high speed, 8-bit, three-state latch in a 20-pin package. After the initial transfer takes place, the latch is updated only after a carry or borrow is generated in the least significant address byte. Four DMA channels are provided when one 8237A is used.

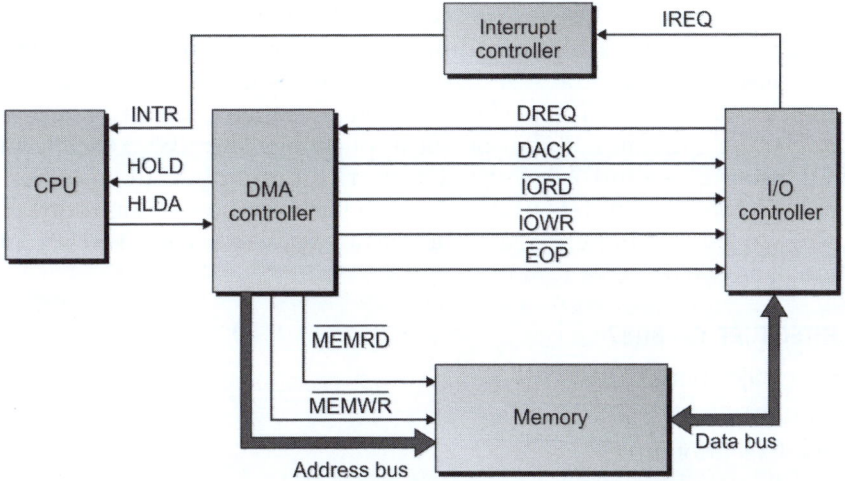

**Fig. 9.10:** Interfacing of 8237A with microprocessor

## PROBLEMS

1. What is direct memory access? Explain it with an example.

2. Describe pin configuration of 8237A DMA controller.

3. Explain the functional block diagram of DMAC 8237A.

4. Explain the operation of DMAC 8237A.

5. Which registers are required for each channel? Explain.

6. Illustrate interfacing of DMA controller to CPU.

7. What is software command? Explain the different types of software command.

8. Explain the cascade mode of DMA controller 8237A.

# 10

# Pentium Processors

## 10.1 INTRODUCTION TO THE 8087 MICROPROCESSOR

The **Intel 8087**, announced in 1980, was the first x87 floating-point co-processor for the 8086 line of microprocessors. The purpose of the 8087 was to speed up computations for floating-point arithmetic, such as addition, subtraction, multiplication, division and square root. It also computed transcendental functions, such as exponential, logarithmic or trigonometric calculations. The performance enhancements were from approximately 20% to over 500%, depending on the specific application. The 8087 could perform about 50,000 FLOPS using around 2.4 watts. Only arithmetic operations benefited from installation of an 8087; computers used only with such applications as word processing, for example, would not benefit from the extra expense (around $150) and power consumption of an 8087.

## 10.2 ARCHITECTURE OF 8087

There are two major parts:

1. Control unit
2. Numeric Execution unit

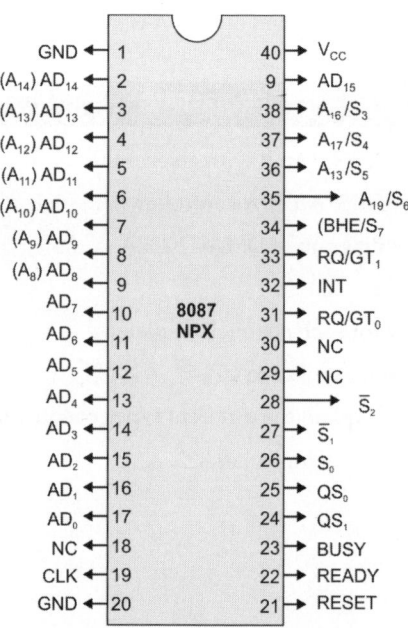

**Fig. 10.1:** Pin diagram of 8087

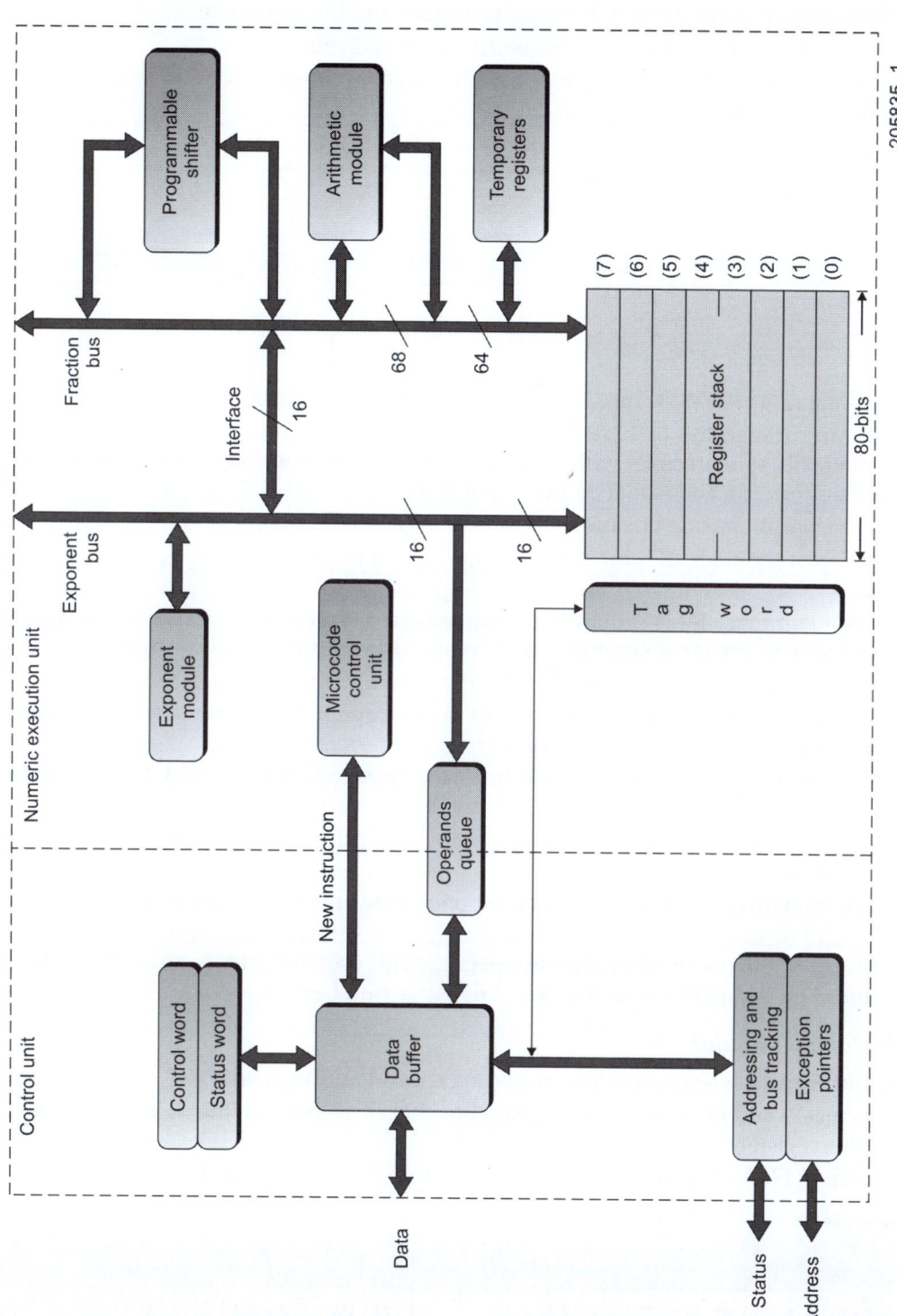

**Fig. 10.2:** Internal Architecture of 8087

### 10.2.1 Control Unit

1. To synchronize the operation of the co-processor and the processor.
2. This unit has a Control word, Status word and Data Buffer.
3. If instruction is an *ESCape* (co-processor) instruction, the co-processor executes it, if not, the microprocessor executes.
4. Status register reflects the overall operation of the co-processor.

#### 10.2.1.1 *Status Register*

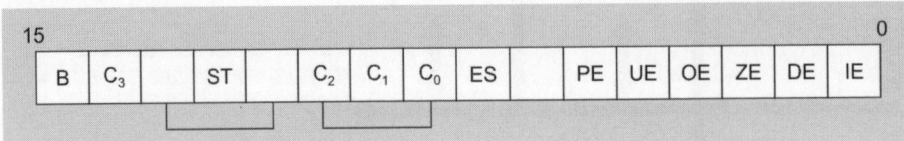

1. B: Busy bit indicates that co-processor is busy executing a task. Busy can be tested by examining the status or by using the FWAIT instruction. Newer co-processor automatically synchronizes with the microprocessor, so busy flag need not be tested before performing additional co-processor tasks.
2. C3–C0: Condition code bits indicate conditions about the co-processor.
3. TOP: Top of the stack (ST) bit indicates the current register address as the top of the stack.
4. ES: Error summary bit is set if any unmasked error bit (PE, UE, OE, ZE, DE or IE) is set. In the 8087, the error summary is also caused as co-processor interrupt.
5. PE: Precision error indicates that the result or operand executes selected precision.
6. UE: Underflow error indicates the result is too large to represent with the current precision selected by the control word.
7. OE: Overflow error indicates a result that is too large to be represented. If this error is masked, the co-processor generates infinity for an overflow error.
8. ZE: A zero error indicates the divisor was zero while the dividend is a non-infinity or non-zero number.
9. DE: Denormalized error indicates at least one of the operands is denormalized.
10. IE: Invalid error indicates a stack overflow or underflow, indeterminate from (0/0, 0, –0, etc.) or the use of a NAN as an operand. This flag indicates error such as those produced by taking the square root of a negative number.

#### 10.2.1.2 *Control Register*

1. Control register selects precision, rounding control, infinity control.
2. It also masks and unmasks the exception bits that correspond to the rightmost six bits of status register.
3. Instruction FLDCW is used to load the value into the control register.

*Control register*

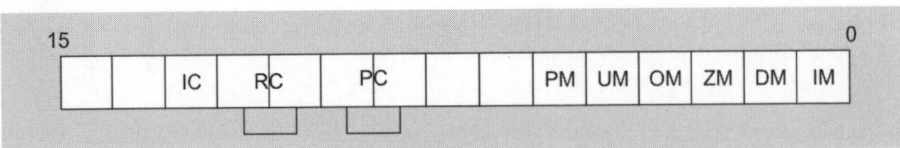

- IC Infinity control
- RC Rounding control
- PC Precision control
- PM Precision mask
- UM Underflow mask
- OM Overflow mask
- ZM Division by zero mask
- DM Denormalized operand mask
- IM Invalid operand mask

1. IC – Infinity control selects either affine or projective infinity. Affine allows positive and negative infinity, while projective assumes infinity is unsigned.

   **INFINITY CONTROL**

   0 = Projective

   1 = Affine

2. RC – Rounding control determines the type of rounding.

   **ROUNDING CONTROL**

   00 = Round to nearest or even

   01 = Round down towards minus infinity

   10 = Round up towards plus infinity

   11 = Chop or truncate towards zero

3. PC – Precision control sets the precision of the result as defined below.

   **PRECISION CONTROL**

   00 = Single precision (short)

   01 = Reserved

   10 = Double precision (long)

   11 = Extended precision (temporary)

4. Exception Masks – It determines whether the error indicated by the exception affects the error bit in the status register. If a logic1 is placed in one of the exception control bits, corresponding status register bit is masked off.

## 10.2.2 Numeric Execution Unit

1. This performs all operations that access and manipulate the numeric data in the co-processor's registers.

2. Numeric registers in NUE are 80 bits wide.

3. NUE is able to perform arithmetic, logical and transcendental operations as well as supply a small number of mathematical constants from its on-chip ROM.

4. Numeric data is routed into two parts—a 64-bit mantissa bus and a 16-bit sign/exponent bus.

## 10.3 CIRCUIT CONNECTION FOR 8086–8087 (Fig. 10.3)

1. Multiplexed address-data bus lines are connected directly from 8086 to 8087.

2. The status lines and the queue status lines are connected directly from 8086 to 8087.

3. The Request/Grant signal RQ/GT0 of 8087 is connected to RQ/GT1 of 8086.

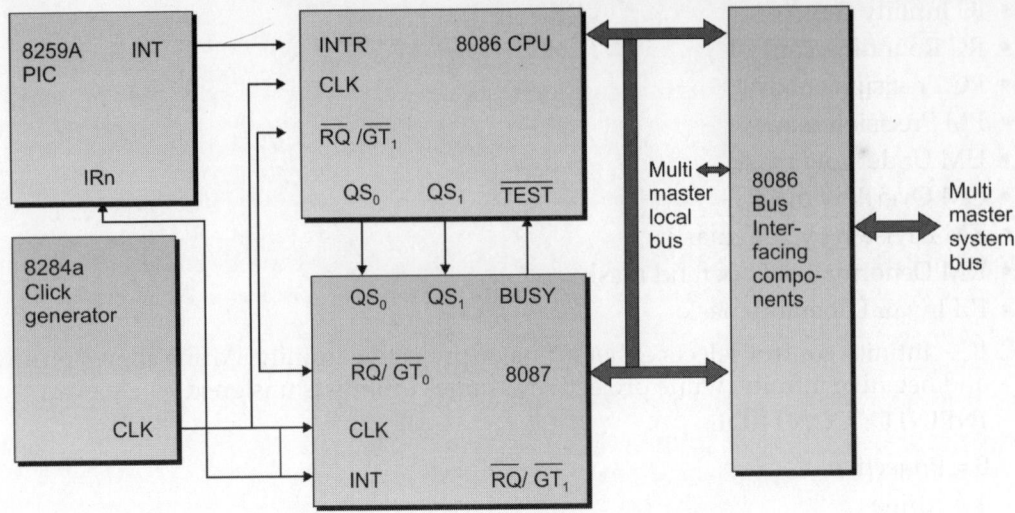

**Fig. 10.3:** Circuit connection for 8086–8087

4. $\overline{BUSY}$ BUSY signal 8087 is connected to TEST pin of 8086.

5. Interrupt output INT of the 8087 to NMI input of 8086. This intimates an error condition.

6. The main purpose of the circuitry between the INT output of 8087 and the NMI input is to make sure that an NMI signal is not present upon reset, to make it possible to mask.

7. NMI input is to make it possible for other devices to cause an NMI interrupt.

8. BHE pin is connected to the system BHE line to enable the upper bank of memory.

9. The RQ/GT1 input is available so that another co-processor such as 8089 I/O processor can be connected and function in parallel with the 8087.

10. One type of Cooperation between the two processors that you need to know about it is how the 8087 transfers data between memory and its internal registers.

11. When 8086 reads an 8087 instruction that needs data from memory or wants to send data to memory, the 8086 sends out the memory address code in the instruction and also the appropriate memory read or memory write signal to transfer a word of data.

12. In the case of memory read, the addressed word will be kept on the data bus by the memory. The 8087 then simply reads the word of data bus. The 8086 ignores this word. If the 8087 only needs this one word of data, it can then go on and execute its instruction.

13. Some 8087 instructions need to read in or write out up to 80-bit word. For these cases, 8086 outputs the address of the first data word on the address bus and outputs the appropriate control signal.

14. The 8087 reads the data word on the data bus by memory or writes a data word to memory on the data bus. The 8087 grabs the 20-bit physical address that was output by the 8086. To transfer additional words, it needs to/from memory, the 8087 then takes over the buses from 8086.

15. To take over the bus, the 8087 sends out a low-going pulse on.

16. $\overline{RQ}/\overline{GT0}$ pin: The 8086 responds to this by sending another low-going pulse back to the $\overline{RQ}/\overline{GT0}$ pin of 8087 and by floating its buses.

17. The 8087 then increments the address it grabbed during the first transfer and outputs the incremented address on the address bus. When the 8087 outputs a memory read or memory write signal, another data word will be transferred to or from the 8087.

18. The 8087 continues the process until it has transferred all the data words required by the instruction to/from memory.

19. When the 8087 is using the buses for its data transfer, it sends another low-going pulse out on its $\overline{RQ}/\overline{GT0}$ pin to 8086 to know it can have the buses back again. The next type of synchronization between the host processor and the co-processor is required to make sure that the 8086 does not attempt to execute the next instruction before the 8087 has completed an instruction.

20. Taking one situation in the case where the 8086 needs the data produced by the execution of an 8087 instruction to carry out its next instruction.

21. In the instruction sequence, for example, the 8087 must complete the **FSTSW STATUS** instruction before the 8086 will have the data it needs to execute the **MOV AX, STATUS** instruction.

22. Without some mechanism to make the 8086 wait until the 8087 completes the FSTSW instruction, the 8086 will go on and execute the **MOV AX, STATUS** with erroneous data.

23. We solve this problem by connecting the 8087 BUSY output to the TEST pin of the 8086 and putting on the WAIT instruction in the program.

24. While 8087 is executing an instruction, it asserts its BUSY pin high. When it is finished with an instruction, the 8087 will drop its BUSY pin low. Since the BUSY pin from 8087 is connected to the TEST pin 8086, the processor can check its pin of 8087 whether it finished its instruction or not.

25. You place the 8086 WAIT instruction in your program after the 8087 FSTSW instruction. When 8086 executes the WAIT instruction, it enters an internal loop where it repeatedly checks the logic level on the TEST input. The 8086 will stay in this loop until it finds the TEST input asserted low, indicating the 8087 has completed its instruction. The 8086 will then exit the internal loop, fetch and execute the next instruction.

## 10.4 INTERFACING

1. Multiplexed address-data bus lines are connected directly from the 8086 to 8087.

2. The status lines and the queue status lines are connected directly from 8086 to 8087.

3. The Request/Grant signal $\overline{RQ}/\overline{GT0}$ of 8087 is connected to $\overline{RQ}/\overline{GT1}$ of 8086.

4. BUSY signal 8087 is connected to $\overline{TEST}$ pin of 8086.

5. Interrupt output INT of the 8087 to NMI input of 8086. This intimates an error condition.

6. A WAIT instruction is passed to keep looking at its $\overline{TEST}$ pin, until it finds pin Low to indicate that the 8087 has completed the computation.

7. SYNCHRONIZATION must be established between the processor and co-processor in two situations.

a. The execution of an ESC instruction that requires the participation of the NUE must not be initiated if the NUE has not completed the execution of the previous instruction.

b. When a processor instruction accesses a memory location that is an operand of a previous co-processor instruction, CPU must synchronize with NPX to ensure that it has completed its instruction. Processor WAIT instruction is provided.

## 10.5 EXCEPTION HANDLING

The 8087 detects an error condition usually called an exception; when it is executing an instruction, it will set the bit in its Status register. The 8087 detects six different types of exception conditions that occur during instruction execution. These will cause an interrupt if unmasked and interrupts are enabled.

1. Invalid operation
2. Overflow
3. Zero divisor
4. Underflow
5. Denormalized operand
6. Inexact result

## 10.6 INSTRUCTION SET

The 8087 instruction mnemonics begins with the letter F which stands for Floating point and distinguishes from 8086. These are grouped into four functional groups:

1. Data transfer instructions
2. Arithmetic instructions
3. Compare instructions
4. Transcendental instructions (Trigonometric and Exponential).

### 10.6.1 Data Transfer Instructions

*Real transfer*

**FLD** Load real
**FST** Store real
**FSTP** Store real and pop
**FXCH** Exchange registers

*Integer transfer*

**FILD** Load integer
**FIST** Store integer
**FISTP** Store integer and pop

*Packed decimal transfer (BCD)*

**FBLD** Load BCD
**FBSTP** Store BCD and pop

**Examples:**

**FLD source:** Decrements the stack pointer by one and copies a real number from a stack element or memory location to the new ST.

FLD ST(3); Copies ST(3) to ST.

FLD LONG_REAL[BX]; Number from memory; copied to ST.

**FLD destination:** Copies ST to a specified stack position or to a specified memory location.

FST ST(2); Copies ST to ST(2), and; increment stack pointer.

FST SHORT_REAL[BX]; Copy ST to a memory at a; SHORT_REAL[BX]

**FXCH destination:** Exchange the contents of ST with the contents of a specified stack element.

FXCH ST(5); Swap ST and ST(5)

**FILD source:** Integer load. Convert integer number from memory to temporary-real format and push on 8087 stack.

FILD DWORD PTR[BX]; Short integer from memory at [BX].

**FIST destination:** Integer store. Convert number from ST to integer and copy to memory.

FIST LONG_INT; ST to memory locations named LONG_INT.

**FISTP destination:** Integer store and pop. Identical to FIST except that stack pointer is incremented after copy.

**FBLD source:** Convert BCD number from memory to temporary-real format and push on top of 8087 stack.

### 10.6.2 Arithmetic Instructions

**Four basic arithmetic functions:** Addition, Subtraction, Multiplication and Division

a. **Addition**

**FADD** Add real

**FADDP** Add real and pop

**FIADD** Add integer

b. **Subtraction**

**FSUB** Subtract real

**FSUBP** Subtract real and pop

**FISUB** Subtract integer

**FSUBR** Subtract real reversed

**FSUBRP** Subtract real and pop

**FISUBR** Subtract integer reversed

c. **Multiplication**

**FMUL** Multiply real

**FMULP** Multiply real and pop

**FIMUL** Multiply integer

d. **Advanced**

**FABS** Absolute value

**FCHS** Change sign

**FPREM** Partial remainder

**FPRNDINT** Round to integer

**FSCALE** Scale

**FSQRT** Square root

**FXTRACT** Extract exponent and mantissa.

**Examples:**

**FADD:** Adding real from specified source to specified destination source can be a stack or memory location. Destination must be a stack element. If no source or destination is specified, then ST is added to ST(1) and stack pointer is incremented so that the result of addition is at ST.

    FADD ST(3), ST; Add ST to ST(3), result in ST(3)

    FADD ST, ST(4); Add ST(4) to ST, result in ST

    FADD; ST + ST(1), pop stack result at ST

    FADDP ST(1); Add ST(1) to ST. Increment stack; pointer so ST(1) becomes ST

    FIADD Car_Sold; Integer number from memory + ST

**FSUB:** Subtract the real number at the specified source from the real number at the specified destination and put the result in the specified destination.

    FSUB ST(2), ST; ST(2) = ST(2) − ST

    FSUB Rate; ST = ST − real no from memory

    FSUB; ST = ( ST(1) − ST)

**FSUBP:** Subtract ST from specified stack element and put result in specified stack element. Then increment the pointer by one.

    FSUBP ST(1); ST(1)-ST. ST(1) becomes new ST

**FISUB:** Integer from memory subtracted from ST, result in ST

    FISUB Cars_Sold; ST becomes ST−integer from memory

## 10.6.3 Comparison Instructions

    **FCOM** Compare real

    **FCOMP** Compare real and pop

    **FCOMPP** Compare real and pop twice

    **FICOM** Compare integer

    **FICOMP** Compare integer and pop

    **FTST** Test ST against + 0.0

    **FXAM** Examine ST

## 10.6.4 Transcendental Instructions

    **FPTAN** Partial tangent

    **FPATAN** Partial arctangent

    **F2XM1** 2x − 1

    **FYL2X** Y log2X

    **FYL2XP1** Y log2(X + 1)

**Examples:**

**FPTAN:** Compute the values for a ratio of Y/X for an angle in ST. The angle must be in radians and the angle must be in the range of $0 <$ angle $< \pi/4$. F2XM1 − Compute Y = 2x-1 for an X value in ST. The result Y replaces X in ST. X must be in the range of $0 \leq X \leq 0.5$.

**FYL2X:** Calculate Y(LOG2X). X must be in the range of $0 < X < \infty$ any Y must be in the range of $-\infty < Y < +\infty$.

**FYL2XP1:** Compute the function Y(LOG2(X + 1)). This instruction is almost identical to FYL2X except that it gives more accurate results when compute log of a number very close to one.

### 10.6.5 Constant Instruction

**FLDZ** Load +0.0

**FLDI** Load+1.0

**FLDPI** Load $\pi$

**FLDL2T** Load log210

**FLDL2E** Load log2e

**FLDLG2** Load log102

**FLDLN2** Load loge2

## 10.7 INTRODUCTION TO THE 80186 MICROPROCESSOR

Figure 10.4 provides the block diagram of the 80186 microprocessor that generically represents all versions except for the enhancements and additional features outlined in Table 10.1. Notice that this microprocessor has a great deal of more internal circuitry than the 8086. The block diagrams of the 80186 and 80188 are identical except for the pre-fetch queue, which is 4 bytes in the 80188 and six bytes in the 80186. Like the 8088, the 80186 contains a bus interface unit (BID) and an execution unit (ED). In addition to the BID and ED, the 80186/80188 family contains a clock generator, a programmable interrupt controller, programmable timers, a programmable DMA controller and a programmable chip selection unit. These enhancements greatly increase the utility of the 80186/80188 and reduce the number of peripheral components required to implement a system. Many popular subsystems for the personal computer use the 80186/80188 as caching disk controllers, local area network (LAN) controllers, etc. The 80186/80188 microprocessors also find application in the cellular telephone network as a switcher.

Software for the 80186/80188 is identical to the 80286 microprocessor without the memory management instructions. This means that the 80286 has features, like instructions—immediate multiplication, immediate shift counts, string I/O, PUSHA, POPA, BOUND, ENTER and LEAVE—all function on the 80186/80188 microprocessors.

## 10.8 ARCHITECTURE OF 80186

*Clock generator:* The internal clock generator replaces the external 8284A clock generator used with the 8086/8088 microprocessors. This reduces the component count in a system. The internal clock generator has three pin connections: Xl, X2 and CLKOUT (or on some versions CLKIN, OSCOUT and CLKOUT). The Xl (CLKIN) and X2 (OSCOUT) pins are connected to a crystal that resonates at twice the operating frequency of the microprocessor. In the 8 MHz version of the 80186/80188, a 16 MHz crystal is attached to Xl (CLKIN) and X2 (OSCOUT). The 80186/80188 are available in 6 MHz, 8 MHz, 12 MHz, 16 MHz or 25 MHz versions. The CLKOUT pin provides a system clock signal that is one-half the crystal frequency with a 50 percent duty cycle. The CLKOUT pin drives other devices in a system and provides a timing source to additional microprocessors in the system. In addition to these external pins, the clock generator provides the internal timing for synchronizing the READY input pin, whereas in the 8086/8088 system, READY synchronization is provided by the 8284A clock generator.

*Programmable interrupt controller:* The programmable interrupt controller (PIC) arbitrates all internal and external interrupts and controls up to two external 8259A PICs. When an external 8259 is attached, the 80186/80188 microprocessors function as the master and the 8259 functions as the slave. The 80C186EC and 80C188EC models contain an 8259A compatible interrupt controller in place of the one described here for the other versions (XL, EA and EB).

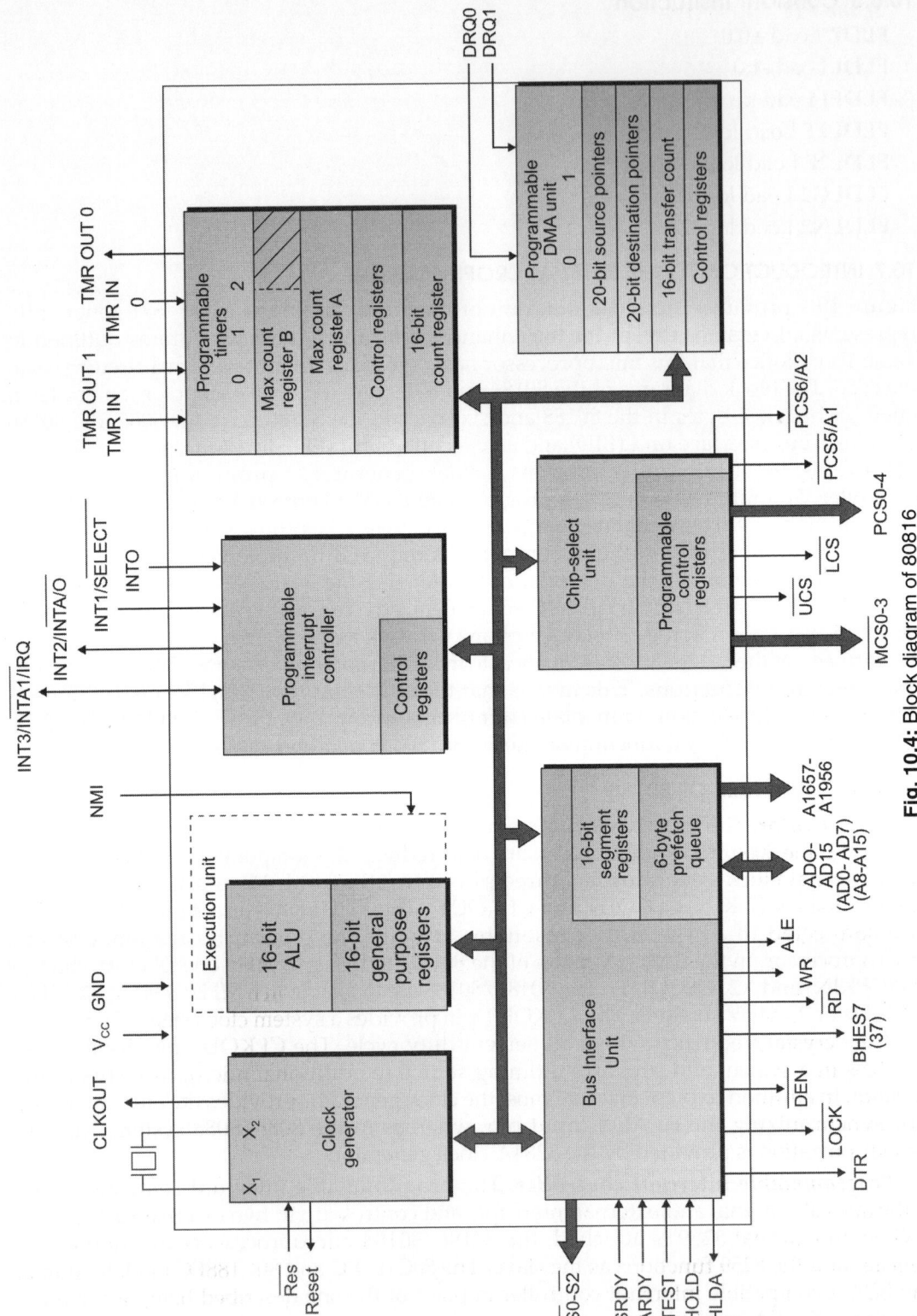

**Fig. 10.4:** Block diagram of 80816

*If the PIC is operated without an 8259, it has five interrupt inputs:* INTO-INT3 and NMI. This is an expansion from the two interrupt inputs available on the 8086/8088 microprocessors. In many systems, the five interrupt inputs are adequate.

*Timers:* The timer section contains three fully programmable 16-bit timers. Timers 0 and 1 generate waveforms for external use and are driven by either the master clock of the 80186/80188 or by an external clock. They are also used to count external events. The third timer, timer 2, is internal and clocked by the master clock. The output of timer 2 generates an interrupt after a specified number of clocks and can also provide a clock to the other timers. Timer 2 can also be used as a watchdog timer, because it can be programmed to interrupt the microprocessor after a certain length of time. The 80C186EC and 80C188EC models have an additional timer called *watchdog*. The watchdog timer is a 32-bit counter that is clocked internally by the CLKOUT signal (one-half the crystal frequency). Each time the counter hits zero, it reloads and generates a pulse on the WDTOUT pin that is four CLKOUT periods wide. This output can be used for any purpose. It can be wired to the reset input to cause a reset or to the NMI input to cause an interrupt. Note that if it is connected to the reset or NMI inputs, it is periodically reprogrammed so that it never counts down to zero. The purpose of a watchdog timer is to reset or interrupt the system if the software goes awry.

*Programmable DMA unit:* The programmable DMA unit contains two DMA channels or four DMA channels in the *80C186EC/80C188EC* models. Each channel can transfer data between memory locations, between memory and I/O or between I/O devices. This DMA controller is similar to the 8237 DMA controller discussed earlier. The main difference is that the 8237 has four channels, as does the EC model.

*Programmable chip selection unit:* The chip selection is a built-in programmable memory and I/O decoder. It has 6 output lines to select memory, 7 lines to select I/O on the XL and EA models and 10 lines that either select memory or I/O on the EB and EC models. On the XL and EA models, the memory selection lines are divided into three groups that select memory for the major sections of the *80186/80188* memory map. The lower memory select signal enables memory for the interrupt vectors, the upper memory select signal enables memory for reset and the middle memory select signals enable up to four middle memory devices. The boundary of the lower memory begins at location OOOOOH and the boundary of the upper memory ends at location FFFFFH. The size of the memory areas are programmable and wait states (0–3 waits) can be automatically inserted with the selection of an area of memory. On the XL and EA models, each programmable I/O selection signal addresses a 128-byte block of I/O space. The programmable I/O area starts at a base I/O address programmed by the user and all seven 128-byte blocks are contiguous. On the EB and EC models, there is an upper and lower memory chip selection pin and eight general-purpose memory or I/O chip selection pins. Another difference is that from 0–15 wait states can be programmed in these two versions of the 80186/80188 embedded controllers.

*Power save/power down feature:* The power save feature allows the system clock to be divided by 4, 8 or 16 to reduce power consumption. The power saving feature is started by software and exited by a hardware event such as an interrupt. The power down feature stops the clock completely, but it is not available on the XL version. The power down mode is entered by executing an HLT instruction and exited by any interrupt.

*Refresh control unit:* The refresh control unit generates the refresh row address at the interval programmed. The refresh control unit does not multiplex the address for the

DRAM—this is still the responsibility of the system designer. The refresh address is provided to the memory system at the end of the programmed refresh interval along with the RFSH control signal. The memory system must run a refresh cycle during the active time of the RFSH control signal. More on memory and refreshing is provided in the section that explains the chip selection unit.

## 10.9 PIN DIAGRAM OF 80186

Figure 10.5 illustrates the pin-out of the 80C188XL microprocessor. Note that the 80C 186XL is packaged in either a 68-pin lead-less chip carrier (LCC) or in a pin grid array (PGA). The LCC package and PGA packages are illustrated in Fig. 10.3.

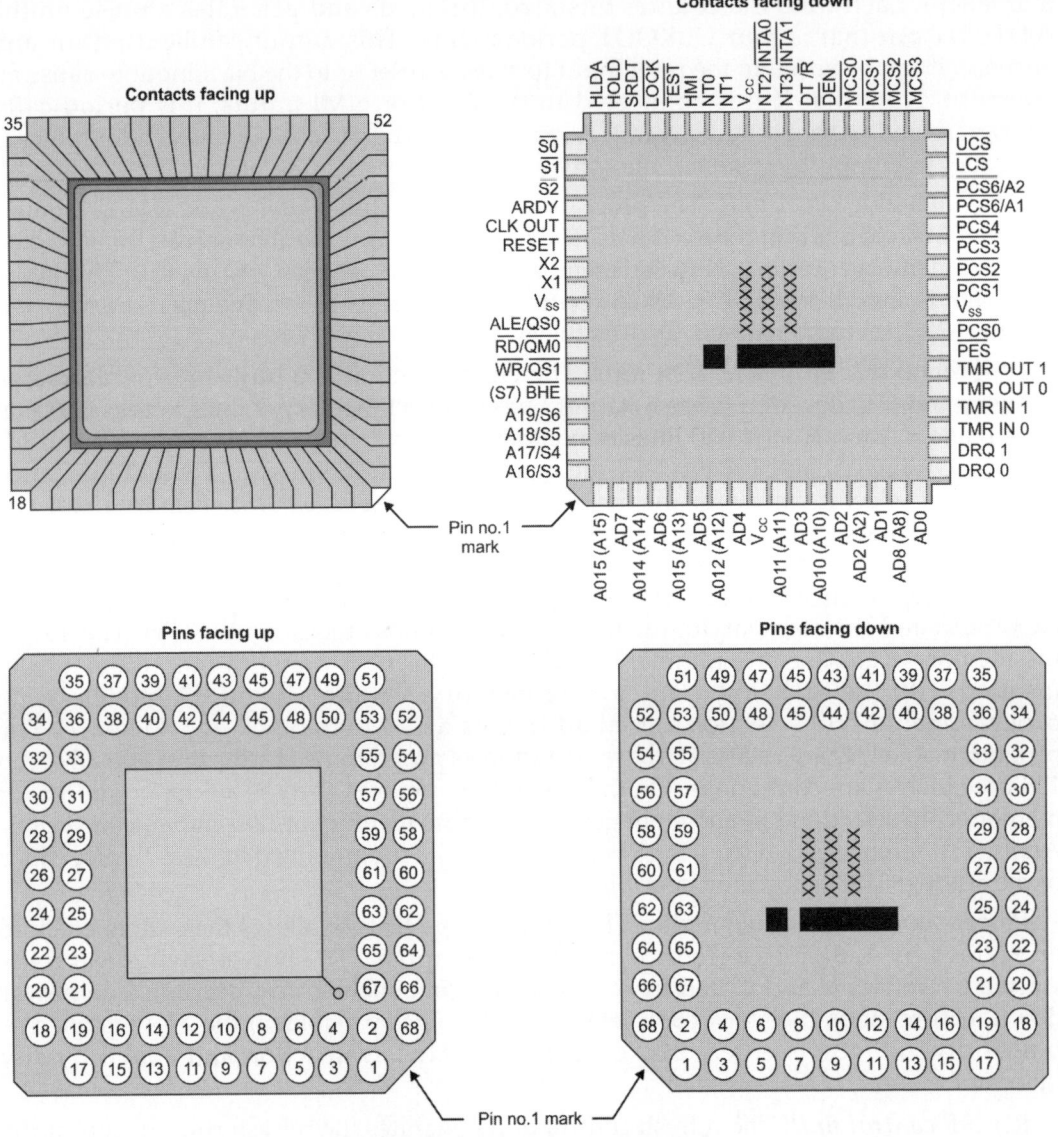

**Fig. 10.5:** Pin diagram of 80186

**Pin definitions:** The following list defines each 80C 188XL pin and notes any differences between the 80C186XL and 80C188XL microprocessors. The enhanced versions are described later in this chapter.

**Vee:** This is the system power supply connection for ±10%, +5.0V.

**Vss:** This is the system ground connection.

**Xl and X2:** These pins generally connect to a fundamental-mode parallel resonant crystal that operates an internal crystal oscillator. An external clock signal may be connected to the Xl pin. The internal master clock operates at one-half the external crystal or clock input signal. Note that these pins are labeled CLKIN (Xl) and OSCOUT (X2) on some versions of the 80186/80188.

**CLKOUT:** This pin provides a timing signal to system peripherals at one-half the clock frequency with a 50 percent duty cycle. The reset pin resets the 80186/80188. For a proper reset, the RES must be held low for at least 50 ms after power is applied. This pin is often connected to an RC circuit that generates a reset signal after power is applied. The reset location is identical to that of the 8086/8088 microprocessors-FFFF0H.

**RESET:** The companion reset output pin (goes high for a reset) connects to system peripherals to initialize them whenever the RES input goes low. This test pin connects to the BUSY output of the 80187 numeric co-processor. The TEST pin is interrogated with the WAIT instruction.

**TMRINO, TMRINI TMROUTO, TMROUTI DRQO and DRQl:** These pins are used as external clocking sources to timers 0 and 1. These pins provide the output signals from timers 0 and 1, which can be programmed to provide square waves or pulses. These pins are active-high level triggered DMA request lines for DMA channels 0 and 1.

**NMI:** This is a non-maskable interrupt input. It is positive edge-triggered and always active. When NMI is activated, it uses interrupt vector 2.

**INTO, INTI, $\overline{INT2}/\overline{INTAO}$ and $\overline{INT3}/\overline{INTAI}$:** These are maskable interrupt inputs. They are active-high and are programmed as either level or edge-triggered. These pins are configured as four interrupt inputs if no external 8259 is present or as two interrupt inputs if 8259s are present.

*i\19/0NCE,* **A18, A17, A16:** These are multiplexed address status connections that provide the address (A 19-A 16) and status (S6-S3). Status bits found on address pins A18-A16 have no system function and are used during manufacturer for testing. The A19 pin is an input for the ONCE function on reset. If ONCE is held low on a reset, the microprocessor enters a testing mode.

**$AD_{15} - AD_0$:** These are multiplexed address/data bus connections. During T1, the 80186 places A15-AO on these pins; during T2, T3 and T4, the 80186 uses these pins as the data bus for signals 015-Do'-Note that the 80188 has pins A07-AOO and A15-Ag.

This pin indicates (when a logic 0) that valid data are transferred through data bus connections 015–08.

**ALE:** This is a multiplexed output pin that contains ALE one-half clock cycle earlier than in the 8086. It is used to de-multiplex the address/data and address/status buses. (Even though the status bits on A19-A16 are not used in the system, they must still be de-multiplexed.)

**WR:** This write pin causes data to be written to memory or I/O.

**RD:** This read pin causes data to be read from memory or I/O.

**ARDY:** The asynchronous READY input informs the 80186/80188 that the memory or I/O is ready for the 80186/80188 to read or write data. If this pin is tied to +5.0Y, the microprocessor functions normally; if it is grounded, the microprocessor enters wait states.

**SRDY:** The synchronous READY input is synchronized with the system clock to provide a relaxed timing for the ready input. As with ARDY, SRDY is tied to +5.0Y for no wait states. This lock pin is an output controlled by the LOCK prefix. If an instruction is prefixed with LOCK, the LOCK pin becomes a logic 0 for the duration of the locked instruction.

**S2, S1 and S0:** These are status bits that provide the system with the type of bus transfer in effect. Refer to Table 10.1 for the states of the status bits. The upper-memory chip select pin selects memory on the upper portion of the memory map. This output is programmable to enable memory sizes of 1 K–256 K bytes ending at location FFFFFH. Note that this pin is programmed differently on the EB and EC versions. The lower-memory chip select pin enables memory beginning at location OOOOOH. This pin is programmed to select memory sizes from l K–256 K bytes. Note that this pin functions differently for the EB and EC versions.

**Table 10.1:** States of the status bus

| S2 | S1 | S0 | Function |
|----|----|----|----------|
| 0 | 0 | 0 | Interrupt acknowledge |
| 0 | 0 | 1 | I/O read |
| 0 | 1 | 0 | I/O write |
| 0 | 1 | 1 | Halt |
| 1 | 0 | 0 | Opcode fetch |
| 1 | 0 | 1 | Memory read |
| 1 | 1 | 0 | Memory write |
| 1 | 1 | 1 | Passive |

$\overline{MCO0} - \overline{MCS3}$ : The middle-memory chip select pins enable four middle memory devices. These pins are programmable to select an 8 K–512 K byte block of memory containing four devices. Note that these pins are not present on the EB and EC versions.

$\overline{PCS0} - \overline{PCS4}$ : These are five different peripheral selection lines. Note that these lines are not present on the EB and EC versions.

**PCS5/Al and PCS6/A2:** These pins are programmed as peripheral selection lines or as internally latched address bits A2 and A1. These lines are not present on the EB and EC versions.

**DT/R:** This pin controls the direction of data bus buffers if attached to the system. This pin enables the external data bus buffers.

## 10.10 INTRODUCTION TO THE 80286 MICROPROCESSOR

The 80286 microprocessor is an advanced version of the 8086 microprocessor that is designed for multi-user and multi-tasking environments. The 80286 addresses 16 M byte of physical memory and 1 G bytes of virtual memory by using its memory-management system. This section of the text introduces the 80286 microprocessor, which finds use in earlier AT-style personal computers that once pervaded the computer market and still finds some application. The 80286 is basically an 8086 that is optimized to execute instructions in fewer clocking periods than the 8086. The 80286 is also an enhanced version of the 8086 because it contains a memory manager. At this time, the 80286 no longer has a place in the personal computer system, but it does find application in control systems as an embedded controller.

## 10.11 HARDWARE FEATURES

Figure 10.6 provides the internal block diagram of the 80286 microprocessor. Notice that like the 80186/80188, the 80286 does not incorporate internal peripherals; instead it contains a memory-management unit (MMU) that is called the *address unit* in the block diagram. As a careful examination of the block diagram reveals that address pins A23-AO~ BUSY, CAP, $\overline{ERROR}$, $\overline{PEREQ}$ and $\overline{PEACK}$ are new or additional pins that do not appear on the 8086 microprocessor. The BUSY, $\overline{ERROR}$, $\overline{PEREQ}$ and $\overline{PEACK}$ signals are used with the microprocessor extension or co-processor of which the 80287 is an example. Note that the TEST pin is now referred to as the BUSY pin. The address bus is now 24-bits wide to accommodate the 16 M bytes of physical memory. The CAP pin is connected to a 0.047 JIF, ±20% capacitor that acts as a 12 V filter and connects to ground. The pin-outs of the 8086 and 80286 are illustrated in Figure 10.6 for comparative purposes. Note that the 80286 does not contain a multiplexed address/data bus. As mentioned in Chapter 1, the 80286 operates in both the real and protected modes. In the real mode, the 80286 addresses a 1 M-byte memory address space and is virtually identical to the 8086. In the protected mode, the 80286 addresses a 16 M byte memory space.

Figure 10.7 illustrates the basic 80286 microprocessor system. Notice that the clock is provided by the 82284 clock generator (similar to the 8284A) and the system control signals are provided by the 82288 system bus controller (similar to the 8288). Also notice the absence of the latch circuits used to de-multiplex the 8086 address/data bus.

## 10.12 ARCHITECTURE OF 80286

The M8086, 88, 186 and 286 CPU family all contain the same basic set of registers, instructions and addressing modes. The 80286 processor is upward compatible with the M8086, M8088 and 80186 CPUs and fully compatible with the HMOS M80286.

### Register Set

The M80C286 base architecture has fifteen registers as shown in Fig. 10.7. These registers are grouped into the following four categories:

**General registers:** Eight 16-bit general purpose registers are used to contain arithmetic and logical operands. Four of these (AX, BX, CX and DX) can be used either in their entirety as 16-bit words or split into pairs of separate 8-bit registers.

**Segment registers:** Four 16-bit special purpose registers select, at any given time, the segments of memory that are immediately addressable for code, stack and data. (For usage, refer to Memory Organization.)

**Base and index registers:** Four of the general purpose registers may also be used to determine offset addresses of operands in memory. These registers may contain base addresses or indexes to particular locations within a segment. The addressing mode determines the specific registers used for operand address calculations.

**Status and control registers:** The 3 16-bit special purpose registers in Fig. 10.7 record or control certain aspects of the M80C286 processor state including the Instruction Pointer, which contains the offset address of the next sequential instruction to be executed.

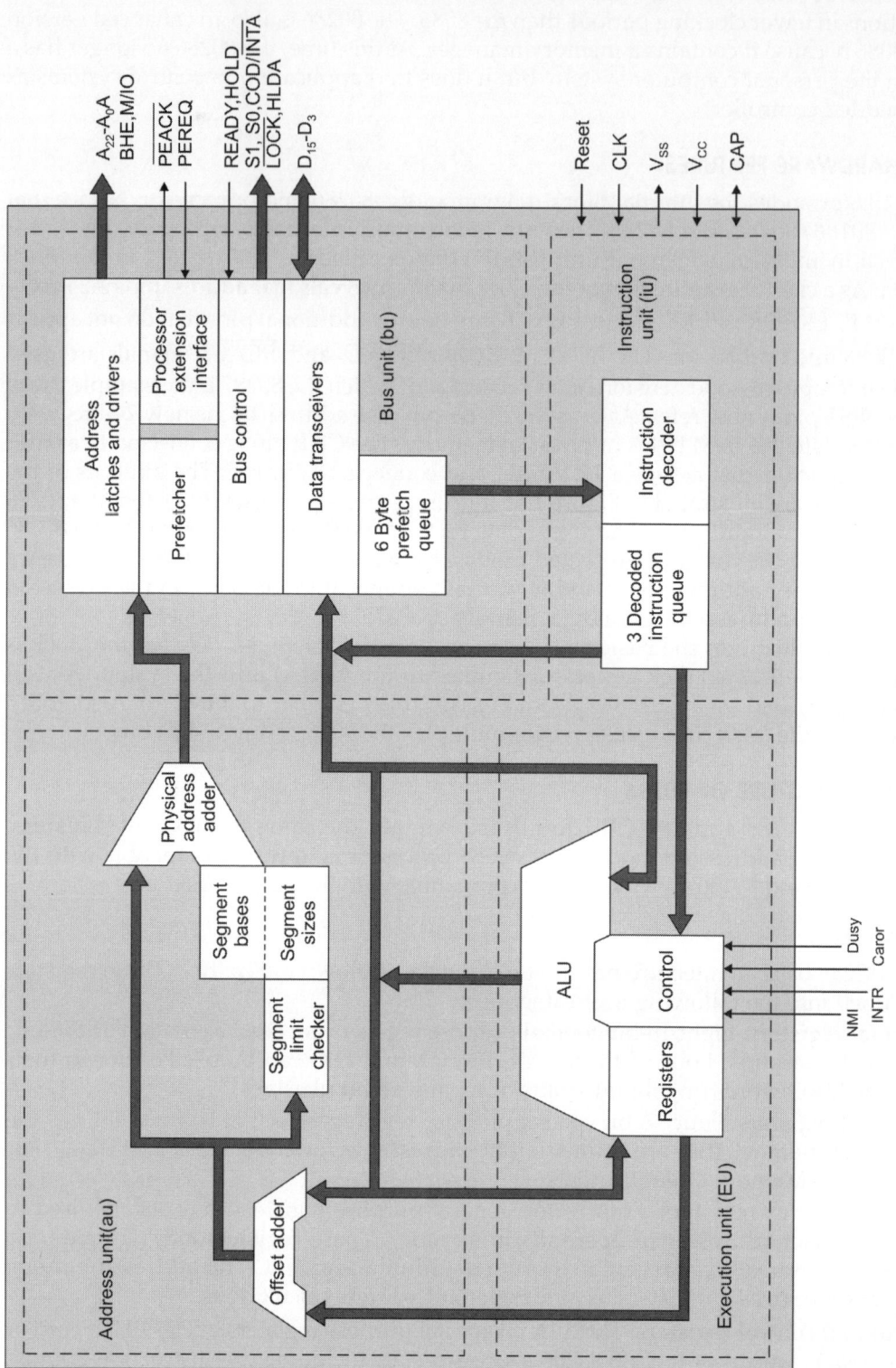

**Fig.10.6:** Architecture of 80286

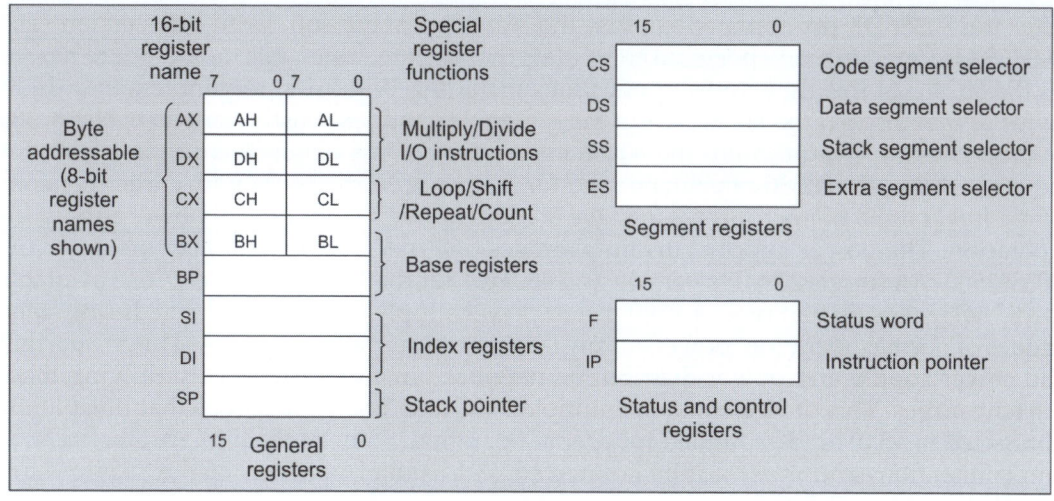

**Fig.10.7:** Register set of 80286

Flags Word Description: The Flags word (Flags) records specific characteristics of the result of logical and arithmetic instructions (bits 0, 2, 4, 6, 7 and 11) and controls the operation of the M80C286 within a given operating mode (bits 8 and 9). Flags is a 16-bit register.

*Flags Word Register*

| X | X | X | X | X | OF | DF | IF | TF | SF | ZF | X | AF | X | X | CF |
|---|---|---|---|---|----|----|----|----|----|----|---|----|---|---|----|

Where

   OF-overflow

   DF-direction flag

   IF-interrupt flag

   TF-trap flag

   SF-sign flag

   ZF-zero flag

   AF-auxiliary carry flag

   CF-carry flag

## 10.13 INTRODUCTION TO THE 80386 MICROPROCESSOR

Before the 80386 or any other microprocessor can be used in a system, the function of each pin must be understood 1. This section of the chapter details the operation of each pin along with the external memory system and I/O structures of the 80386. Figure 10.8 illustrates the pin-out of the 80386DX microprocessor, which is packaged in a 132-pin PGA (pin grid array). Two versions of the 80386 are commonly available: the 80386DX, illustrated and described in this chapter, is the full version, and the 80386SX is a reduced bus version of the 80386. A new version of the 80386-the 80386EX-incorporates the AT bus system, dynamic RAM controller, programmable chip selection logic, 26 address pins, 16 data pins and 24 I/O pins. The 80386DX addresses 4 G bytes of memory through its 32-bit data bus and 32-bit address. The 80386SX, more like the 80286, addresses 16 M bytes of memory with its 24-bit address bus via its 16-bit data bus. The 80386SX was developed after the 80386DX for applications that did not require the full 32-bit bus version. The 80386SX is found in many personal computers that use the same basic motherboard design as the 80286. At this time, most applications including Windows, require less than 16 M bytes of memory, so the 80386SX is a fairly popular and less costly version of the 80386 microprocessor. Even though the 80486 has become a less expensive upgrade path for newer systems, the 80386 can still be used for many applications. As with earlier versions of the Intel family of microprocessors, the 80386 requires a single +5.0V power supply for operation. The power supply current averages 550 rnA for the 25 MHz version of the 80386, 500 rnA for the 20 MHz version and 450 rnA for the 16 MHz version. Also available is a 33 MHz version that requires 600 rnA of power supply current. Note that during some modes of normal operation, power supply current can surge to over 1.0 A. This means that the power supply and power distribution network must be capable of supplying these current surges. This device contains multiple VCC and VSS connections that must all be connected to +5.0V and grounded for proper operation. Some of the pins are labeled N/C (no connection) and must not be connected. Additional versions of the 80386SX are available with a +3.3V power supply. These are often found in portable notebook or laptop computers and are usually packaged in a surface mount device. Each 80386 output pin is capable of providing 4.0 rnA (address and data connections) or 5.0 rnA (other connections). This represents an increase in drive current as compared to the 2.0 rnA available on earlier 8086, 8088 and 80286 output pins. Each input pin represents a small load requiring only ±IO J.lA of current. In some systems except the smallest, these current levels require bus buffers.

## 10.14 PIN DIAGRAM OF 80386 (Fig. 10.8)

**A31-A2:** Address bus connections address any of the 1 G x 32 memory locations found in the 80386 memory system. Note that AO and Al are encoded in the bus enable (BE3-BEO) to select any or all of the four bytes in a 32-bit wide memory location. Also note that because the 80386SX contains a 16-bit data bus in place of the 32-bit data bus found on the 80386DX, A1 is present on the 80386SX and the bank selection signals are replaced with BHE and BLE. The BHE signal enables the upper data bus half and the BLE signal the lower.

**D31-DO:** Data bus connections transfer data between the microprocessor and its memory and I/O system. Note that the 80386SX contains DiS-Do.

**$\overline{BE3}$ – $\overline{BE0}$:** Bank enable signals select the access of a byte, word, or doubleword of data. These signals are generated internally by the microprocessor from address bits Al and AO. On the 80386SX, these pins are replaced by BHE, BLE and AI.

**$\overline{M/IO}$:** (Memory) It selects a memory device during logic I or an I/O device during logic O. During the I/O operation, the address bus contains a 16-bit I/O address on address connections AIS-A2.

**$\overline{W/R}$:** Write/read indicates that the current bus cycle is a write during logic 1 or a read during logic O. The address data strobe becomes active whenever the 80386 has issued a valid memory or I/O address. This signal is combined with the W/R signal to generate the separate read and write signals present in the earlier 8086–80286 microprocessor-based systems.

**RESET:** Reset initializes the 80386, causing it to begin executing software at memory location FFFFFFFOH. The 80386 is reset to the real mode and the left-most 12 address connections remain logic I's (FFFH) until a far jump or far call is executed. This allows compatibility with earlier microprocessors.

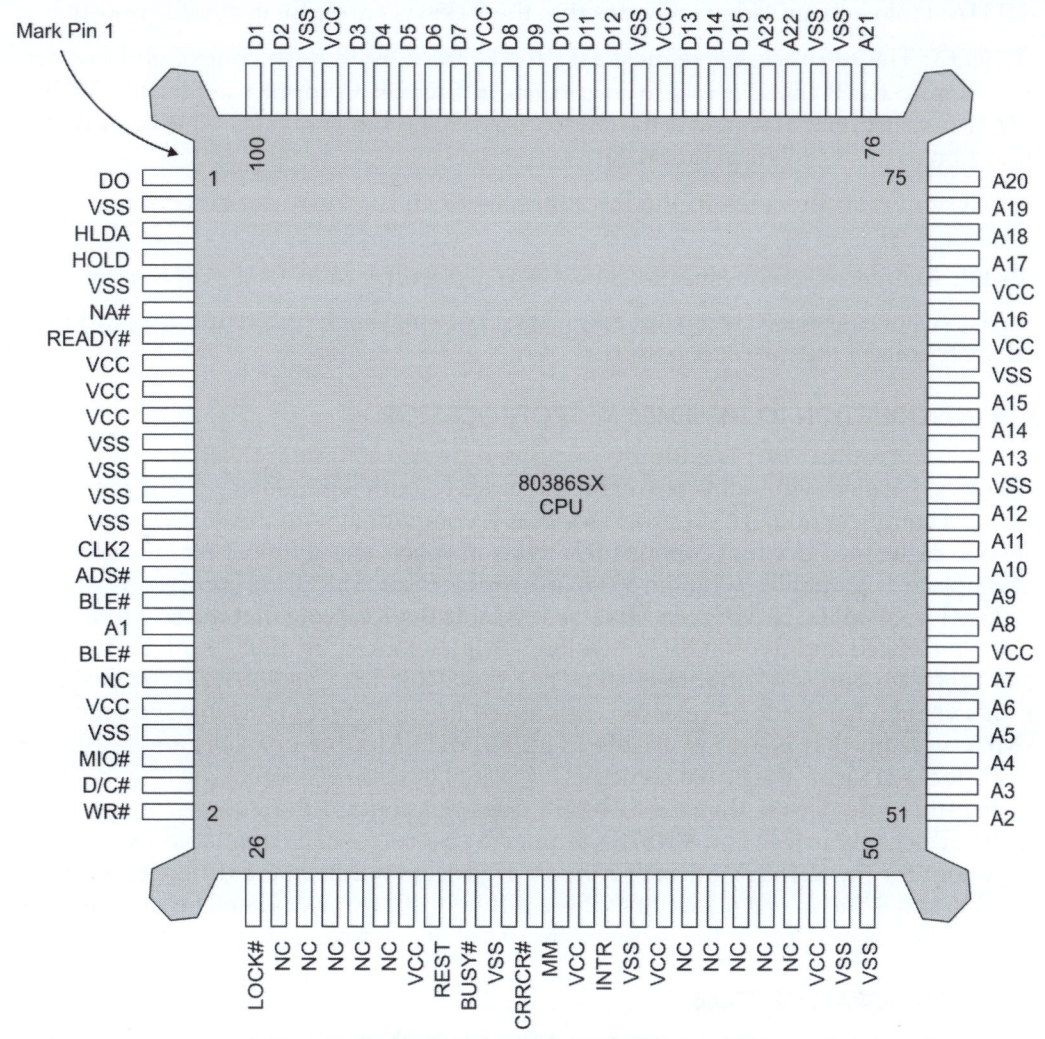

**Fig.10.8:** Pin diagram of 80386

**CLK2:** Clock times 2 is driven by a clock signal that is twice the operating frequency of the 80386. For example, to operate the 80386 at 16 MHz, we apply a 32 MHz clock to this pin.

$\overline{\text{READY}}$ : Ready controls the number of wait states inserted into the timing to lengthen memory accesses. Lock becomes a logic 0 whenever an instruction is prefixed with the LOCK: prefix. This is most often used during DMA accesses.

$\overline{\text{D/C}}$ : Data/control indicates that the data bus contains data for or from memory or I/O during logic 1. If *DIC* is a logic 0, the microprocessor is halted or executes an interrupt acknowledge. Bus size 16 selects either a 32-bit data bus (BS 16 = 1) or a 16-bit data bus (BS 16 = 0). In most cases, if an 80386DX is operated on a 16-bit data bus, we use the 80386SX that has a 16-bit data bus. Next address causes the 80386 to output the address of the next instruction or data in the current bus cycle. This pin is often used for pipelining the address.

**HOLD:** Hold requests a DMA action.

**HLDA:** Hold acknowledge indicates that the 80386 is currently in a hold condition.

**PEREQ :** The co-processor request asks the 80386 to relinquish control and is a direct connection to the 80387 arithmetic co-processor. Busy is an input used by the WAIT or F\VAIT instruction that waits for the co-processor to become not busy. This is also a direct connection to the 80387 from the 80386.

**ERROR :** Error indicates to the microprocessor that an error is detected by the co-processor.

**INTR:** An interrupt request is used by external circuitry to request an interrupt.

**NMI:** A non-maskable interrupt requests a non-maskable interrupt as it did on the earlier versions of the microprocessor.

## 10.15 INTRODUCTION TO THE 80486 MICROPROCESSOR

The 80486 microprocessor is a highly integrated device containing well over 1.2 million transistors. Located within this powerful integrated circuit is a memory-management unit (MMU); a complete numeric co-processor that is compatible with the 80387; a high-speed level one cache memory that contains 8 K bytes of space; and a full 32-bit microprocessor that is upward compatible with the 80386 microprocessor. The 80486 is currently available as a 25 MHz, 33 MHz, 50 MHz, 66 MHz or 100 MHz device. Note that the 66 MHz version is double-clocked and the 100 MHz version is triple-clocked. In 1990, Intel demonstrated a 100 MHz version (not double-clocked) of the 80486 for *Computer Design* magazine, but this version has yet to be released. Advanced Micro Devices (AMD) have produced a 40 MHz version that is also available in an 80 MHz (double-clocked) and a 120 MHz (triple-clocked) form. The 80486 comes as an 80486DX or an 80486SX. The only difference between these devices is that the 80486SX does not contain the numeric co-processor, which reduces its price. The 80487SX numeric co-processor is available as a separate component for the 80486SX microprocessor. This section details the differences between the 80486 and 80386 microprocessors. These differences are few, as shall be seen. The most notable differences apply to the cache memory system and parity generator.

## 10.16 PIN DIAGRAM OF 80486

Figure 10.9 1*Cr28* illustrates the pin-out of the 80486DX microprocessor, a 168-pin PGA. The 80486SX, also packaged in a 168-pin PGA, is not illustrated because only a few differences exist. Note that pin HIS is NMI on the 80486DX and pin A15 is NMI on the 80486SX. The only other differences are that pin A15 is IGNNE on the 80486DX (not present o~ the 80486SX), pin C14 is FERR on the 80486DX and pins HIS and C14 on the 80486SX are not connected. When connecting the 80486 microprocessor, all Vee and Vss pins must be connected to the power supply for proper operation. The power supply must be capable of supplying 5.0V ±10 % with up to 1.2 A of surge current for the 33 MHz version. The average supply current is 650 rnA for the 33 MHz version. Intel has also produced a 3.3V version that requires an average of 500 rnA at a triple-clock speed of 100 MHz. Logic 0 outputs allow up to 4.0 rnA of current and logic 1 outputs allow up to 1.0 rnA. If larger currents are required, as they often are, then the 80486 must be buffered. Figure 10.9 1*Cr29* shows a buffered 80486DX system. In the circuit shown, only the address, data and parity signals are buffered.

**A31-A2:** Address outputs A31-A2 provide the memory and I/O with the address during normal operation; during a cache line invalidation, A31-A4 are used to drive the microprocessor.

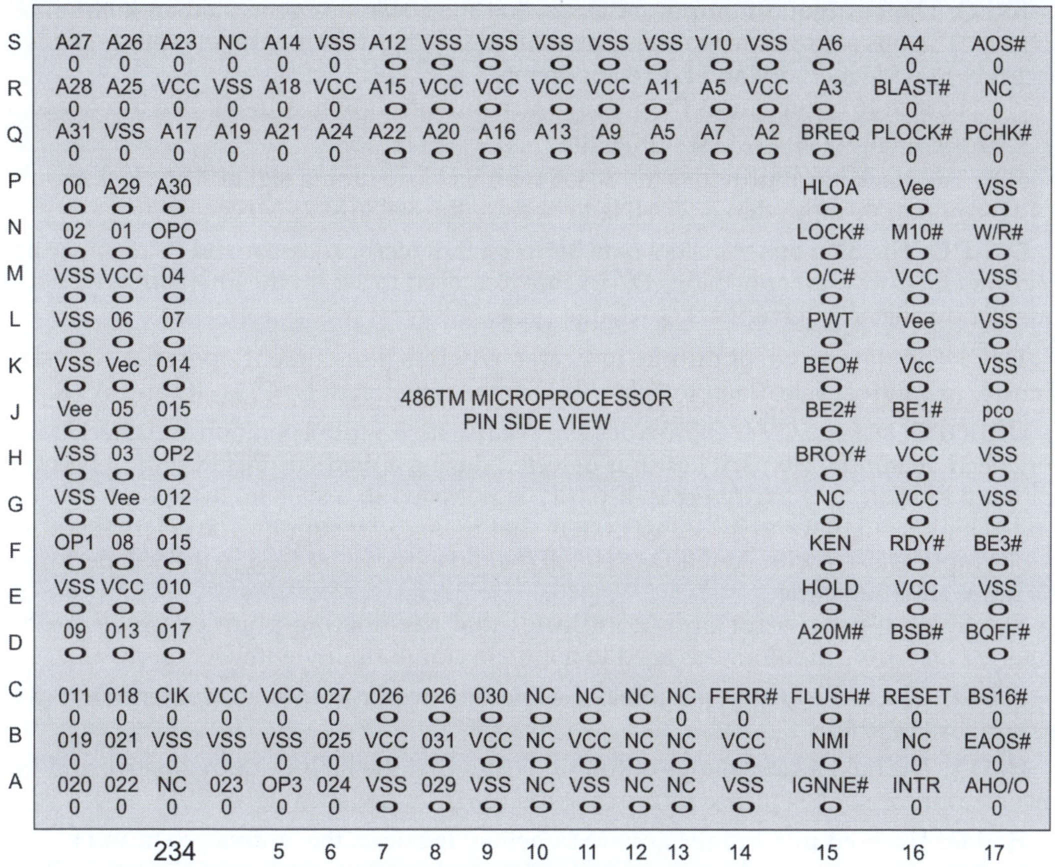

| | 234 | | | 5 | 6 | 7 | 8 | 9 | 10 | 11 | 12 | 13 | 14 | 15 | 16 | 17 |
|---|---|---|---|---|---|---|---|---|---|---|---|---|---|---|---|---|
| S | A27 | A26 | A23 | NC | A14 | VSS | A12 | VSS | VSS | VSS | VSS | VSS | V10 | VSS | A6 | A4 | AOS# |
| R | A28 | A25 | VCC | VSS | A18 | VCC | A15 | VCC | VCC | VCC | VCC | A11 | A5 | VCC | A3 | BLAST# | NC |
| Q | A31 | VSS | A17 | A19 | A21 | A24 | A22 | A20 | A16 | A13 | A9 | A5 | A7 | A2 | BREQ | PLOCK# | PCHK# |
| P | 00 | A29 | A30 | | | | | | | | | | | | HLOA | Vee | VSS |
| N | 02 | 01 | OPO | | | | | | | | | | | | LOCK# | M10# | W/R# |
| M | VSS | VCC | 04 | | | | | | | | | | | | O/C# | VCC | VSS |
| L | VSS | 06 | 07 | | | | | | | | | | | | PWT | Vee | VSS |
| K | VSS | Vec | 014 | | | | | | | | | | | | BEO# | Vcc | VSS |
| J | Vee | 05 | 015 | | | | | | | | | | | | BE2# | BE1# | pco |
| H | VSS | 03 | OP2 | | | | | | | | | | | | BROY# | VCC | VSS |
| G | VSS | Vee | 012 | | | | | | | | | | | | NC | VCC | VSS |
| F | OP1 | 08 | 015 | | | | | | | | | | | | KEN | RDY# | BE3# |
| E | VSS | VCC | 010 | | | | | | | | | | | | HOLD | VCC | VSS |
| D | 09 | 013 | 017 | | | | | | | | | | | | A20M# | BSB# | BQFF# |
| C | 011 | 018 | CIK | VCC | VCC | 027 | 026 | 026 | 030 | NC | NC | NC | NC | FERR# | FLUSH# | RESET | BS16# |
| B | 019 | 021 | VSS | VSS | VSS | 025 | VCC | 031 | VCC | NC | VCC | NC | NC | VCC | NMI | NC | EAOS# |
| A | 020 | 022 | NC | 023 | OP3 | 024 | VSS | 029 | VSS | NC | VSS | NC | NC | VSS | IGNNE# | INTR | AHO/O |

486TM MICROPROCESSOR
PIN SIDE VIEW

**Fig.10.9:** Pin diagram of 80486

**20M:** The address bit 20 mask causes the 80486 to wrap its address around from location OOOFFFFFH to OOOOOOOOH, as does the 8086 microprocessor. This provides a memory system that functions like the 1 M-byte real memory system in the 8086 microprocessor.

**ADS:** The address data strobe becomes a logic zero to indicate that the address bus contains a valid memory address.

**AHOLD:** The address hold input causes the microprocessor to place its address bus connections at their high-impedance state, with the remainder of the buses staying active. It is often used by another bus master to gain access for a cache invalidation cycle.

**BE3-BEO:** Byte enable outputs select a bank of the memory system when information is transferred between the microprocessor and its memory and I/O space. The BE3 signal enables D31-D24, BE2 enables D23-D16, BEl enables DIS-D8 and BEO enables D7-DO.

**BLAST:** The burst last output shows that the burst bus cycle is complete on the next activation of the BRDY signal.

**BOFF:** The back-off input causes the microprocessor to place its buses at their high-impedance state during the next clock cycle. The microprocessor remains in the bus hold state until the BOFF pin is placed at a logic 1 level. The burst ready input is used to signal the microprocessor that a burst cycle is complete.

**BREQ:** The bus request output indicates that the 80486 has generated an internal bus request. The bus size 8 input causes the 80486 to structure itself with an 8-bit data bus to access byte-wide memory and I/O components.

The bus size 16 input causes the 80486 to structure itself with a 16-bit data bus to access word-wide memory and I/O components.

**eLK:** The clock input provides the 80486 with its basic timing signal. The clock input is a TTL-compatible input that is 25 MHz to operate the 80486 at 25 MHz.

**D31-DO:** The data bus transfers data between the microprocessor and its memory and I/O system. Data bus connections D7-DO are also used to accept the interrupt vector type number during an interrupt acknowledge cycle.

**D/$\overline{C}$:** The data/control output indicates whether the current operation is a data transfer or control cycle. Refer to Table 10.2 for the function of D/C, M/IO and W/R.

**DP3-DPO:** Data parity I/O provides even parity for a write operation and check parity for a read operation. If a parity error is detected during a read, the PCHK output becomes a logic 0 to indicate a parity error. If parity is not used in a system, these lines must be pulled high to +5.0V or to 3.3 V in a system that uses a 3.3V supply. The external address strobe input is used with AHOLD to signal that an external address is used to perform a cache invalidation cycle.

The floating-point error output indicates that the floating-point co-processor has detected an error condition. It is used to maintain compatibility with DOS software.

**FLUSH:** The cache flush input forces the microprocessor to erase the contents of its 8 K-byte internal cache.

**HLDA:** The hold acknowledge output indicates that the HOLD input is active and that the microprocessor has placed its buses at their high-impedance state.

**HOLD:** The hold input requests a DMA action. It causes the address, data and control buses to be placed at their high-impedance state and also, once recognized, causes HLDA to become a logic O.

**IGNNE:** The ignore numeric error input causes the co-processor to ignore floating-point errors and to continue processing data. This signal does not affect the state of the FERR pin.

**INTR:** The interrupt request input requests a maskable interrupt as it does in all other family members.

The cache enable input causes the current bus to be stored in the internal cache. The lock output becomes a logic 0 for any instruction that is prefixed with the lock prefix.

**Table 10.2:** Function of M/IO, D/C and W/R

| M/IO | D/C | W/R | Bus cycle type |
|------|-----|-----|----------------|
| 0 | 0 | 0 | Interrupt acknowledge |
| 0 | 0 | 1 | Halt/special |
| 0 | 1 | 0 | I/O read |
| 0 | 1 | 1 | I/O write |
| 1 | 0 | 0 | Opcode fetch |
| 1 | 0 | 1 | Reserved |
| 1 | 1 | 0 | Memory read |
| 1 | 1 | 1 | Memory write |

**M/IO:** Memory/IO defines whether the address bus contains a memory address or an I/O port number. It is also combined with the W/R signal to generate memory and I/O read and write control signals.

**NMI:** The non-maskable interrupt input requests a type 2 interrupt.

**PCD:** The page cache disable output reflects the state of the peD attribute bit in the page table entry or the page directory entry. The parity check output indicates that a parity error was detected during a read operation on the DP3-DPO pins.

**PLOCK:** The pseudo-lock output indicates that the current operation requires more than one bus cycle to perform. This signal becomes a logic 0 for arithmetic co-processor operations that access 64- or 80-bit memory data.

**PWT:** The page write through output indicates the state of the PWT attribute bit in the page table entry or the page directory entry. The ready input indicates that a non-burst bus cycle is complete. The RDY signal must be returned or the microprocessor places wait states into its timing until RDY is asserted.

**RESET:** The reset input initializes the 80486 as it does in other family members. Table 10.2 shows the effect of the RESET input on the 80486 microprocessor.

**W/R:** Write/read signals that the current bus cycle is either a read or a write.

## 10.17 OVERALL ARCHITECTURAL COMPARISON OF THE PENTIUM FAMILY OF MICRO-PROCESSORS

The Pentium (P54) was first shipped in 1993 and had 3.1 million transistors. It used a 5 Volt to power its core and I/O logic, PGA on Socket 4, had a 2 × 8kb L1 cache, and operated at 50, 60 and 66 MHz. The system bus also operated at these speeds. The Pentium (P54C) was released in 1994 and had PGA on Socket 5 and 7, 3.3 Volts supply for core and I/O logic. It was also the first to use a multiplier to give processor speeds of 75, 90, 100, 120, 133, 150, 166 and 200 MHz. The last version of the first member of this sub-generation was the Pentium MMX (P55C). This had 4.1 million transistors, fit Socket 7 and had a 2 × 16 KB L1 cache with improved branch prediction logic. It operated at 2.8 V for its core logic and 3.3 V for I/O logic. Its 60 and 66 MHz system clock speed was multiplied on board the CPU to give between 120 and 300 MHz CPU clock speeds. Overall features included:

1. Superscalar architecture: Two integers [U (slow) and V (fast)] and one floating-point pipelines. The U and V pipelines contained 5 stages of instruction execution, while the floating-point pipeline had 8 stages. The U and V pipelines were served by two 32 byte prefetch buffers. This allowed overlapping execution of instructions in the pipelines.

2. Dynamic branch prediction used the Branch Target Buffer. The Pentium's branch prediction logic helped speed up program execution by anticipating branches and ensuring that branched-to-code was available in cache.

3. An Instruction and a Data Cache each of 8 K byte capacity

4. A 64-bit system data bus and 32-bit address bus

5. Dual processing capability

6. On-board Advanced Programmable Interrupt Controller

7. The Pentium MMX version contained an additional MMX unit that speeds up multimedia and 3D applications. Processing multimedia data involved instructions operating on large volumes of packetized data. Intel proposed a new approach: *single instruction multiple data*, which could operate on video pixels or Internet audio streams.

The MMX unit contained eight new 64-bit registers and 57 'simple' hardwired MMX instructions that operate on 4 new data types. To leverage the features of the MMX unit, applications were programmed to include the new instructions.

## 10.18 PENTIUM PRO: ARCHITECTURE FOR 6TH GENERATION PROCESSORS

The Pentium Pro was designed around the 6th generation P6 architecture, which was optimized for 32-bit instructions and 32-bit operating systems, such as Windows NT and Linux. It was the first of the P6 family, which included the Pentium II, the Celeron variants and the Pentium III. The physical package was also significant advance, as was the incorporation of additional RISC features. However, aimed as it was at the server market, the Pentium Pro did not incorporate MMX technology. It was expensive to produce as it included the L2 cache on its substrate (but on a separate die) and had 5.5 million transistors at its core and over 8 million in its L2 cache. Its core logic operated at 3.3 Volts. The microprocessor was still, however, chiefly CISC in design and optimized for 32-bit operation. The chief features of the Pentium Pro were:

1. A partly integrated L2 cache of up to 512 KB (on a specially manufactured SRAM separate die) that was connected via a dedicated 'backside' bus that ran at full CPU speed.
2. Three 12 staged pipelines
3. Speculative execution of instructions
4. Out-of-order completion of instructions
5. 40 renamed registers
6. Dynamic branch prediction
7. Multi-processing with up to 4 Pentium Pros
8. An increased bus size to 36-bits (from 32) to enable up to 64 Gb of memory to be used. (Please note that the 4 extra bits can address up to 16 memory locations; this gives 4 Gb × 16 = 64 Gb of memory.)

The following description is taken from Intel's introduction to its microprocessor architecture which is relevant to all members of the P6 family including the Celeron, Pentium II and III. The Intel Pentium Pro processor had three-way superscalar architecture. The term "three-way superscalar" means that using parallel processing techniques, the processor is able on average to decode, dispatch and complete execution of (retire) three instructions per clock cycle. To handle this level of instruction throughput, the Pentium Pro processor used a decoupled, 12-stage super pipeline that supports out-of-order instruction execution. It did this by incorporating even more parallelism than the Pentium processor. The Pentium Pro processor provided Dynamic Execution (micro-data-flow analysis, out-of-order execution, superior branch prediction and speculative execution) in a superscalar implementation.

The centerpiece of the Pentium Pro processor architecture was an innovative out-of-order execution mechanism called "dynamic execution." Dynamic execution incorporates three data-processing concepts:

a. Deep branch prediction
b. Dynamic data-flow analysis
c. Speculative execution.

Branch prediction is a concept found in most mainframe and high-speed RISC microprocessor architectures. It allows the processor to decode instructions beyond branches to keep the instruction pipeline full. In the Pentium Pro processor, the instruction

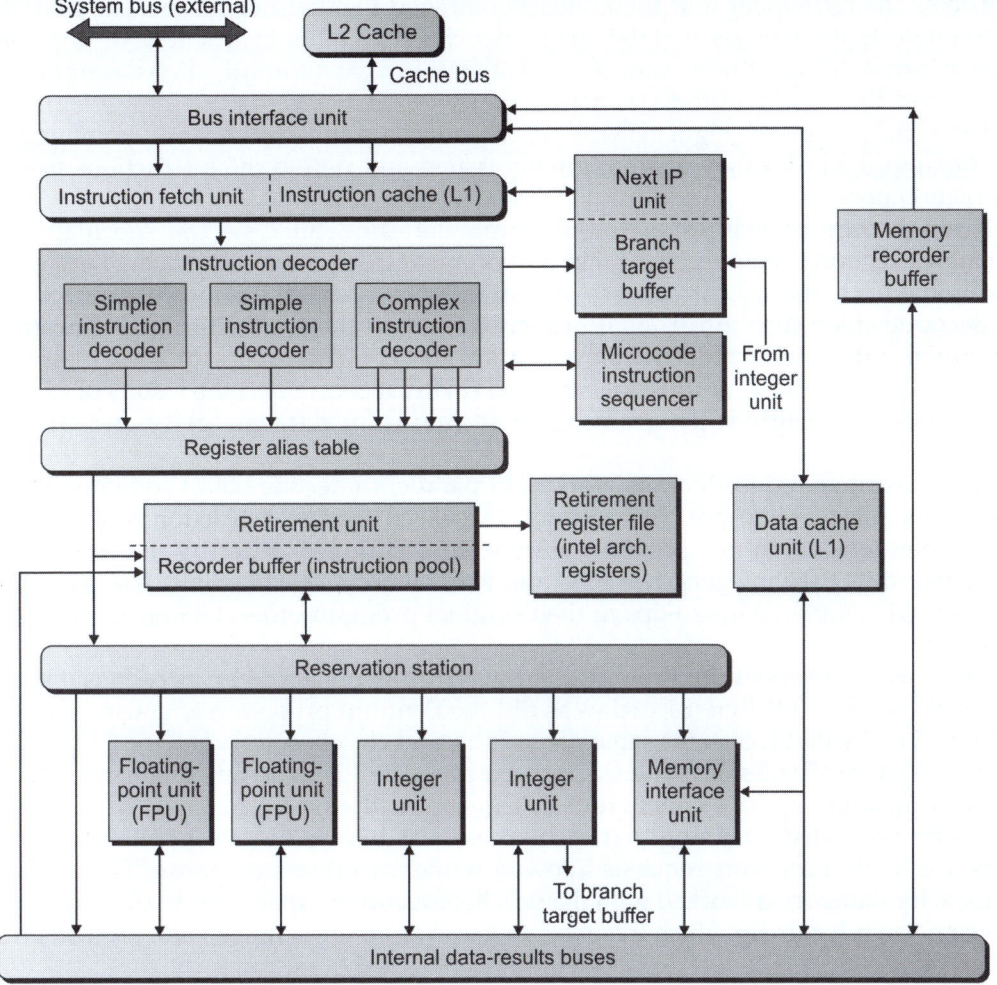

**Fig.10.10:** Functional block diagram of the pentium pro processor micro-architecture

fetch/decode unit used a highly optimized branch prediction algorithm to predict the direction of the instruction stream through multiple levels of branches, procedure calls and returns.

Dynamic data-flow analysis involves real-time analysis of the flow of data through the processor to determine data and register dependencies and to detect opportunities for out-of-order instruction execution. The Pentium Pro processor dispatch/execute unit can simultaneously monitor many instructions and execute these instructions in the order that optimizes the use of the processor's multiple execution units, while maintaining the integrity of the data being operated on. This out-of-order execution keeps the execution units busy even when cache misses and data dependencies among instructions occur.

Speculative execution refers to the processor's ability to execute instructions ahead of the program counter, but ultimately to commit the results in the order of the original instruction stream. To make speculative execution possible, the Pentium Pro processor micro-architecture decoupled the dispatching and executing of instructions from the commitment of results. The processor's dispatch/execute unit used data-flow analysis to execute all available instructions in the instruction pool and store the results in temporary registers. The retirement unit then linearly searched the instruction pool for completed instructions that no longer had data dependencies with other instructions or unresolved branch predictions. When completed instructions were found, the retirement unit committed the results of these instructions to memory and/or the Intel Architecture registers (the processor's eight general-purpose registers and eight floating-point unit data registers) in the order they were originally issued and retired the instructions from the instruction pool.

Through deep branch prediction, dynamic data-flow analysis and speculative and dynamic execution removed the constraint of linear instruction sequencing between the traditional fetch and execute phases of instruction execution. It allowed instructions to be decoded deep into multi-level branches to keep the instruction pipeline full. It promoted out-of-order instruction execution to keep the processor's six instruction execution units running at full capacity. And finally it committed the results of executed instructions in original program order to maintain data integrity and program coherency.

Three instruction decode units worked in parallel to decode object code into smaller operations called "micro-ops" (microcode). These went into an instruction pool and (when interdependencies do not prevent) were executed out-of-order by the five parallel execution units (two integers, two FPU and one memory interface unit). The Retirement Unit retired completed micro-ops in their original program order, taking account of any branches.

The power of the Pentium Pro processor was further enhanced by its caches: it had the same two on-chip 8-K Byte L1 caches as did the Pentium processor, and also had a 256–512 KB L2 cache that was in the same package as, and closely coupled to, the CPU, using a dedicated 64-bit ("backside") full clock speed bus. The L1 cache was dual ported, the L2 cache supported up to 4 concurrent accesses and the 64-bit external data bus was transaction-oriented, meaning that each access was handled as a separate request and response, with numerous requests allowed while awaiting a response. These parallel features for data access worked with the parallel execution capabilities to provide a "non-blocking" architecture in which the processor was more fully utilized and performance is enhanced.

### 10.18.1 Pentium Pro Modes of Operation

The Intel I-32 Architecture supports three operating modes: protected mode, real-address mode and system management mode. The operating mode determines which instructions and architectural features are accessible.

1. **Protected mode:** It is the native state of the processor. In this mode, all instructions and architectural features are available, providing the highest performance and capability. This is the recommended mode for all new applications and operating systems. Among the capabilities of protected mode is the ability to directly execute "real-address mode" 8086 software in a protected, multi-tasking environment. This feature is called **virtual-8086 mode**, although it is not actually a processor mode. Virtual-8086 mode is actually a protected mode attribute that can be enabled for any task.

2. **Real-address mode:** This mode provides the programming environment of the Intel 8086 processor with a few extensions (such as the ability to switch to protected or system management mode). The processor is placed in real-address mode following power-up or a reset.

3. **System management mode:** This mode has a standard architectural feature unique to all Intel processors, beginning with the Intel 386 SL processor. This mode provides

an operating system or executive with a transparent mechanism for implementing platform-specific functions, such as power management and system security. The processor enters SMM when the external SMM interrupt pin (SMI#) is activated or an SMI is received from the advanced programmable interrupt controller (APIC). In SMM, the processor switches to a separate address space while saving the entire context of the currently running program or task. SMM-specific code may then be executed transparently. Upon returning from SMM, the processor is placed back into its state prior to the system management interrupt.

The basic execution environment is the same for each of these operating modes.

## 10.19 PENTIUM II

The Pentium II incorporated many of the salient features of the Pentium Pro and Pentium MMX; however, its physical package was based on the SECC/Slot 1 interface and its 512 KB L2 cache ran at only half the processor internal clock rate. First generation Pentium II Klamath CPUs operated at 233, 266, 300 and 333 MHz with an FSB of 66 MHz and a core voltage of 2.8 Volts. In 1998, Intel introduced the Pentium II Deschutes that operated at a speed of 350, 400 and 450 MHz with a 100 MHz and later 66 MHz, FSB and at 2.0 Volts at the core. Its major improvements were:

1. 16 KB L1 instruction and data caches
2. L2 cache with non-proprietary commercially available SRAM
3. Improved 16-bit capability through segment register caches
4. MMX unit
5. Standard Pentium II could only be used in dual multi-processor configurations; however, Pentium Xeon Processors had up to 2 MB of L2 cache and could be used in multi-processor configurations of up to 4 processors.

## 10.20 CELERON

The Celeron began as a scaled down version of the Pentium II and was designed to compete against similar offerings from Intel's competitors. The Klamath-based Celeron Covington core ran at 266 and 300 MHz and was constructed without an L2 cache. However, adverse market reaction saw the Deschutes-based Mendocino core introduced with an 128 Kb L2 cache and ran at 300, 333, 400, 433, 466, 500 and 533 MHz. Celerons had the same L1 cache as their bigger brothers—Pentium II and III. The important distinction is that the L2 cache operated at full CPU clock rates, unlike the Pentium II and the SECC packaged Pentium III. Later variants of the Pentium III had an on-die L2 cache which ran at full CPU clock rate. The Celeron III (Coppermine128 core) had the same internal features as the Pentium III, but has reduced functionality: 66 MHz clock rate, no error correction codes for the data bus, and parity creation for the address bus, and a maximum of 4 GB of address space. Celeron III Coppermine128s with a 1.6 V core and a 100 MHz were produced in 2001 and operated at core speeds of up to 1.1 MHz. Tualatin-core Celerons were put on the market in late 2001 and ran at 1.2 GHz. 2002 saw the final versions produced running at 1.3 and 1.4 MHz.

## 10.21 PENTIUM III

The only significant difference between the Pentium III and its predecessor was the inclusion of 72 MMX instructions, known as the Internet Streaming Single Instruction Multiple Data Extensions (ISSE), they include integer and floating-point operations.

However, like the original MMX instructions, application programmers must include the corresponding extensions if any use is to be made of these instructions. The most controversial and short-lived addition was the CPU ID number which could be used for software licensing and e-commerce. After protest from various sources, Intel disabled it as default, but did not remove it. Depending on the BIOS and motherboard manufacturer, it may remain as such but it can be enabled via the BIOS. In reality, Pentium III performance was based. The three variants of Pentium III were the Katami, Coppermine and Tualatin. Katami first introduced the ISSE (MMX/2) as described with an FSB of 100 MHZ. The Coppermine also introduced Advanced Transfer Cache (ATC) for the L2 cache which reduced cache capacity to 256 KB but saw the cache run at full processor speed. Also the 64-bit Katami cache bus was quadrupled to 256-bits. Coppermine also uses an 8-way set associative cache, rather than the 4-way set associative cache in the Katami and older Pentiums. Bringing the cache on-die also increased the transistor count to 30 million, from the 10 million on the Katami. Another advance in the Coppermine was Advanced System Buffering (ASB), which simply increased the number of buffers to account for the increased FSB speed of 133 MHz. The Pentium III Tualatin had a reduced die size that allowed it to run at higher speeds. Tualatins use a 133 MHz FSB and have ATC and ASB.

## 10.22 PENTIUM IV: THE NEXT GENERATION

The release of the Pentium IV in 2000 heralded the seventh generation of Intel microprocessors. The release was premature, however, due to the outperformance of the Pentium III Coppermine, with its 1 GHz performance threshold, by Intel's major competitor in the microprocessor market, the AMD Athlon. Intel was not ready to answer the competition through the early release of the next member of its Pentium III family, the Pentium III Tualatins, which were designed to break the 1 GHz barrier. Previous attempts to do so with the Pentium III Coppermine 1.13 GHz met with failure due to design flaws. Paradoxically, however, Intel was in a position to release the first of the Pentium IV family, the Willamette, which ran at 1.3, 1.4 and 1.5 MHz, using a FC-PGA package on the short-lived Socket 423, which was a design dead-end for motherboard manufacturers and consumers. Worse still, the only Intel chipset available for the Pentium IV could only house the highly expensive RAMBus DRAM. In addition, the early versions of Pentium IV CPU were outperformed by slower AMD Athlons. Nevertheless, the core capability of Intel's seventh generation processors is that they can run at ever-higher speeds. For example, Intel's sixth generation Pentiums began at 120 MHz with the Pentium Pro and ended at over 1.2 GHz, a tenfold increase. The bottom line here is that Intel's seventh generation chips could end up running at speeds of 10 GHz or more. How has Intel achieved this? Through a radical redesign of the Pentium's core architecture. The following sections illustrate the major advances, while a general overview of all Intel processors may be found at http://www.intel.com/products/roadmap/.

The most visible feature seen of the new Pentium IV is the Front Side Bus (FSB) which initially operated at equivalent speed of 400 MHz as compared to 100 MHz on the Pentium III. The Pentium III has a 64-bit data bus that delivered a data throughput of 1.066 GB (64* 133 = 1.066). The Pentium IV FSB bus is also 64-bit wide, however, in the earlier model, the 100 MHz bus speed was 'quad-pumped' giving an effective bus speed of 400 MHz and a data transfer rate of 3.2 Gbps. In late 2002, Pentium IV and associated chipsets operated at 133 MHz core bus speed, which was 'quad-pumped' to 533 MHz, delivering a throughput of 4.2 Gbps. Thus, the first Pentium IV versions exchanged data with the i845 and i850 chipsets faster than any other processor, thereby removing the Pentium III's most significant bottleneck. For example, the Intel 850 chipset for the Pentium IV used two RAMBus channels to 2–4 RDRAM RIMMs. Together, these two RDRAM channels are able to deliver the same data bandwidth as the Pentium IV FSB. As the earlier discussion on DRAM indicates, similar transfer rates are delivered using the i845 chipset and DDR DRAM. In 2004, the speed of the FSB on the Pentium IV was increased to 800 MHz, with a throughput of 6.4 Gbps. This is accomplished through a physical signaling scheme of 'quad-pumping' the data transfers over a 200 MHz clocked system bus and a buffering scheme allowing for sustained 800 MHz data transfers.

## PROBLEMS

1. What are the basic differences between 8086 and 8087 microprocessors?

2. Explain the internal architecture of 8087 in detail.

3. Explain data transfer instructions of 8087 with examples.

4. Explain arithmetic instructions of 8087 with examples.

5. Draw the pin diagram of 80186 and explain its pin functions.

6. Explain the different pin functions of 80286 microprocessor.

7. Explain the internal architecture of 80286.

8. Draw the pin diagram of 80386 and explain its pin functions.

9. Draw the pin diagram of 80486 and explain its pin functions.

10. Explain the internal architecture of Pentium pro processor.

11. Give an overview of Celeron processor.

12. Give an overview of Pentium II processor.

13. Give an overview of Pentium III processor.

14. Give an overview of Pentium IV processor.

# Appendix 1

**Table A1.1:** Instruction set summary

| *Mnemonic and description* | *Instruction code* | | | |
|---|---|---|---|---|
| **DATA TRANSFER** | | | | |
| **MOV = Move:** | 76543210 | 76543210 | 76543210 | 76543210 |
| Register/memory to/from register | 100010 dw | mod reg r/m | | |
| Immediate to register/memory | 1100011 w | mod 000 r/m | data | data if w = 1 |
| Immediate to register | 1011 w reg | data | data if w = 1 | |
| Memory to accumulator | 1010000 w | addr-low | addr-high | |
| Accumulator to memory | 1010001 w | addr-low | addr-high | |
| Register/memory to segment register | 10001110 | mod 0 reg r/m | | |
| Segment register to register/memory | 10001100 | mod 0 reg r/m | | |
| **PUSH = Push:** | | | | |
| Register/memory | 11111111 | mod 110 r/m | | |
| Register | 01010 reg | | | |
| Segment register | 000 reg 110 | | | |
| **POP = Pop:** | | | | |
| Register/memory | 10001111 | mod 000 r/m | | |
| Register | 01011 reg | | | |
| Segment register | 000 reg 111 | | | |
| **XCHG = Exchange:** | | | | |
| Register/memory with register | 1000011 w | mod reg r/m | | |
| Register with accumulator | 10010 reg | | | |
| **IN = Input from:** | | | | |
| Fixed port | 1110010 w | port | | |
| Variable port | 1110110 w | | | |
| **OUT = Output to:** | | | | |
| Fixed port | 1110011 w | port | | |
| Variable port | 1110111 w | | | |
| **XLAT** = Translate byte to AL | 11010111 | | | |
| **LEA** = Load EA to register | 10001101 | mod reg r/m | | |
| **LDS** = Load pointer to DS | 11000101 | mod reg r/m | | |
| **LES** = Load pointer to ES | 11000100 | mod reg r/m | | |

*(Contd...)*

**Table A1.1:** Instruction set summary *(Contd...)*

| *Mnemonic and description* | *Instruction code* | | | |
|---|---|---|---|---|
| **LAHF** = Load AH with flags | 10011111 | | | |
| **SAHF** = Store AH into flags | 10011110 | | | |
| **PUSHF** = Push flags | 10011100 | | | |
| **POPF** = Pop flags | 10011101 | | | |
| **ARITHMETIC** | 76543210 | 76543210 | 7654321 0 | 76543210 |
| **ADD = Add:** | | | | |
| Reg./memory with register to either | 000000 dw | mod reg r/m | | |
| Immediate to register/memory | 100000 sw | mod 000 r/m | data | data if s: w = 01 |
| Immediate to accumulator | 0000010 w | data | data if w = 1 | |
| **ADC = Add with carry:** | | | | |
| Reg./memory with register to either | 000100 dw | mod reg r/m | | |
| Immediate to register/memory | 100000 sw | mod 010 r/m | data | data if s: w = 01 |
| Immediate to accumulator | 0001010 w | data | data if w = 1 | |
| **INC = Increment:** | | | | |
| Register/memory | 1111111 w | mod 000 r/m | | |
| Register | 01000 reg | | | |
| **AAA** = ASCII adjust for add | 00110111 | | | |
| **BAA** = Decimal adjust for add | 00100111 | | | |
| **SUB = Subtract:** | | | | |
| Reg./memory and register to either | 001010 dw | mod reg r/m | | |
| Immediate from register/memory | 100000 sw | mod 101 r/m | data | data if s w = 01 |
| Immediate from accumulator | 0010110 w | data | data if w = 1 | |
| **SSB = Subtract with borrow** | | | | |
| Reg./memory and register to either | 000110 dw | mod reg r/m | | |
| Immediate from register/memory | 100000 sw | mod 011 r/m | data | data if s w = 01 |
| Immediate from accumulator | 000111 w | data | data if w = 1 | |
| **DEC = Decrement:** | | | | |
| Register/memory | 1111111 w | mod 001 r/m | | |
| Register | 01001 reg | | | |
| **NEG** = Change sign | 1111011 w | mod 011 r/m | | |
| **CMP = Compare:** | | | | |
| Register/memory and register | 001110 dw | mod reg r/m | | |
| Immediate with register/memory | 100000 sw | mod 111 r/m | data | data if s w = 01 |
| Immediate with accumulator | 0011110 w | data | data if w = 1 | |
| **AAS** = ASCII adjust for subtract | 00111111 | | | |
| **DAS** = Decimal adjust for subtract | 00101111 | | | |
| **MUL** = Multiply (unsigned) | 1111011 w | mod 100 r/m | | |
| **IMUL** = Integer multiply (signed) | 1111011 w | mod 101 r/m | | |
| **AAM** = ASCII adjust for multiply | 11010100 | 00001010 | | |

*(Contd...)*

**Table A1.1:** Instruction set summary *(Contd...)*

| *Mnemonic and description* | *Instruction code* | | | |
|---|---|---|---|---|
| **DIV** = Divide (unsigned) | 1111011 w | mod 110 r/m | | |
| **IDIV** = Integer divide (signed) | 1111011 w | mod 111 r/m | | |
| **AAD** = ASCII adjust for divide | 11010101 | 00001010 | | |
| **CBW** = Convert byte to word | 10011000 | | | |
| **CWD** = Convert word to double word | 10011001 | | | |
| **LOGIC** | 76543210 | 76543210 | 76543210 | 76543210 |
| **NOT** = Invert | 1111011 w | mod 010 r/m | | |
| **SHL/SAL** = Shift logical/arithmetic left | 110100 vw | mod 100 r/m | | |
| **SHR** = Shift logical right | 110100 vw | mod 101 r/m | | |
| **SAR** = Shift arithmetic right | 110100 vw | mod 111 r/m | | |
| **ROL** = Rotate left | 110100 vw | mod 000 r/m | | |
| **ROR** = Rotate right | 110100 vw | mod 001 r/m | | |
| **RCL** = Rotate through carry flag left | 110100 vw | mod 010 r/m | | |
| **RCR** = Rotate through carry right | 110100 vw | mod 011 r/m | | |
| **AND** = **And:** | | | | |
| Reg./memory and register to either | 001000 dw | mod reg r/m | | |
| Immediate to register/memory | 1000000 w | mod 100 r/m | data | data if w = 1 |
| Immediate to accumulator | 0010010 w | data | data if w = 1 | |
| **TEST** = **And function to flags, no result:** | | | | |
| Register/memory and register | 1000010 w | mod reg r/m | | |
| Immediate data and register/memory | 1111011 w | mod 000 r/m | data | data if w = 1 |
| Immediate data and accumulator | 1010100 w | data | data if w = 1 | |
| **OR** = **Or:** | | | | |
| Reg./memory and register to either | 000010 dw | mod reg r/m | | |
| Immediate to register/memory | 1000000 w | mod 001 r/m | data | data if w = 1 |
| Immediate to accumulator | 0000110 w | data | data if w = 1 | |
| **XOR** = **Exclusive or:** | | | | |
| Reg./memory and register to either | 001100 dw | mod reg r/m | | |
| Immediate to register/memory | 1000000 w | mod 110 r/m | data | data if w = 1 |
| Immediate to accumulator | 0011010 w | data | data if w = 1 | |
| **STRING MANIPULATION** | | | | |
| **REP** = Repeat | 1111001 z | | | |
| **MOVS** = Move byte/word | 1010010 w | | | |
| **CMPS** = Compare byte/word | 1010011 w | | | |
| **SCAS** = Scan byte/word | 1010111 w | | | |
| **LODS** = Load byte/wd to AL/AX | 1010110 w | | | |
| **STOS** = Store byte/wd from AL/A | 1010101 w | | | |
| **CONTROL TRANSFER** | | | | |
| **CALL** = **Call:** | | | | |
| Direct within segment | 11101000 | disp-low | disp-high | |
| Indirect within segment | 11111111 | mod 010 r/m | | |

*(Contd...)*

**Table A1.1:** Instruction set summary *(Contd...)*

| *Mnemonic and description* | | *Instruction code* | |
|---|---|---|---|
| Direct intersegment | 10011010 | offset-low<br>seg-low | offset-high<br>seg-high |
| Indirect intersegment | 11111111 | mod 011 r/m | |
| **JMP = Unconditional jump:** | **76543210** | **76543210** | **76543210** |
| Direct within segment | 11101001 | disp-low | disp-high |
| Direct within segment-short | 11101011 | disp | |
| Indirect within segment | 11111111 | mod 100 r/m | |
| Direct intersegment | 11101010 | offset-low<br>seg-low | offset-high<br>seg-high |
| Indirect intersegment | 11111111 | mod 101 r/m | |
| **RET = Return from CALL:** | | | |
| Within segment | 11000011 | | |
| Within seg adding immed to SP | 11000010 | data-low | data-high |
| Intersegment | 11001011 | | |
| Intersegment adding immediate to SP | 11001010 | data-low | data-high |
| **JE/JZ** = Jump on equal/zero | 01110100 | disp | |
| **JL/JNGE** = Jump on less/not greater or equal | 01111100 | disp | |
| **JLE/JNG** = Jump on less or equal/ not greater | 01111110 | disp | |
| **JB/JNAE** = Jump on below/not above or equal | 01110010 | disp | |
| **JBE/JNA** = Jump on below or equal/ not above | 01110110 | disp | |
| **JP/JPE** = Jump on parity/parity even | 01111010 | disp | |
| **JO** = Jump on overflow | 01110000 | disp | |
| **JS** = Jump on sign | 01111000 | disp | |
| **JNE/JNZ** = Jump on not equal/not zero | 01110101 | disp | |
| **JNL/JGE** = Jump on not less/greater or equal | 01111101 | disp | |
| **JNLE/JG** = Jump on not less or equal/ greater | 01111111 | disp | |
| **JNB/JAE** = Jump on not below/above or equal | 01110011 | disp | |
| **JNBE/JA** = Jump on not below or equal/above | 01110111 | disp | |
| **JNP/JPO** = Jump on not par/par odd | 01111011 | disp | |
| **JNO** = Jump on not overflow | 01110001 | disp | |
| **JNS** = Jump on not sign | 01111001 | disp | |
| **LOOP** = Loop CX times | 11100010 | disp | |
| **LOOPZ/LOOPE** = Loop while zero/equal | 11100001 | disp | |
| **LOOPNZ/LOOPNE** = Loop while not zero/equal | 11100000 | disp | |

*(Contd...)*

**Table A1.1:** Instruction set summary *(Contd...)*

| Mnemonic and description | Instruction code | |
|---|---|---|
| JCXZ = Jump on CX zero | 11100011 | disp |
| INT = Interrupt | | |
| Type specified | 11001101 | type |
| Type 3 | 11001100 | |
| INTO = Interrupt on overflow | 11001110 | |
| IRET = Interrupt return | 11001111 | |
| **PROCESSOR CONTROL** | **76543210** | **76543210** |
| **CLC** = Clear carry | 11111000 | |
| **CMC** = Complement carry | 11110101 | |
| **STC** = Set carry | 11111001 | |
| **CLD** = Clear direction | 11111100 | |
| **STD** = Set direction | 11111101 | |
| **CLI** = Clear interrupt | 11111010 | |
| **STI** = Set interrupt | 11111011 | |
| **HLT** = Halt | 11110100 | |
| **WAIT** = Wait | 10011011 | |
| **ESC** = Escape (to external device) | 11011xxx | mod xxx r/m |
| **LOCK** = Bus lock prefix | 11110000 | |

*Notes:*

AL = 8-bit accumulator

AX = 16-bit accumulator

CX = Count register

DS = Data segment

ES = Extra segment

Above/below refers to unsigned value

Greater = More positive;

Less = Less positive (more negative) signed values

if d = 1 then "to" reg; if d = 0 then "from" reg

if w = 1 then word instruction; if w = 0 then byte instruction

if mod = 11 then r/m is treated as a REG field

if mod = 00 then DISP = 0*, disp-low and disp-high are absent

if mod = 01 then DISP = disp-low sign-extended to 16-bits, disp-high is absent

if mod = 10 then DISP = disp-high; disp-low

if r/m = 000 then EA = (BX) + (SI) + DISP

if r/m = 001 then EA = (BX) + (DI) + DISP

if r/m = 010 then EA = (BP) + (SI) + DISP

if r/m = 011 then EA = (BP) + (DI) + DISP

if r/m = 100 then EA = (SI) + DISP

if r/m = 101 then EA = (DI) + DISP

if r/m = 110 then EA = (BP) + DISP*

if r/m = 111 then EA = (BX) + DISP

DISP follows 2nd byte of instruction (before data if required)

*except if mod = 00 and r/m = 110 then EA = disp-high; disp-low.

Mnemonics © Intel, 1978

## DATA SHEET REVISION REVIEW

The following list represents key differences between this and the –004 data sheet. Please review this summary carefully.

1. The Intel 8086 implementation technology (HMOS) has been changed to (HMOS-III).
2. Delete all "changes from 1985 Handbook Specification" sentences.
   If s w = 01 then 16-bits of immediate data form the operand
   If s w = 11 then an immediate data byte is sign extended to form the 16-bit operand
   If v = 0 then "count" = 1; if v = 1 then "count" in (CL)
   x = Don't care
   z is used for string primitives for comparison with ZF FLAG.

## Segment Override Prefix

**0 0 1 reg 1 1 0**

REG is assigned according to Table A1.2:

**Table A1.2:** Segment override prefix

| 16-Bit (w = 1) | | 8-Bit (w = 0) | | Segment | |
|---|---|---|---|---|---|
| 000 | AX | 000 | AL | 00 | ES |
| 001 | CX | 001 | CL | 01 | CS |
| 010 | DX | 010 | DL | 10 | SS |
| 011 | BX | 011 | BL | 11 | DS |
| 100 | SP | 100 | AH | | |
| 101 | BP | 101 | CH | | |
| 110 | SI | 110 | DH | | |
| 111 | DI | 111 | BH | | |

Instructions which reference the flag register file as a 16-bit object use the symbol FLAGS to represent the file:

FLAGS = X:X:X:X:(OF):(DF):(IF):(TF):(SF):(ZF):X:(AF):X:(PF):X:(CF)

# Appendix 2

| S. No. | Mnemonics, operand | Opcode | Bytes |
|:---:|:---:|:---:|:---:|
| | **Opcode table of INTEL 8085 in alphabetical order** | | |
| 1. | ACI Data | CE | 2 |
| 2. | ADC A | 8F | 1 |
| 3. | ADC B | 88 | 1 |
| 4. | ADC C | 89 | 1 |
| 5. | ADC D | 8A | 1 |
| 6. | ADC E | 8B | 1 |
| 7. | ADC H | 8C | 1 |
| 8. | ADC L | 8D | 1 |
| 9. | ADC M | 8E | 1 |
| 10. | ADD A | 87 | 1 |
| 11. | ADD B | 80 | 1 |
| 12. | ADD C | 81 | 1 |
| 13. | ADD D | 82 | 1 |
| 14. | ADD E | 83 | 1 |
| 15. | ADD H | 84 | 1 |
| 16. | ADD L | 85 | 1 |
| 17. | ADD M | 86 | 1 |
| 18. | ADI Data | C6 | 2 |
| 19. | ANA A | A7 | 1 |
| 20. | ANA B | A0 | 1 |
| 21. | ANA C | A1 | 1 |
| 22. | ANA D | A2 | 1 |
| 23. | ANA E | A3 | 1 |
| 24. | ANA H | A4 | 1 |
| 25. | ANA L | A5 | 1 |
| 26. | ANA M | A6 | 1 |
| 27. | ANI Data | E6 | 2 |
| 28. | CALL Label | CD | 3 |
| 29. | CC Label | DC | 3 |
| 30. | CM Label | FC | 3 |
| 31. | CMA | 2F | 1 |
| 32. | CMC | 3F | 1 |

*(Contd...)*

## Opcode table of INTEL 8085 in alphabetical order *(Contd...)*

| S. No. | Mnemonics, operand | Opcode | Bytes |
|--------|---------------------|--------|-------|
| 33. | CMP A | BF | 1 |
| 34. | CMP B | B8 | 1 |
| 35. | CMP C | B9 | 1 |
| 36. | CMP D | BA | 1 |
| 37. | CMP E | BB | 1 |
| 38. | CMP H | BC | 1 |
| 39. | CMP L | BD | 1 |
| 40. | CMP M | BD | 1 |
| 41. | CNC Label | D4 | 3 |
| 42. | CNZ Label | C4 | 3 |
| 43. | CP Label | F4 | 3 |
| 44. | CPE Label | EC | 3 |
| 45. | CPI Data | FE | 2 |
| 46. | CPO Label | E4 | 3 |
| 47. | CZ Label | CC | 3 |
| 48. | DAA | 27 | 1 |
| 49. | DAD B | 09 | 1 |
| 50. | DAD D | 19 | 1 |
| 51. | DAD H | 29 | 1 |
| 52. | DAD SP | 39 | 1 |
| 53. | DCR A | 3D | 1 |
| 54. | DCR B | 05 | 1 |
| 55. | DCR C | 0D | 1 |
| 56. | DCR D | 15 | 1 |
| 57. | DCR E | 1D | 1 |
| 58. | DCR H | 25 | 1 |
| 59. | DCR L | 2D | 1 |
| 60. | DCR M | 35 | 1 |
| 61. | DCX B | 0B | 1 |
| 62. | DCX D | 1B | 1 |
| 63. | DCX H | 2B | 1 |
| 64. | DCX SP | 3B | 1 |
| 65. | DI | F3 | 1 |
| 66. | EI | FB | 1 |
| 67. | HLT | 76 | 1 |
| 68. | IN Port-address | DB | 2 |
| 69. | INR A | 3C | 1 |
| 70. | INR B | 04 | 1 |
| 71. | INR C | 0C | 1 |
| 72. | INR D | 14 | 1 |
| 73. | INR E | 1C | 1 |
| 74. | INR H | 24 | 1 |

*(Contd...)*

**Opcode table of INTEL 8085 in alphabetical order** *(Contd...)*

| S. No. | Mnemonics, operand | Opcode | Bytes |
|--------|--------------------|--------|-------|
| 75. | INR L | 2C | 1 |
| 76. | INR M | 34 | 1 |
| 77. | INX B | 03 | 1 |
| 78. | INX D | 13 | 1 |
| 79. | INX H | 23 | 1 |
| 80. | INX SP | 33 | 1 |
| 81. | JC Label | DA | 3 |
| 82. | JM Label | FA | 3 |
| 83. | JMP Label | C3 | 3 |
| 84. | JNC Label | D2 | 3 |
| 85. | JNZ Label | C2 | 3 |
| 86. | JP Label | F2 | 3 |
| 87. | JPE Label | EA | 3 |
| 88. | JPO Label | E2 | 3 |
| 89. | JZ Label | CA | 3 |
| 90. | LDA Address | 3A | 3 |
| 91. | LDAX B | 0A | 1 |
| 92. | LDAX D | 1A | 1 |
| 93. | LHLD Address | 2A | 3 |
| 94. | LXI B | 01 | 3 |
| 95. | LXI D | 11 | 3 |
| 96. | LXI H | 21 | 3 |
| 97. | LXI SP | 31 | 3 |
| 98. | MOV A, A | 7F | 1 |
| 99. | MOV A, B | 78 | 1 |
| 100. | MOV A, C | 79 | 1 |
| 101. | MOV A, D | 7A | 1 |
| 102. | MOV A, E | 7B | 1 |
| 103. | MOV A, H | 7C | 1 |
| 104. | MOV A, L | 7D | 1 |
| 105. | MOV A, M | 7E | 1 |
| 106. | MOV B, A | 47 | 1 |
| 107. | MOV B, B | 40 | 1 |
| 108. | MOV B, C | 41 | 1 |
| 109. | MOV B, D | 42 | 1 |
| 110. | MOV B, E | 43 | 1 |
| 111. | MOV B, H | 44 | 1 |
| 112. | MOV B, L | 45 | 1 |
| 113. | MOV B, M | 46 | 1 |
| 114. | MOV C, A | 4F | 1 |
| 115. | MOV C, B | 48 | 1 |
| 116. | MOV C, C | 49 | 1 |

*(Contd...)*

## Opcode table of INTEL 8085 in alphabetical order *(Contd...)*

| S. No. | Mnemonics, operand | Opcode | Bytes |
|--------|--------------------|--------|-------|
| 117. | MOV C, D | 4A | 1 |
| 118. | MOV C, E | 4B | 1 |
| 119. | MOV C, H | 4C | 1 |
| 120. | MOV C, L | 4D | 1 |
| 121. | MOV C, M | 4E | 1 |
| 122. | MOV D, A | 57 | 1 |
| 123. | MOV D, B | 50 | 1 |
| 124. | MOV D, C | 51 | 1 |
| 125. | MOV D, D | 52 | 1 |
| 126. | MOV D, E | 53 | 1 |
| 127. | MOV D, H | 54 | 1 |
| 128. | MOV D, L | 55 | 1 |
| 129. | MOV D, M | 56 | 1 |
| 130. | MOV E, A | 5F | 1 |
| 131. | MOV E, B | 58 | 1 |
| 132. | MOV E, C | 59 | 1 |
| 133. | MOV E, D | 5A | 1 |
| 134. | MOV E, E | 5B | 1 |
| 135. | MOV E, H | 5C | 1 |
| 136. | MOV E, L | 5D | 1 |
| 137. | MOV E, M | 5E | 1 |
| 138. | MOV H, A | 67 | 1 |
| 139. | MOV H, B | 60 | 1 |
| 140. | MOV H, C | 61 | 1 |
| 141. | MOV H, D | 62 | 1 |
| 142. | MOV H, E | 63 | 1 |
| 143. | MOV H, H | 64 | 1 |
| 144. | MOV H, L | 65 | 1 |
| 145. | MOV H, M | 66 | 1 |
| 146. | MOV L, A | 6F | 1 |
| 147. | MOV L, B | 68 | 1 |
| 148. | MOV L, C | 69 | 1 |
| 149. | MOV L, D | 6A | 1 |
| 150. | MOV L, E | 6B | 1 |
| 151. | MOV L, H | 6C | 1 |
| 152. | MOV L, L | 6D | 1 |
| 153. | MOV L, M | 6E | 1 |
| 154. | MOV M, A | 77 | 1 |
| 155. | MOV M, B | 70 | 1 |
| 156. | MOV M, C | 71 | 1 |
| 157. | MOV M, D | 72 | 1 |
| 158. | MOV M, E | 73 | 1 |

*(Contd...)*

**Opcode table of INTEL 8085 in alphabetical order** *(Contd...)*

| S. No. | Mnemonics, operand | Opcode | Bytes |
|--------|--------------------|--------|-------|
| 159. | MOV M, H | 74 | 1 |
| 160. | MOV M, L | 75 | 1 |
| 161. | MVI A, Data | 3E | 2 |
| 162. | MVI B, Data | 06 | 2 |
| 163. | MVI C, Data | 0E | 2 |
| 164. | MVI D, Data | 16 | 2 |
| 165. | MVI E, Data | 1E | 2 |
| 166. | MVI H, Data | 26 | 2 |
| 167. | MVI L, Data | 2E | 2 |
| 168. | MVI M, Data | 36 | 2 |
| 169. | NOP | 00 | 1 |
| 170. | ORA A | B7 | 1 |
| 171. | ORA B | B0 | 1 |
| 172. | ORA C | B1 | 1 |
| 173. | ORA D | B2 | 1 |
| 174. | ORA E | B3 | 1 |
| 175. | ORA H | B4 | 1 |
| 176. | ORA L | B5 | 1 |
| 177. | ORA M | B6 | 1 |
| 178. | ORI Data | F6 | 2 |
| 179. | OUT Port-Address | D3 | 2 |
| 180. | PCHL | E9 | 1 |
| 181. | POP B | C1 | 1 |
| 182. | POP D | D1 | 1 |
| 183. | POP H | E1 | 1 |
| 184. | POP PSW | F1 | 1 |
| 185. | PUSH B | C5 | 1 |
| 186. | PUSH D | D5 | 1 |
| 187. | PUSH H | E5 | 1 |
| 188. | PUSH PSW | F5 | 1 |
| 189. | RAL | 17 | 1 |
| 190. | RAR | 1F | 1 |
| 191. | RC | D8 | 1 |
| 192. | RET | C9 | 1 |
| 193. | RIM | 20 | 1 |
| 194. | RLC | 07 | 1 |
| 195. | RM | F8 | 1 |
| 196. | RNC | D0 | 1 |
| 197. | RNZ | C0 | 1 |
| 198. | RP | F0 | 1 |
| 199. | RPE | E8 | 1 |
| 200. | RPO | E0 | 1 |

*(Contd...)*

| S. No. | Mnemonics, operand | Opcode | Bytes |
|---|---|---|---|
| | **Opcode table of INTEL 8085 in alphabetical order** *(Contd...)* | | |
| 201. | RRC | 0F | 1 |
| 202. | RST 0 | C7 | 1 |
| 203. | RST 1 | CF | 1 |
| 204. | RST 2 | D7 | 1 |
| 205. | RST 3 | DF | 1 |
| 206. | RST 4 | E7 | 1 |
| 207. | RST 5 | EF | 1 |
| 208. | RST 6 | F7 | 1 |
| 209. | RST 7 | FF | 1 |
| 210. | RZ | C8 | 1 |
| 211. | SBB A | 9F | 1 |
| 212. | SBB B | 98 | 1 |
| 213. | SBB C | 99 | 1 |
| 214. | SBB D | 9A | 1 |
| 215. | SBB E | 9B | 1 |
| 216. | SBB H | 9C | 1 |
| 217. | SBB L | 9D | 1 |
| 218. | SBB M | 9E | 1 |
| 219. | SBI Data | DE | 2 |
| 220. | SHLD Address | 22 | 3 |
| 221. | SIM | 30 | 1 |
| 222. | SPHL | F9 | 1 |
| 223. | STA Address | 32 | 3 |
| 224. | STAX B | 02 | 1 |
| 225. | STAX D | 12 | 1 |
| 226. | STC | 37 | 1 |
| 227. | SUB A | 97 | 1 |
| 228. | SUB B | 90 | 1 |
| 229. | SUB C | 91 | 1 |
| 230. | SUB D | 92 | 1 |
| 231. | SUB E | 93 | 1 |
| 232. | SUB H | 94 | 1 |
| 233. | SUB L | 95 | 1 |
| 234. | SUB M | 96 | 1 |
| 235. | SUI Data | D6 | 2 |
| 236. | XCHG | EB | 1 |
| 237. | XRA A | AF | 1 |
| 238. | XRA B | A8 | 1 |
| 239. | XRA C | A9 | 1 |
| 240. | XRA D | AA | 1 |
| 241. | XRA E | AB | 1 |
| 242. | XRA H | AC | 1 |
| 243. | XRA L | AD | 1 |
| 244. | XRA M | AE | 1 |
| 245. | XRI Data | EE | 2 |
| 246. | XTHL | E3 | 1 |

# Viva-Voce Questions

**Q 1. What is a Microprocessor?**

**Ans:** Microprocessor is a program-controlled device, which fetches the instructions from memory and then decodes and executes them. Most microprocessors are single-chip devices.

**Q 2. What is the difference between 8086 and 8088?**

**Ans:** The BIU in 8088 is an 8-bit data bus and 16-bit in 8086. Instruction queue is 4 byte long in 8088 and 6 byte in 8086.

**Q 3. What are the functional units in 8086?**

**Ans:** 8086 has two independent functional units due to which the processor speed is more. The Bus Interface and Execution Units are the two functional units.

**Q 4. What is the maximum clock frequency in 8086?**

**Ans:** 5 MHz is the maximum clock frequency in 8086.

**Q 5. In which type of registers logic calculations are done?**

**Ans:** Accumulator is the register in which Arithmetic and Logic calculations are done.

**Q 6. How is 8086 faster than 8085?**

**Ans:** Due to pipelining concept, 8086 BIU fetches the next instruction when EU is busy in executing another instruction.

**Q 7. What does EU do?**

**Ans:** Execution Unit receives program instruction codes and data from BIU, executes these instructions and stores the result in general registers.

**Q 8. Which segment is used to store interrupt and subroutine return address registers?**

**Ans:** Stack segment in segment register is used to store interrupt and subroutine return address registers.

**Q 9. What does microprocessor speed depend on?**

**Ans:** The processing speed depends on Data Bus Width.

**Q 10. What is the size of data bus and address bus in 8086?**

**Ans:** 8086 has 16-bit data bus and 20-bit address bus.

**Q 11. What is the maximum memory addressing capability of 8086?**

**Ans:** The maximum memory capability of 8086 is 1 MB.

**Q 12. What is flag?**

**Ans:** Flag is a flip-flop used to store information about the status of a processor and the status of the instruction executed most recently.

**Q 13. Which flags can be set or reset by the programmer and also used to control the operation of the processor?**

**Ans:** Trace Flag, Interrupt Flag, Direction Flag.

**Q 14. In how many modes 8086 can be operated and how?**

**Ans:** 8086 can be operated in two modes: Minimum and Maximum modes. They are minimum modes if MN/MX pin is active high and maximum modes if MN/MX pin is ground.

**Q 15. What is the difference between min mode and max mode of 8086?**

**Ans:** Minimum mode operation is the least expensive way to operate the 8086 microprocessor because all the control signals for the memory and I/O are generated by the microprocessor. In the maximum mode which some of the control signals must be externally generated requires the addition of an external bus controller. It is used only when the system contains external co-processors such as 8087 arithmetic co-processor.

**Q 16. Which bus controller is used in maximum mode of 8086?**

**Ans:** 8288 bus controller is used to provide the signals eliminated from the 8086 by the maximum mode operation.

**Q 17. What is stack?**

**Ans:** Stack is a portion of RAM used for saving the content of Program Counter and general purpose registers.

**Q 18. Which stack is used in 8086?**

**Ans:** FIFO (First In First Out) stack is used in 8086. In this type of Stack, the first stored information is retrieved first.

**Q 19. What is the position of the Stack Pointer after the PUSH instruction?**

**Ans:** The address line is 02 less than the earlier value.

**Q 20. What is the position of the Stack Pointer after the POP instruction?**

**Ans:** The address line is 02 greater than the earlier value.

**Q 21. What is interrupt?**

**Ans:** Interrupt is a signal sent by external device to the processor so as to request the processor to perform a particular work.

**Q 22. What are the various interrupts in 8086?**

**Ans:** Maskable interrupts, non-maskable interrupts.

**Q 23. What is meant by maskable interrupts?**

**Ans:** An interrupt that can be turned off by the programmer is known as maskable interrupt.

**Q 24. What is non-maskable interrupts?**

**Ans:** An interrupt which can never be turned off (i.e. disabled) is known as non-maskable interrupt.

**Q 25. Which interrupts are generally used for critical events?**

**Ans:** Non-maskable interrupts are used in critical events, such as power failure, emergency, shut off, etc.

**Q 26. Give an example of non-maskable interrupts.**

**Ans:** Trap is known as non-maskable interrupt which is used in emergency condition.

**Q 27. Give examples of maskable interrupts.**

**Ans:** RST 7.5, RST 6.5, RST 5.5 are Maskable interrupts. When RST 5.5 interrupt is received, the processor saves the contents of the PC register into stack and branches to 2Ch (hexadecimal) address.

When RST 6.5 interrupt is received, the processor saves the contents of the PC register into stack and branches to 34h (hexadecimal) address.

When RST 7.5 interrupt is received, the processor saves the contents of the PC register into stack and branches to 3Ch (hexadecimal) address.

**Q 28. What are SIM and RIM instructions?**

**Ans:** SIM is Set Interrupt Mask used to mask the hardware interrupts. RIM is Read Interrupt Mask used to check whether the interrupt is Masked or not.

**Q 29. What is macro?**

**Ans:** Macro is a set of instructions that performs a task and all the instructions defined in it are inserted in the program at the point of usage.

**Q 30. What is the difference between macro and procedure?**

**Ans:** A procedure is accessed via a CALL instruction and a macro is inserted in the program at the point of execution.

**Q 31. What is meant by LATCH?**

**Ans:** Latch is a D-type flip-flop used as a temporary storage device controlled by a timing signal, which can store 0 or 1. The primary function of a Latch is data storage. It is used in output devices such as LED, to hold the data for display.

**Q 32. What is a compiler?**

**Ans:** Compiler is used to translate the high-level language program into machine code at a time. It does not require special instruction to store in a memory, it stores automatically. The execution time is less as compared to interpreter.

**Q 33. What is the disadvantage of microprocessor?**

**Ans:** It has limitations on the size of data. Most microprocessors do not support floating-point operations.

**Q 34. What is the 82C55A device?**

**Ans:** The 8255A/82C55A interfaces peripheral I/O devices to the microcomputer system bus. It is programmable by the system software. It has a three-state bi-directional 8-bit buffer which interfaces the 8255A/82C55A to the system data bus.

**Q 35. What kind of input/output interface does a PPI implement?**

**Ans:** It provides a parallel interface which includes features, such as single-bit, 4-bit, and byte-wide input and output ports; level-sensitive inputs; latched outputs; strobed inputs or outputs; and strobed bi-directional input/outputs.

**Q 36. How many I/O lines are available on the 82C55A?**

**Ans:** 82C55A has a total of 24 I/O lines.

**Q 37.** **Describe the mode 0, mode 1 and mode 2 operations of the 82C55A.**

**Ans:** **MODE 0:** Simple I/O mode: In this mode, any of the ports A, B and C can be programmed as input or output. In this mode, all the bits are out or in.
**MODE 1:** Ports A and B can be used as input or output ports with handshaking capabilities. Handshaking signals are provided by the bits of port C.
**MODE 2:** Port A can be used as a bi-directional I/O port with handshaking capabilities whose signals are provided by port C. Port B can be used either in simple I/O mode or handshaking mode 1.

**Q 38.** **What is the mode and I/O configuration for ports A, B and C of an 82C55A after its control register is loaded with 82H?**

**Ans:** If control register is loaded with 82H, then the port B is configured as an input port, port A and port C are configured as output ports and in mode 0.

**Q 39.** **What is a bus?**

**Ans:** Bus is a group of conducting lines that carries data, address and control signals.

**Q 40.** **Why is data bus bi-directional?**

**Ans:** The microprocessor has to fetch (read) the data from memory or input device for processing and after processing, it has to store (write) the data to memory or output device, hence, the data bus is bi-directional.

**Q 41.** **Why is address bus unidirectional?**

**Ans:** The address is an identification number used by the microprocessor to identify or access a memory location or I/O device. It is an output signal from the processor, hence, the address bus is unidirectional.

**Q 42.** **What is the function of microprocessor in a system?**

**Ans:** The microprocessor is the master in the system which controls all the activity of the system. It issues address, controls signals and fetches the instruction and data from memory. Then it executes the instruction to take appropriate action.

**Q 43.** **What are the modes in which 8086 can operate?**

**Ans:** The 8086 can operate in two modes and they are minimum (or uniprocessor) mode and maximum (or multiprocessor) mode.

**Q 44.** **What is the data and address size in 8086?**

**Ans:** The 8086 can operate on either 8-bit or 16-bit data. The 8086 uses 20-bit address to access memory and 16-bit address to access I/O devices.

**Q 45.** **Explain the function of M/IO in 8086.**

**Ans:** The signal M/IO is used to differentiate memory address and I/O address. When the processor is accessing memory locations, M/IO is asserted high and when it is accessing I/O mapped devices it is asserted low.

**Q 46.** **Write the flags of 8086.**

**Ans:** The 8086 has nine flags and they are as follows:

1. Carry flag (CF)
2. Parity flag (PF)
3. Auxiliary carry flag (AF)
4. Zero flag (ZF)
5. Sign flag (SF)

    6. Overflow flag (OF)

    7. Trace flag (TF)

    8. Interrupt flag (IF)

    9. Direction flag (DF)

**Q 47. What are the interrupts of 8086?**

**Ans:** The interrupts of 8086 are INTR and NMI. The INTR is general maskable interrupt and NMI is non-maskable interrupt.

**Q 48. How clock signal is generated in 8086? What is the maximum internal clock frequency of 8086?**

**Ans:** The 8086 does not have on-chip clock generation circuit, hence, the clock generator chip, 8284 is connected to the CLK pin of 8086. The clock signal supplied by 8284 is divided by three for internal use. The maximum internal clock frequency of 8086 is 5 MHz.

**Q 49. Write the special functions carried by the general purpose registers of 8086.**

**Ans:** The special functions carried by the general purpose registers of 8086 are the following:

    1. AX 16-bit accumulator

    2. AL 8-bit accumulator

    3. BX Base register

    4. CX Count register

    5. DX Data register

**Q 50. What is pipelined architecture?**

**Ans:** In pipelined architecture, the processor will have number of functional units and the execution time of functional units is overlapped. Each functional unit works independently most of the time.

**Q 51. List the segment registers of 8086.**

**Ans:** The segment registers of 8086 are: Code segment, Data segment, Stack segment and Extra segment registers.

**Q 52. Define machine cycle.**

**Ans:** Machine cycle is defined as the time required to complete one operation of accessing memory, I/O or acknowledging an external request. This cycle may consist of three to six T-states.

**Q 53. Define T-State.**

**Ans:** T-State is defined as one subdivision of the operation performed in one clock period. These subdivisions are internal states synchronized with the system clock and each T-State is precisely equal to one clock period.

**Q 54. List the components of microprocessor (single board microcomputer) based system.**

**Ans:** The microprocessor based system consists of microprocessor as CPU, semiconductor memories like EPROM and RAM, input device, output device and interfacing devices.

**Q 55. Why interfacing is needed for I/O devices?**

**Ans:** Generally I/O devices are slow devices, therefore, the speed of I/O devices does not match with the speed of microprocessor, and so an interface is provided between system bus and I/O devices.

**Q 56. What is the difference between CPU bus and system bus?**

**Ans:** The CPU bus has multiplexed lines, whereas the system bus has separate lines for each signal. (The multiplexed CPU lines are de-multiplexed by the CPU interface circuit to form system bus.)

**Q 57. What does memory-mapping mean?**

**Ans:** The memory mapping is the process of interfacing memories to microprocessor and allocating addresses to each memory locations.

**Q 58. Why is EPROM mapped at the beginning of memory space in 8085 system?**

**Ans:** In 8085 microprocessor, after a reset, the program counter will have OOOOH address. If the monitor program is stored from this address then after a reset, it will be executed automatically. The monitor program is a permanent program and stored in EPROM memory. If EPROM memory is mapped at the beginning of memory space, i.e. at OOOOH, then the monitor program will be executed automatically after a reset.

**Q 59. What is the need for system clock and how is it generated in 8085?**

**Ans:** The system clock is necessary for synchronizing various internal operations or devices in the microprocessor and to synchronize the microprocessor with other peripherals in the system.

**Q 60. What is DMA?**

**Ans:** The direct data transfer between I/O device and memory is called DMA.

**Q 61. What is the need for port?**

**Ans:** The I/O devices are generally slow devices and their timing characteristics do not match with processor timings. Hence, the I/O devices are connected to system bus through the ports.

**Q 62. What is a port?**

**Ans:** The port is a buffered I/O, which is used to hold the data transmitted from the microprocessor to I/O device or vice-versa.

**Q 63. Give some examples of port devices used in 8085 microprocessor based system.**

**Ans:** The various INTEL I/O port devices used in 8085 microprocessor based system are 8212, 8155, 8156, 8255, 8355 and 8755.

**Q 64. Write a short note on INTEL 8255.**

**Ans:** The INTEL 8255 is an I/O port device consisting of three numbers of 8-bit parallel I/O ports. The ports can be programmed to function either as an input port or as an output port in different operating modes. It requires four internal addresses and has one logic LOW chip select pin.

**Q 65. What is the drawback in memory mapped I/O?**

**Ans:** When I/O devices are memory mapped, some of the addresses are allotted to I/O devices and so the full address space cannot be used for addressing memory (i.e. physical memory address space will be reduced). Hence, memory mapping is useful only for small systems, where the memory requirement is less.

**Q 66. How DMA is initiated?**

**Ans:** When the I/O device needs a DMA transfer, it sends a DMA request signal to DMA controller. The DMA controller in turn sends a HOLD request to the processor. When the processor receives a HOLD request, it drives its tri-stated pins to high-impedance state at the end of current instruction execution and sends an acknowledge signal to DMA controller. Now the DMA controller performs DMA transfer.

**Q 67. What is processor cycle (machine cycle)?**

**Ans:** The processor cycle or machine cycle is the basic operation performed by the processor. To execute an instruction, the processor will run one or more machine cycles in a particular order.

**Q 68. What is instruction cycle?**

**Ans:** The sequence of operations that a processor has to carry out while executing the instruction is called Instruction cycle. Each instruction cycle of a processor indium consists of a number of machine cycles.

**Q 69. What is fetch and execute cycle?**

**Ans:** In general, the instruction cycle of an instruction can be divided into fetch and execute cycles. The fetch cycle is executed to fetch the opcode from memory. The execute cycle is executed to decode the instruction and to perform the work instructed by the instruction.

**Q 70. What is block and demand transfer mode DMA?**

**Ans:** In block transfer mode, the DMA controller transfers a block of data and relieves the bus for processor. After sometime, another block of data is transferred by DMA and so on. In demand transfer modes the DMA controller completes the entire data transfer at a stretch and then relieves the bus to processor.

**Q 71. What is the need for timing diagram?**

**Ans:** The timing diagram provides information regarding the status of various signals, when a machine cycle is executed. The knowledge of timing diagram is essential for system designer to select matched peripheral devices like memories, latches, ports, etc. to form a microprocessor system.

**Q 72. How many machine cycles constitute one instruction cycle in 8085?**

**Ans:** Each instruction of the 8085 processor consists of one to five machine cycles.

**Q 73. Define opcode and operand.**

**Ans:** Opcode (Operation code) is the part of an instruction/directive that identifies a specific operation.

Operand is a part of an instruction/directive that represents a value on which the instruction acts.

**Q 74. What is opcode fetch cycle?**

**Ans:** The opcode fetch cycle is a machine cycle executed to fetch the opcode of an instruction stored in memory. Every instruction starts with opcode fetch machine cycle.

**Q 75. What operation is performed during first T-state of every machine cycle in 8085?**

**Ans:** In 8085, during the first T-state of every machine cycle, the low byte address is latched into an external latch using ALE signal.

**Q 76. Why are status signals provided in microprocessor?**

**Ans:** The status signals can be used by the system designer to track the internal operations of the processor. Also, it can be used for memory expansion (by providing separate memory banks for program and data and selecting the bank using status signals).

**Q 77. What are the Control and Status available in 8085?**

**Ans:** There are four Control and Status signals available in 8085. They are as follows:

    a. Address Latch Enable (ALE) enables the address latches so that the lower half of an address is stored in the latches which is available throughout the machine cycle.

    b. RD and WR are basically used to control the direction of the data flow between processor and memory or I/O device/port.

    c. IO/M indicates whether I/O operation or memory operation is being carried out. S0 and S1 indicate the type of machine cycle in progress.

    d. Ready is used by the microprocessor to sense whether a peripheral is ready or not for data transfer, if not, the processor waits. It is thus used to synchronize slower peripherals to the microprocessor.

**Q 78. When does the 8085 processor check for an interrupt?**

**Ans:** In the second T-state of the last machine cycle of every instruction, the 8085 processor checks whether an interrupt request is made or not.

**Q 79. What is interrupt acknowledge cycle?**

**Ans:** The interrupt acknowledge cycle is a machine cycle executed by 8085 processor to get the address of the interrupt service routine in-order to service the interrupt device.

**Q 80. State if HOLD has higher priority than TRAP or not.**

**Ans:** The interrupts including mAP are recognized only if the HOLD is not valid, hence TRAP has lower priority than HOLD.

**Q 81. What is masking and why is it required?**

**Ans:** Masking is preventing the interrupt from disturbing the current program execution. When the processor is performing an important job (process) and if the process should not be interrupted then all the interrupts should be masked or disabled. In processor with multiple interrupts, the lower priority interrupt can be masked so as to prevent it from interrupting the execution of interrupt service routine of higher priority interrupt.

**Q 82. When does the 8085 processor accept hardware interrupt?**

**Ans:** The processor keeps on checking the interrupt pins at the second T-state of last machine cycle of every instruction. If the processor finds a valid interrupt signal and if the interrupt is unmasked and enabled then the processor accepts the interrupt. The acceptance of the interrupt is acknowledged by sending an OOA signal to the interrupted device.

**Q 83. When will the 8085 processor disable the interrupt system?**

**Ans:** The interrupts of 8085 except TRAP are disabled after anyone of the following operations:

    1. Executing El instruction.

    2. System or processor reset.

    3. After reorganization (acceptance) of an interrupt.

**Q 84. What is the function performed by Dl instruction?**

**Ans:** The function of Dl instruction is to enable the disabled interrupt system.

**Q 85. What is the function performed by El instruction?**

**Ans:** The El instruction can be used to enable the interrupts after disabling.

**Q 86. How is the vector address generated for the INTR interrupt of 8085?**

**Ans:** For the interrupt INTR, the interrupting device has to place either RST opcode or CALL opcode followed by 16-bit address. I~RST opcode is placed then the corresponding vector address is generated by the processor. In case of CALL opcode, the given l6-bit address will be the vector address.

**Q 87. How are clock signals generated in 8085 and what is the frequency of the internal clock?**

**Ans:** The 8085 has the clock generation circuit on the chip but an external quartz crystal or LC circuit or RC circuit should be connected at the pins XI and X2. The maximum internal clock frequency of 8085A is 3.03 MHz.

**Q 88. What happens to the 8085 processor when it resets?**

**Ans:** When the 8085 processor resets, it executes the first instruction at the OOOOH location. The 8085 resets (clears) instruction register, interrupt mask bits and other registers.

**Q 89. What are the operations performed by ALU of 8085?**

**Ans:** The operations performed by ALU of 8085 are: Addition, Subtraction, Logical AND, OR, Exclusive OR, Compare Complement, Increment, Decrement and Left I Right shift.

**Q 90. What are the hardware interrupts in 8085?**

**Ans:** The hardware interrupts in 8085 are TRAP, RST 7.5, RST 6.5 and RST 5.5.

**Q 91. Which interrupt has the highest priority in 8085? What is the priority of other interrupts?**

**Ans:** The TRAP has the highest priority, followed by RST 7.5, RST 6.5, RST 5.5 and INTR.

**Q 92. What is ALE?**

**Ans:** The ALE (Address Latch Enable) is a signal used to de-multiplex the address and data lines, using an external latch. It is used to enable the external latch.

**Q 93. Explain the function of IO/M in 8085.**

**Ans:** The IO/M is used to differentiate memory access and I/O access. For IN and OUT instructions, it is high. For memory reference instructions, it is low.

**Q 94. Where is the READY signal used?**

**Ans:** READY is an input signal to the processor, used by the memory or I/O devices to get extra time for data transfer or to introduce wait states in the bus cycles.

**Q 95. What are HOLD and HLDA and how are they used?**

**Ans:** HOLD and HLDA acknowledge signals are used for the Direct Memory Access (DMA) type of data transfer. The DMA controller places a high on HOLD pin in order to take control of the system bus. The HOLD request is acknowledged by the 8085 by driving all its tri-stated pins to high-impedance state and asserting HLDA signal high.

**Q 96. What is polling?**

**Ans:** Polling is a scheme or an algorithm to identify the devices interrupting the processor. Polling is employed when multiple devices interrupt the processor through one interrupt pin of the processor.

**Q 97. What are the different types of polling?**

**Ans:** The polling can be classified into software and hardware polling. In software polling, the entire polling process is governed by a program. In hardware polling, the hardware takes care of checking the status of interrupting devices and allowing one-by-one to the processor.

**Q 98. What is the need for interrupt controller?**

**Ans:** The interrupt controller is employed to expand the interrupt inputs. It can handle the interrupt request from various devices and allow one-by-one to the processor.

**Q 99. List some of the features of INTEL 8259 (Programmable Interrupt Controller).**

1. It manages eight interrupt requests.
2. The interrupt vector addresses are programmable.
3. The priorities of interrupts are programmable.
4. The interrupts can be masked or unmasked individually.

**Q 100. What is a programmable peripheral device?**

**Ans:** If the functions performed by a peripheral device can be altered or changed by a program instruction then the peripheral device is called programmable device. Usually the programmable devices will have control registers. The device can be programmed by sending control word in the prescribed format to the control register.

**Q 101. What is synchronous data transfer scheme?**

**Ans:** For synchronous data transfer scheme, the processor does not check the readiness of the device after a command has been issued for read/write operation. For this scheme, the processor will request the device to get ready and then read/Write to the device immediately after the request. In some synchronous schemes, a small delay is allowed after the request.

**Q 102. What is asynchronous data transfer scheme?**

**Ans:** In asynchronous data transfer scheme, first the processor sends a request to the device for read/write operation. Then the processor keeps on polling the status of the device. Once the device is ready, the processor executes a data transfer instruction to complete the process.

**Q 103. What are the operating modes of 8212?**

**Ans:** The 8212 can be hardwired to work either as a latch or tri-state buffer. If mode (MD) pin is tied HIGH then it will work as a latch and so it can be used as output port. If mode (MD) pin is tied LOW then it will work as tri-state buffer and so it can be used as input port.

**Q 104. Explain the working of a handshake output port.**

**Ans:** In handshake output operation, the processor will load a data to port. When the port receives the data, it will inform the output device to collect the data. Once the output device accepts the data, the port will inform the processor that it is empty. Now the processor can load another data to port and the above process is repeated.

**Q 105. What are the internal devices of 8255?**

Ans: The internal devices of 8255 are port-A, port-B and port-C. The ports can be programmed for either input or output function in different operating modes.

**Q 106. What is baud rate?**

Ans: The baud rate is the rate at which the serial data are transmitted. Baud rate is defined as 1/(The time for a bit cell). In some systems, one bit cell has one data bit, then the baud rate and bits/sec are same.

**Q 107. What is USART?**

Ans: The device which can be programmed to perform Synchronous or Asynchronous serial communication is called USART (Universal Synchronous Asynchronous Receiver Transmitter). The INTEL 8251A is an example of USART.

**Q 108. What are the functions performed by INTEL 8251A?**

Ans: The INTEL 8251A is used for converting parallel data to serial or vice-versa. The data transmission or reception can be either asynchronous or synchronous. The 8251A can be used to interface MODEM and establish serial communication through MODEM over telephone lines.

**Q 109. What are the control words of 8251A and what are its functions?**

Ans: The control words of 8251A are Mode word and Command word. The mode word informs 8251 about the baud rate, character length, parity and stop bits. The command word can be sent to enable the data transmission and reception.

**Q 110. What is the information that can be obtained from the status word of 8251?**

Ans: The status word can be read by the CPU to check the readiness of the transmitter or receiver and to check the character synchronization in synchronous reception. It also provides information regarding various errors in the data received. The various error conditions that can be checked from the status word are parity error, overrun error and framing error.

**Q 111. Give some examples of input devices to microprocessor-based system.**

Ans: The input devices used in the microprocessor-based system are keyboards, DIP switches, ADC, Floppy disc, etc.

**Q 112. What are the tasks involved in keyboard interface?**

Ans: The tasks involved in keyboard interfacing are sensing a key actuation, debouncing the key and generating key codes (Decoding the key). These tasks are performed by software if the keyboard is interfaced through ports and they are performed by hardware if the keyboard is interfaced through 8279.

**Q 113. How is a keyboard matrix formed in keyboard interface using 8279?**

Ans: The return lines, RL0 to RL7 of 8279 are used to form the columns of keyboard matrix. In decoded scan, the scan lines SL0 to SL3 of 8279 are used to form the rows of keyboard matrix. In encoded scan mode, the output lines of external decoder are used as rows of keyboard matrix.

**Q 114. What is scanning in keyboard and what is scan time?**

Ans: The process of sending a zero to each row of a keyboard matrix and reading the columns for key actuation is called scanning. The scan time is the time taken by the processor to scan all the rows one-by-one starting from first row and coming back to the first row again.

**Q 115. What is scanning in display and what is the scan time?**

Ans: In display devices, the process of sending display codes to 7-segment LEDs to display the LEDs one-by-one is called scanning (or multiplexed display). The scan time is the time taken to display all the 7-segment LEDs one by one, starting from the first LED and coming back to the first LED again.

**Q 116. What are the internal devices of a typical DAC?**

Ans: The internal devices of a DAC are R/2R resistive network, an internal latch and current to voltage converting amplifier.

**Q 117. What is settling or conversion time in DAC?**

Ans: The time taken by the DAC to convert a given digital data to corresponding analog signal is called conversion time.

**Q 118. What are the different types of ADC?**

Ans: The different types of ADC are successive approximation ADC, counter type ADC, flash type ADC, integrator converters and voltage to frequency converters.

**Q 119. Define stack.**

Ans: Stack is a sequence of RAM memory locations defined by the programmer.

**Q 120. What is program counter? How is it useful in program execution?**

Ans: The program counter keeps track of program execution. To execute a program, the starting address of the program is loaded in program counter. The PC sends out an address to fetch a byte of instruction from memory and increments its content automatically.

**Q 121. How is the microprocessor synchronized with peripherals?**

Ans: The timing and control unit synchronizes all the microprocessor operations with clock and generates control signals necessary for communication between the microprocessor and peripherals.

**Q 122. What is a minimum system and how is it formed in 8085?**

Ans: A minimum system is one which is formed using minimum number of IC chips. The 8085 based minimum system is formed using 8155, 8355 and 8755.

# Index